Spectroscopy

VOLUME ONE

SPECTROSCOPY

VOLUME ONE

Atomic spectra; Nuclear magnetic resonance spectroscopy; Nuclear quadrupole resonance spectroscopy; Electron spin resonance spectroscopy; Mössbauer spectroscopy.

VOLUME TWO

Molecular spectra; Symmetry and group theory; Microwave spectroscopy; Infrared and raman spectroscopy; Far-infrared spectroscopy; Force constants; Thermodynamic functions.

VOLUME THREE

Molecular quantum numbers; Electronic spectra of diatomic molecules; Dissociation energies of diatomic molecules; Electronic spectra of polyatomic molecules; Fluorescence and phosphorescence spectroscopy; Astrochemistry; Photoelectron spectroscopy.

Spectroscopy
VOLUME ONE

EDITED BY

B.P. STRAUGHAN Ph.D., F.R.I.C.

Department of Chemistry,
University of Newcastle upon Tyne,
England

AND

S. WALKER, M.A., D.PHIL., D.Sc.

Chairman, Department of Chemistry,
Lakehead University,
Ontario, Canada

LONDON
CHAPMAN AND HALL

A HALSTED PRESS BOOK
JOHN WILEY & SONS INC., NEW YORK

First published 1976
by Chapman and Hall Ltd
11 New Fetter Lane, London EC4P 4EE

© 1976 Chapman and Hall Ltd

Typeset by EWC Wilkins Ltd., London and Northampton
and printed in Great Britain by
Butler and Tanner Ltd., Frome and London

ISBN 0 412 13340 7 (cased edition)
ISBN 0 412 13350 4 (Science Paperback edition)

Distributed in the U.S.A. by Halsted Press,
a Division of John Wiley & Sons, Inc., New York

Library of Congress Cataloging in Publication Data
Main entry under title:

Spectroscopy.

    Previous editions by S. Walker and H. Straw published
in 1962 and 1967, entered under: Walker, Stanley.
    Includes bibliographies.
    1. Spectrum analysis. I. Straughan, B.P.
II. Walker, Stanley. Spectroscopy.
QC451.W33 1976   535'.84   75—45328
ISBN 0—470—15031—9 (v. 1)
    0—470—15032—7 (v. 2)
    0—470—15033—5 (v. 3)

Preface

It is fifteen years since Walker and Straw wrote the first edition of 'Spectroscopy' and considerable developments have taken place during that time in all fields of this expanding subject. In atomic spectroscopy, for example, where the principles required in a student text have been laid down for many years, there have been advances in optical pumping and double resonance which cannot be neglected at undergraduate level. In addition, nuclear quadrupole resonance (n.q.r.) and far infrared spectroscopy now merit separate chapters while additional chapters dealing with Mössbauer spectroscopy, photoelectron spectroscopy and group theory are an essential requisite for any modern spectroscopy textbook.

When the idea for a new edition of Spectroscopy was first discussed it quickly became clear that the task of revision would be an impossible one for two authors working alone. Consequently it was decided that the new edition be planned and co-ordinated by two editors who were to invite specialists, each of whom had experience of presenting their subject at an undergraduate level, to contribute a new chapter or to revise extensively an existing chapter. In this manner a proper perspective of each topic has been provided without any sacrifice of the essential character and unity of the first edition.

The expansion of subject matter has necessitated the division of the complete work into three self contained volumes.

Volume 1 includes atomic, n.m.r., n.q.r., e.s.r. and Mössbauer spectroscopy.

Volume 2 contains chapters on molecular symmetry and group theory, microwave, infrared and Raman, far-infrared spectroscopy, force constants, evaluation of thermodynamic functions.

Volume 3 centres on the information which results when a valence electron(s) is excited or removed from the parent molecule. It includes electronic spectroscopy, quantum numbers, dissociation energies, fluorescence and phosphorescence spectroscopy, astrochemistry, photoelectron spectroscopy.

The complete work now provides a single source of reference for all the spectroscopy that a student of chemistry will normally encounter as an undergraduate. Furthermore, the depth of coverage should ensure the books' use on graduate courses and for those starting research work in one of the main branches of spectroscopy.

A continued source of confusion in the spectroscopic literature is the duplication of symbols and the use of the same symbol by different authors to represent different factors. The literature use of both SI and non SI units further complicates the picture. In this book we have tried to use SI units throughout or units such as the electron volt which are recognised for continued use in conjunction with SI units. The symbols and recognised values of physical constants are those published by the Symbols Committee of the Royal Society 1975.

B.P. Straughan
S. Walker

October, 1975

Acknowledgements

Although not involved in the production of this second edition, we would like to express our sincere thanks to Mr. H. Straw whose vital contribution to the first edition of Spectroscopy helped to ensure its widespread success and hence the demand for a new edition. One of us (S.W.) wishes to thank his wife, Kathleen, without whose help at many stages part of this work could not have gone forward.

Terms and symbols describing magnetic phenomena

When discussing magnetic phenomena, the chemist has been accustomed to write H instead of B when H is the *magnetic field strength* and B is the *magnetic flux density* (or *magnetic induction field*). When using electromagnetic units (e.m.u.) H and B have the same magnitude but in SI units they differ by $4\pi \times 10^{-7}$. Consequently, B has been used throughout this book and in imprecise contexts it has been referred to as the *magnetic field*.

Contributors to Volume One

Contents

1 Atomic spectroscopy

1.1 INTRODUCTION TO ATOMIC SPECTROSCOPY

Atomic spectra are principally concerned with the interchange of energy between the atom and electromagnetic radiation where the exchange in energy may be associated on the simplest pictorial model with a valence electron changing its orbit. Energy may be absorbed from the radiation field (absorption spectra) or may be added to it (emission spectra). The actual change in energy (ΔE) between two energy levels in an atom is related to the frequency of the radiation absorbed or emitted by the equation:

$$E' - E'' = \Delta E = h\nu \tag{1.1}$$

where h is Planck's constant and ν is the frequency in hertz (previously 'cycles per second' was employed). E' is the total energy of the atom in its higher energy state, and E'' is the value for the lower state. The quantity $h\nu$ is really the photon of energy, either absorbed or emitted.

Even for the hydrogen atom there are many possible energy levels. On the basis of the simple circular orbit model a few of the various orbits in which the electron may exist are represented as in Fig. 1.1, the proton being at the centre of the innermost orbit. The values of energy levels corresponding to the principal quantum numbers (n) (see later) 1, 2, 3, 4, 5, are E_1, E_2, E_3, E_4, and E_5, respectively. If hydrogen is present in an electric discharge, some of the modecules split up into atoms, and some of these hydrogen atoms may reach an excited electronic state, that is have n values greater than 1. Some of the excited atoms lose the whole of their excess energy, returning to the lowest level (that is, $n = 1$, the ground state) and emit radiation with frequencies $\nu_1, \nu_2, \nu_3, \nu_4 \ldots$, for example:

$$E_2 - E_1 = h\nu_1 \tag{1.2}$$

$$E_3 - E_1 = h\nu_2 \tag{1.3}$$

1

$$E_4 - E_1 = h\nu_3 \quad \text{and so on} \tag{1.4}$$

Some electrons return to the $n = 2^\dagger$ state, and these produce another set of frequencies ν_1', ν_2', ν_3' given by:

$$E_3 - E_2 = h\nu_1' \tag{1.5}$$

$$E_4 - E_2 = h\nu_2' \tag{1.6}$$

$$E_5 - E_2 = h\nu_3' \quad \text{and so on} \tag{1.7}$$

These different frequencies are sorted out either by passage through a prism or by means of a diffraction grating; the separation of the radiation into its component frequencies gives *spectrum of the element*. If, however, it is impossible to separate the frequencies—and this applies over a fairly wide frequency region—the spectrum is said to be continuous; an example of a continuous spectrum is exhibited by filament lamps.

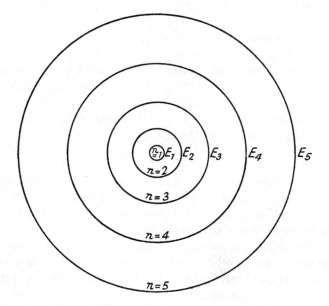

Fig. 1.1 Circular orbit model for the hydrogen atom

An apparatus for producing the hydrogen atom emission frequencies is shown in Fig. 1.2. To the end of a quartz tube an optical quartz window is attached, and two metal electrodes are sealed through the quartz in the side arms A and B. The gas to be examined (in this case hydrogen) is introduced into the tube, and a discharge is produced by applying a voltage of $\sim 3000\,\text{V}$ across the electrodes from a transformer. The swiftly moving electrons in the discharge

† This is, of course, only an intermediate stop because they must finally return to the lowest energy state ($n = 1$).

bombard the hydrogen molecules; this bombardment results in the formation of some excited hydrogen atoms which are capable of emitting light. Light from this source is lined up with a *spectroscope* (i.e. an instrument which splits the light up into a spectrum). This in principle consists of a slit, a collimating lens producing a parallel beam of light which is then passed through the prism, and a telescope lens. The latter brings the frequencies, separated by the prism, to a focus. For example, if the frequencies $\nu_1, \nu_2, \nu_3 \ldots$ are directed on to the slit, as in Fig. 1.3, a narrow beam of light is allowed to enter; the lens B collimates the light on to the prism which disperses the light into its respective frequencies. The lens C then focuses these separated frequencies (which are monochromatic images of the slit) on to a photographic plate. When the plate is developed, sharp black lines are observed corresponding to each of the emitted frequencies $\nu_1, \nu_2, \nu_3 \ldots$, and this photographic record of the spectrum is termed a *spectrogram*. When the instrument gives a photographic record it is called a *spectrograph*; the term *spectroscope* is usually reserved for instruments where the spectrum is viewed by the eye.

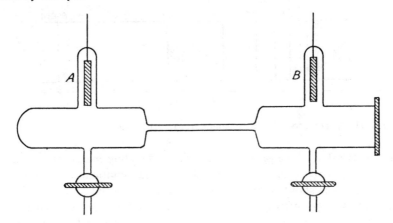

Fig. 1.2 One type of discharge tube

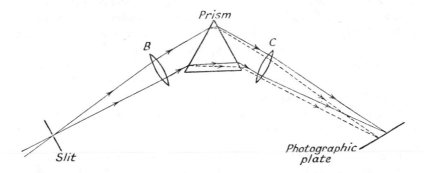

Fig. 1.3 Optical arrangement of a medium quartz type of spectrograph

3

In Fig. 1.4 the frequencies resulting from transitions from the $n > 2$ levels to the $n = 2$ level are given for the hydrogen atom. This gives a series of lines known as the *Balmer series*. For transitions to the $n = 1$ level another series (*Lyman*) is obtained.

In Fig. 1.3 the simplest kind of spectrograph has been shown, and this type would not be able to distinguish between two lines only a fraction of an Ångström unit apart in the visible region of the spectrum ($1 \text{Å} = 10^{-10}$ m). A most important function of a spectrograph is to separate wavelengths of approximately the same value. This ability is measured by the *resolving power*, which is defined as $\lambda/d\lambda$, where $d\lambda = \lambda_1 - \lambda_2$. This is the wavelength separation between the two closest lines λ_1 and λ_2 of similar intensity which can just be separated (resolved) by the spectrograph at the average wavelength, λ, where $\lambda = (\lambda_1 + \lambda_2)/2$. The resolving power may vary considerably, depending on the type of spectrograph used. Much fundamental information is derived from a study of spectral 'lines' with a spectrograph which has a high resolving power.

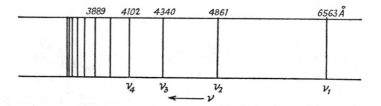
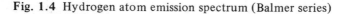

Fig. 1.4 Hydrogen atom emission spectrum (Balmer series)

The absorption of energy may be studied by directing energy from a filament lamp (e.g. a tungsten filament lamp which provides a continuous range of wavelengths from about 3000 Å to 10 000 Å) through a cell containing the atoms[†] whose absorption spectrum is required. The frequencies which are required to accomplish the permitted energy transitions are selected by the atoms from the continuous source of radiation, and when the photographic plate is developed these appear as light lines on a dark background. When a transition between E' and E'' levels occurs in both emission and absorption for separate experiments on a particular atom, the corresponding frequencies are in agreement within the limits of experimental error.

The complexity of the spectrum varies quite considerably with the atom concerned. The emission spectrum for mercury is given in Fig. 1.5 for the 2400 to 6000 Å region and does not appear overwhelmingly complex. The spectrum for iron, however, consists of an extremely large number of lines which are distributed right across the ultraviolet and visible regions, and, in fact,

† The atoms could be produced, for example, by the dissociation of a compound, if a high-temperature furnace were employed as the absorption cell.

these lines are often used for interpolating the value of the wavelengths for the lines of other elements. In Fig. 1.6 the 2400 to 6000 Å region of the iron emission spectrum may be observed, while in Fig. 1.7 part of the absorption spectrum for the principal series of sodium is given (see later).

The primary standard of wavelength which is now employed is the orange emission line of krypton (Kr_{36}^{86}) resulting from a cathode discharge tube situated in a cryostat at the triple point of nitrogen (63 K), a standard which is claimed to be reproducible to 1 part in 10^8. This line is employed to define the basic SI unit (see later) the metre of which is 1 650 763.73 vacuum wavelengths of this radiation. Many other lines have been measured relative to this and defined as secondary standards, although to be classed as such the measured wavelengths from several laboratories have to be in agreement.

It will be observed from Figs. 1.5–1.7 that the intensity of the different lines within a spectrum varies considerably, and in addition to determining the frequencies accurately, quantitative studies can be made of the intensities of these lines, where the intensity of the lines is dependent on the concentration of the element producing it. The study of both frequency and intensity is known as *spectrophotometry*.

1.1.1. Units and fundamental constants

In recent years there has been a trend to adopt a more logical system of units termed the *Système International d'Unitès* (SI units) which was the ultimate outcome of a resolution of the 9th General Conference of Weights and Measures in 1948. It was adopted in 1960 and was based on the seven units listed in (A) below.

SI units may be divided into the following three categories:
(A) *Base units.* These are the standards of:
    (a) length, the metre (m);
    (b) time, the second (s);
    (c) mass, the kilogram (kg);
    (d) electric current, the ampere (A);
    (e) temperature, the kelvin (K);
    (f) amount of substance, the mole (mol), where the mole is the amount of substance which contains the same number of molecules (atoms) as there are atoms of carbon in 0.012 kg of carbon-12;
    (g) luminosity intensity, the candela (cd).

The initial practice in electrical and magnetic quantities was to define them in terms of the centimetre, gramme, and second which were then the units employed in mechanics. Later it became apparent that if the metre, kilogramme, and second were made the fundamental units, a consistent system of electrical and magnetic units could be built up. Known as the mks system, this with the ampere as the electrical base unit has been adopted for SI units.
(B) *Derived units.* These result from combining the above base units as is

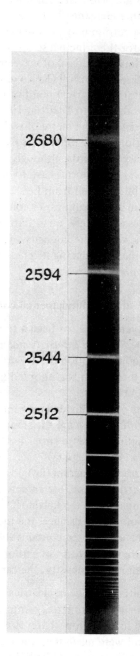

Fig. 1.5 Emission spectrum of the mercury atom between 2400 and 6000 Å (The marked scale is in Ångström units). (After Pearse and Gaydon [1.1]).

Fig. 1.6 Iron arc spectrum between 2400 and 6000 Å (The marked scale is in Ångström units). After Pearse and Gaydon [1.1]).

Fig. 1.7 Part of the absorption spectrum of the principal series of sodium. (After Kuhn [1.2]. Courtesy of Springer Verlag, Heidelberg).

indicated in the two examples below.

For energy (E) the derived unit is the joule, and thus it is the practice now to consider energy changes in joules whereas it was previously considered in terms of ergs or calories. The joule may be defined by inserting the SI units in the Einstein equation $E = mc^2 = \text{kg m}^2\text{s}^{-2}$.

The derived unit of force is the newton (N), and since energy = force × distance, the newton may be defined as follows:

$$\text{force} = \text{J m}^{-1} = \text{kg m s}^{-2}$$

The derived unit for pressure is the pascal (Pa) and may be defined as newton per square metre, that is from the formula force/unit area.

Some derived units without special names are those for:

(i) volume, which has the unit of cubic metre (m^3);

(ii) density, which has the unit of kilogramme per cubic metre (kg m^{-3});

(iii) molar mass, which has the unit of kilogramme per mole (kg mol^{-1}).

The derived units in electric and magnetic systems of spectroscopic application are:

(a) the charge (Q), where the quantity of unit charge is the coulomb (C) and may be defined as the quantity of electricity carried per second by a current of one ampere, i.e., $\text{C} = \text{A s}$;

(b) the capacitance (C) the unit of which is the farad (F) which may be defined as the capacitance of a capacitor of which there is a difference of one volt potential between the plates when they hold one coulomb of charge;

(c) the unit of magnetic flux which is the weber (Wb). This may be defined as the amount of flux which produces an electromotive force of one volt in a one-turn conductor as it reduces uniformly to zero in one second $(1 \text{ Wb} = 1 \text{ kg m}^2 \text{ s}^{-2} \text{ A}^{-1} = 1 \text{ V s} = 10^8 \text{ lines of flux})$;

(d) the magnetic flux density or induction $(\boldsymbol{B})$ may be defined by means of the equation for the force on a current element placed in a magnetic field $\boldsymbol{B}$, is measured in tesla units where a tesla unit can be identified with $\text{kg s}^{-2} \text{ A}^{-1}$ $(= \text{V s m}^{-2} = \text{Wb m}^{-2})$.

(e) the magnetic moment (μ) is the couple experienced by a magnet placed at right angles to a uniform field with unit flux density and is measured in ampere metre^2;

(f) the electrical intensity (E) at a point in an electric field is the force experienced by a unit charge of one coulomb situated at that point and is measured in volt metre^{-1} $(= \text{newton coulomb}^{-1})$;

(g) the electric potential difference is the volt (V) and may be defined as $\text{J A}^{-1}\text{ s}^{-1}$.

As a result of common usage certain multiples and sub-multiples of length are treated as derived units although none is recognized as being strictly an SI unit, e.g. micron $(\mu) = 10^{-6}$ m, ångström $(\text{Å}) = 10^{-10}$ m, and fermi (fm) =

10^{-15} m. The ångström unit is particularly useful since internuclear distances are of that order, and the literature of physical chemistry abounds with its use. Consequently, it will be retained here. The conversion to metres is a straight-forward one since 1 Å is now precisely 10^{-10} m based on the Kr_{36}^{86} standard line (see Section 1.1); this was not the case before 1960 when the previous standard of wavelength was the red line of cadmium which made 1 Å = 1.0000002 × 10^{-10} m.

(C) *Supplementary units.* At present these are the radian (rad) which is the unit for a plane angle while the term steradian (sr) is employed for solid angles.

Multiples and submultiples of a particular unit. These are obtained by employing such prefixes as those listed in Table 1.1 and have been agreed upon internationally for use with SI units.

Table 1.1 Multiplication factors used in conjunction with SI units

Prefix	Multiplication factor	Symbol
giga	10^9	G
mega	10^6	M
kilo	10^3	k
centi	10^{-2}	c
milli	10^{-3}	m
micro	10^{-6}	μ
nano	10^{-9}	n
pico	10^{-12}	p

The application of such prefixes may be seen as follows. It is common practice in spectroscopy to write a wavelength of 10×10^{-8} cm as 10 Å. However, in SI units this would be 10^{-9} m, and in terms of the appropriate prefix this would be written as 1 nm. In SI units the first line of the Balmer series would be 656.28 nm whereas it has normally been quoted as 6562.8 Å. In spectroscopic studies of dissociation energies and barriers to internal rotation the energies are best expressed in terms of kJ (i.e. units of 1000 × the basic unit). The use of prefixes allows some flexibility in the expression of results. For example, the van der Walls radius of the chlorine atom is 1.80×10^{-8} cm, and this in SI units could be written as 1.80×10^{-10} m or as 180 pm; the former procedure would be preferable for use in calculations whereas the latter would seem more appropriate when results were being tabulated.

To summarize, the main changes which have been brought about by the adoption of SI units are:

(a) to replace the centimetre and gram by the metre and kilogram respectively, although the first pair are still retained as sub-multiples;

(b) to make the newton (kg m s^{-2}) the unit of force and the joule (kg m^2 s^{-2}) the unit of energy;

(c) to replace the electrostatic and electromagnetic units by SI electrical units.

A comparison of some SI and previously employed units is made in Table 1.2 for some quantities of spectroscopic interest, and includes some of the derived units.

Table 1.2 Comparison for some quantities of spectroscopic interest of SI and the previously employed units

Quantity	Symbol	SI unit	Previously employed units
Dissociation energy	D	joule mol^{-1}	cal mole^{-1}
Energy	E	joule (J)	erg
Frequency	ν	hertz (Hz)	s^{-1}
Moment of inertia	I	kg m^2	g cm^2
Force constant (stretching)	f	N m^{-1}	dyne cm^{-1}
Force constant (bending)	f	joules rad^{-1}	ergs rad^{-1}

A number of physical constants are of frequent use in spectroscopic work, and it is useful to have the values of these quantities in cgs and SI units. Some such data are listed in Table 1.3

Most textbooks on spectroscopy have not yet adopted SI units, and, in addition, the bulk of previous literature has not employed these units. It often becomes necessary to make use of conversion factors from data such as:

1 e.v./molecule = 8065.7 cm^{-1} = 23.061 kcal mol^{-1} = 1.6021 x 10^{-12} ergs/molecule

$$= 1.6021 \times 10^{-19} \text{ joules/molecule}$$

1.1.2 Frequency, wavenumber, and wavelength

In spectroscopic studies which take place in the absence of an applied electric or magnetic field the absorption or emission of radiation is recorded at particular frequencies (ν), wavelengths (λ), or wavenumbers (σ), where:

$$\nu \times \lambda = \text{velocity of light} = c \tag{1.8}$$

where the values of λ and c depend slightly on whether the measurements are made in vacuum or in air; the frequency, however, in each case, is given by:

$$\nu = \frac{c_{air}}{\lambda_{air}} = \frac{c_{vac}}{\lambda_{vac}} \tag{1.9}$$

Table 1.3 The values of some physical constants expressed in SI and cgs units

Quantity	Symbol	SI Units	cgs Units
Speed of light in vacuo	c	$2.997924580 \times 10^8 \, ms^{-1}$	$2.997924580 \times 10^{10} \, cm \, s^{-1}$
Electric charge of a proton	e	$1.6021892 \times 10^{-19} \, C$	$1.6021892 \times 10^{-20} \, e.m.u.$
Planck's constant	h	$6.626176 \times 10^{-34} \, J \, s$	$6.626176 \times 10^{-27} \, ergs$
Mass of electron at rest	m_e	$9.109534 \times 10^{-31} \, kg$	$9.109534 \times 10^{-28} \, g$
Mass of proton at rest	m_p	$1.6726485 \times 10^{-27} \, kg$	$1.6726485 \times 10^{-24} \, g$
Avogadro constant	N_a	$6.022045 \times 10^{23} \, mol^{-1}$	$6.022045 \times 10^{23} \, mol^{-1}$
Faraday constant	F	$9.648456 \times 10^4 \, C \, mol^{-1}$	$9.648456 \times 10^3 \, e.m.u. \, mol^{-1}$
Gas constant $(=N_A k)$	R	$8.31441 \, JK^{-1} \, mol^{-1}$	$8.31441 \times 10^7 \, erg \, K^{-1} \, mol^{-1}$
Boltzmann constant	k	$1.380662 \times 10^{-23} \, KJ^{-1}$	$1.380662 \times 10^{-16} \, erg \, K^{-1}$
Rydberg constant (nucleus infinite mass)	R_∞	$1.097373177 \times 10^7 \, m^{-1}$	$1.0973731 \times 10^5 \, cm^{-1}$
Bohr magneton $(=e\hbar/2m_e)$	μ_B	$9.274078 \times 10^{-2} \, J \, T^{-1}$	$9.050824 \times 10^{-24} \, e.m.u.$
Nuclear magneton $(=e\hbar/2m_p)$	μ_N	$5.050824 \times 10^{-27} \, J \, T^{-1}$	$5.050824 \times 10^{-24} \, e.m.u.$

The spectroscopist measures the wavelength in air, and this may be corrected to λ_{vac} by $\lambda_{vac} = n\lambda_{air}$, where n is the refractive index of air at that particular wavelength. With a view to the subsequent analysis of the spectral lines it is often more satisfactory to employ frequency than wavelength. In addition, frequency is more fundamental than the wavelength, since the frequency of monochromatic light does not alter in different media. The wavelength, however, does alter as is obvious from Equation (1.9). A value proportional to frequency may be chosen ($1/\lambda_{vac}$); this is known as the *wavenumber* (σ). It has been the practice to use different units in different spectral regions. For example, Ångström units (Å) are normally used for atomic spectra, where:

$$1 \, \text{Å} = 10^{-8} \, cm = 10^{-10} \, m = 10 \, nm$$

Other procedures employed for recording wavelength in other regions are the micron (μm), and the nanometre (nm). Hence:

$$1 \, \mu m = 10^{-4} \, cm = 10^{-6} \, m$$

$$1 \, nm = 10^{-7} \, cm = 10^{-9} \, m = 1 \, nm$$

$$1 \, \mu m = 10\,000 \, \text{Å} = 1000 \, nm = 10^{-6} \, m = 1 \, \mu m$$

Here the μ employed as a prefix has the value of 10^{-6} and the symbol n has the value of 10^{-9}.

In the infrared region wavelengths are recorded in microns so that the results may be expressed in terms of simple numbers between ~ 1 to $1000\,\mu m$.

No one spectrometer can separate all the radiation ranging in wavelength from $\sim 10\,\text{Å}$ to $\sim 10^4$ cm into a spectrum; instead, several different types of instruments have to be used. The type of spectral technique employed sometimes lends its name to a certain region of electromagnetic radiation; for example, the region between $\sim 10\,\text{Å}$ and $2000\,\text{Å}$ is known as the *vacuum ultraviolet*. Since the oxygen in the air and many other gases absorb in this region and would interfere with the spectrum, a vacuum spectrograph is used; that is why this region is called the vacuum ultraviolet.

The various spectral regions are listed in Fig. 1.8. By different types of spectrographs it is possible to study the entire region from long wavelength X-radiation to radio wavelengths, although in the case of atomic spectra we shall

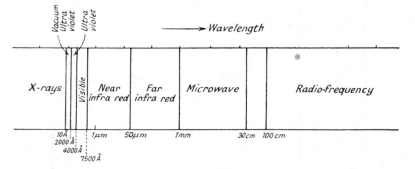

Fig. 1.8 Spectroscopic regions of the electromagnetic spectrum

be mainly concerned with the ultraviolet and visible, and occasionally the vacuum ultraviolet and the infrared regions. However, this division into regions in terms of a particular type of energy change in the atoms or molecules is to some extent arbitrary.

1.2   THE HYDROGEN ATOM AND THE THREE QUANTUM NUMBERS (n, l and m_l)

1.2.1   Introduction

As might be expected, the hydrogen atom yields the simplest analysable spectrum. It consists of five series of lines which can be represented by the equation:

$$\sigma = R_H (1/n_2^2 - 1/n_1^2) \qquad (1.10)$$

where R_H is its Rydberg constant. When $n_2 = 1$, then $n_1 = 2, 3, 4 \ldots$, and each of the lines calculated from Equation (1.10) corresponds to the wavenumber in the Lyman series. For the case where $n_2 = 2$, lines in the Balmer series are obtained corresponding to n values of $3, 4, 5 \ldots$. In general, for these five series, if the appropriate values of n_1 and n_2 are inserted in Equation (1.10),

11

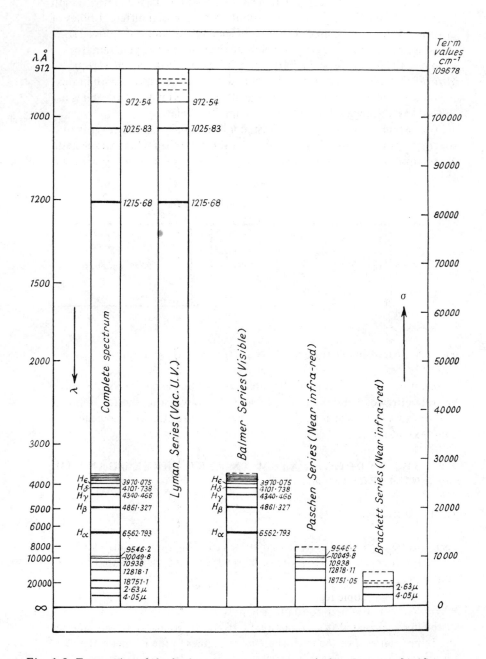

Fig. 1.9 Four series of the hydrogen atom spectrum. (After Grotrian [1.3]. Courtesy of Springer-Verlag, Heidelberg).

the corresponding wavenumbers may be calculated, and these agree almost exactly with the measured values. In Fig. 1.9 four of the five series of the hydrogen atom are given and identified with the Lyman, Balmer, Paschen, and Brackett series. In each case the series limit is represented by a broken line; the wavenumber of this line is obtained by placing n_1 equal to a very large value, and then σ approaches a limit given by R/n_2^2 where the series has approached a limit beyond which there are no lines.

The emission transitions for the five line series of the hydrogen atom as based on the Bohr circular orbit model are given in Fig. 1.10.

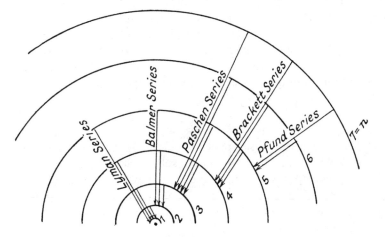

Fig. 1.10 Emission transitions for the five series of the hydrogen atom

The n_1 and n_2 values were later identified by Bohr in terms of values of the principal number and spectral changes in the hydrogen atom of the orbit of the electron in the hydrogen atom. Thus, for the hydrogen atom Bohr visualized an electron of mass m moving with velocity σ describing a circular orbit of radius r around charge $+e$. The orbital angular momentum ($m\sigma r$) was taken to be quantised and could have values governed by the equation:

$$m\sigma r = nh/2\pi \tag{1.11}$$

where $n = 1, 2, 3, \ldots$. The spectral change between the energy levels E' and E'' is then governed by the equation:

$$h\nu = E' - E'' = R_H c\,(1/n_2^2 - 1/n_1^2) \tag{1.12}$$

Equations (1.11) and (1.12) were key ones in the Bohr theory (1913) of the hydrogen atom which then appeared to account for its observed electronic spectrum. The theory was built up by (a) the application of Planck's quantum theory, and (b) the use of Newton's laws of motion.

Later it was appreciated that it is inadequate to apply the laws of classical mechanics to systems on an atomic scale. New concepts were developed to

13

express the behaviour of particles on such a scale, and the approach became that of quantum mechanics.

We shall now list a few key features in this approach, although ultimately our aim is to indicate how quantum mechanics leads to the emergence of quantum numbers. It then becomes possible to characterize spectral changes in terms of quantum number changes. In addition, quantum mechanics often permits only certain quantum number changes and helps to account for the fact that certain lines appear and others are absent.

Key features in the quantum mechanical approach are:

(1) The dual nature of electrons and photons which may be regarded as possessing both wave and particle properties. For example, the phenomena of diffraction and interference can best be explained in terms of wave theory whereas the photoelectric effect is best accounted for in terms of particle theory. A key equation is that suggested by de Broglie in 1924 which relates the momentum (p) of the particle to the wave property (λ):

$$p = h/\lambda \tag{1.13}$$

where h is Planck's constant.

(2) Heisenberg in 1927 made a most important postulate in quantum mechanics which indicated that it is not possible to determine both conjugate variables (e.g. energy, position, time, and momentum) simultaneously. For example, as applied to energy and time there must be uncertainty in the energy (ΔE) and in the time (Δt), and the two are related by

$$\Delta E \times \Delta t \approx h/2\pi \tag{1.14}$$

(3) Schrödinger in 1926 developed an equation of motion which incorporated both particle and wave properties. The wave property of a particle of mass m which has a total energy E and a potential energy V as expressed by the time-independent Schrödinger equation is:

$$\nabla^2 \psi + \frac{8\pi^2 m}{h^2}(E - V)\psi = 0 \tag{1.15}$$

where ψ is the wave function which describes the wave property of the particle, and ∇^2 is the Laplacian operator:

$$\nabla^2 = \frac{\partial^2}{\partial x^2} + \frac{\partial^2}{\partial y^2} + \frac{\partial^2}{\partial z^2} \tag{1.16}$$

On rearrangement of Equation (1.15) we get:

$$\left(-\frac{h^2}{8\pi^2 m}\nabla^2 + V\right)\psi = E\psi \tag{1.17}$$

or

$$\mathcal{H}\psi = E\psi \tag{1.18}$$

where

$$\mathcal{H} = -\frac{h^2}{8\pi^2 m}\nabla^2 + V$$

$\mathcal{H}$ is known as the Hamiltonian operator and will be employed in later chapters.

Equation (1.15) is applicable to a variety of spectroscopic problems. Those of particular interest are:

(1) Solution of the equation when applied to the hydrogen atom. From one such study the quantum numbers n (the principal quantum number), l (the azimuthal quantum number), and m_l (magnetic quantum number) emerge.

(2) In the study of type (1) a necessary condition from quantum mechanics for an electronic transition is that l may change by only $+1$ or -1. This gives what is known as the selection rule:

$$\Delta l = \pm 1$$

which limits the number of permissible transitions.

(3) An equation for the rotational energy of the molecule in terms of a rotational quantum number (J). Changes in J are limited by the selection rule:

$$\Delta J = \pm 1$$

(4) An equation for the vibrational energy of a molecule in terms of the vibrational quantum number (v) and the corresponding selection rule. In addition, the criterion for a transition between two vibrational states may also be deduced.

(5) Equations for the intensity of a spectral line in both emission and absorption.

Cases such as (1) to (5) are vital in the proper appreciation of spectroscopic phenomena. Cases (3) to (5) will be considered in Vol. 2, and at present we shall limit our consideration to case (1).

1.2.2 Quantum-mechanical considerations of the hydrogen atom and the emergence of the quantum numbers n, l, and m_l

The behaviour of the electron in the hydrogen atom can be represented by means of the Schrödinger wave equation which in Cartesian coordinates is:

$$\frac{\partial^2 \psi}{\partial x^2} + \frac{\partial^2 \psi}{\partial y^2} + \frac{\partial^2 \psi}{\partial z^2} + \frac{8\pi^2 u}{h^2}(E - V)\psi = 0 \tag{1.19}$$

where E is the total energy, V the potential energy, and u the reduced mass of the hydrogen atom. $\psi \equiv \psi(x, y, z)$ and may be interpreted in terms of the amplitude of the wave. The function $|\psi|^2 dx.\,dy.\,dz$ measures the probability that the electron can be found in the small element of volume $dx.\,dy.\,dz$.

For the hydrogen atom with a nuclear charge $+e$ and an electron with charge $-e$ at a distance r from the nucleus the potential energy is:

$$V = -e^2/r \tag{1.20}$$

and this is the value of V to be substituted into Equation (1.19).

From a mathematical point of view the solution of Equation (1.19) is simplified if the equation is transformed from Cartesian coordinates into spherical polar coordinates. Figure 1.11 may be employed to illustrate the relationship between Cartesian (x, y, z) and polar coordinates (r, θ, ϕ). From the geometry of this figure we obtain:

$$r^2 = x^2 + y^2 + z^2$$
$$\tan \theta = \frac{\sqrt{(x^2 + y^2)}}{z} \tag{1.21}$$
$$\tan \phi = y/x$$

where

$$x = r \sin \theta \cos \phi$$
$$y = r \sin \theta \sin \phi$$
$$z = r \cos \theta$$

In polar coordinates Equation (1.19) becomes:

$$\frac{1}{r^2} \cdot \frac{\partial}{\partial r}\left(r^2 \frac{\partial \psi}{\partial r}\right) + \frac{1}{r^2 \sin^2 \theta} \cdot \frac{\partial^2 \psi}{\partial \phi^2} + \frac{1}{r^2 \sin\theta} \cdot \frac{\partial}{\partial \theta}\left(\sin \theta \cdot \frac{\partial \psi}{\partial \theta}\right)$$
$$+ \frac{8\pi^2 u}{h^2}\left(E + \frac{e^2}{r}\right)\psi = 0 \tag{1.22}$$

The only acceptable solutions for ψ are those where ψ and its first derivative are everywhere finite, single-valued, and continuous. Such solutions which satisfy these conditions are termed eigenfunctions.

Inspection of Equation (1.22) shows that it can be resolved into three differential equations by making the substitution:

$$\psi(r\theta\phi) = R(r) \cdot \Theta(\theta) \cdot \Phi(\phi) \tag{1.23}$$

where $R(r)$ is a function of r only, $\Theta(\theta)$ of θ only, and $\Phi(\phi)$ of ϕ only. Taking the first and second derivatives of ψ, substituting into Equation (1.22) and multiplying throughout by $(r^2 \sin^2 \theta)/R\Theta\Phi$ abbreviating $R(r)$ to R, $\Theta(\theta)$ to Θ, and $\Phi(\phi)$ to Φ, we obtain:

$$\frac{\sin^2 \theta}{R} \cdot \frac{\partial}{\partial r}\left(r^2 \frac{\partial R}{\partial r}\right) + \frac{\sin \theta}{\Theta} \frac{\partial}{\partial \theta}\left(\sin \theta \frac{\partial \Theta}{\partial \theta}\right) + \frac{1}{\Phi} \cdot \frac{\partial^2 \Phi}{\partial \phi^2}$$
$$+ r^2 \sin^2 \theta \frac{8\pi^2 u}{h^2}\left(E + \frac{e^2}{r}\right) = 0 \tag{1.24}$$

It now becomes possible to separate out three differential equations, the first

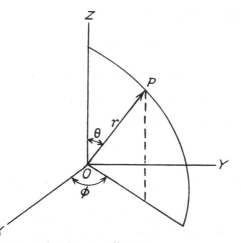

Fig. 1.11 Representation of polar coordinates

involving ϕ only, the second and third involving only θ and r respectively. This enables each of the three resulting equations to be solved independently.

(a) Differential equation involving ϕ only.

Since only the third term in Equation (1.24) involves ϕ, this ϕ term may be placed equal to a constant $(-m^2)$, because only this term is a function of the independent variable ϕ. Thus:

$$\frac{1}{\Phi} \cdot \frac{\partial^2 \Phi}{\partial \phi^2} = \text{constant} = -m^2 \qquad (1.25)$$

The reason for choosing $-m^2$ as the constant is that it facilitates the solution of this differential equation.

Equation (1.25) may be readily solved from a knowledge of differential equations.

Let

$$D = \partial/\partial\phi \qquad (1.26)$$

then

$$D^2 \Phi = -m^2 \Phi \qquad (1.27)$$

$$(D^2 + m^2)\Phi = 0 \qquad (1.28)$$

This equation may be factorized:

$$(D + im)(D - im)\Phi = 0 \qquad (1.29)$$

where $i = \sqrt{-1}$, and the general solution of the equation is:

$$\Phi = Ae^{im\phi} + Be^{-im\phi} \qquad (1.30)$$

The above particular solution may be obtained by taking $B = 0$ so that

$$\Phi = Ae^{im\phi} = A(\cos m\phi + i \sin m\phi) \qquad (1.31)$$

17

and then normalizing Φ, we get:

$$\int_0^{2\pi} \Phi(\phi) \times \Phi^*(\phi)d\phi = 1 \tag{1.32}$$

where $\Phi^*(\phi)$ is the complex conjugate of $\Phi(\phi)$, that is:

$$\Phi^*(\phi) = Ae^{-im\phi} = A(\cos m\phi - i \sin m\phi) \tag{1.33}$$

From substitution for $\Phi(\phi)$ and $\Phi^*(\phi)$:

$$A = 1/\sqrt{(2\pi)} \tag{1.34}$$

Hence,

$$\Phi = \frac{1}{\sqrt{(2\pi)}}(\cos m\phi + i \sin m\phi) \tag{1.35}$$

As Φ is a wave-function it has to be single-valued, and it follows therefore that m must be an integer. If m were not an integer, an increase in ϕ by 2π would not repeat the same value of Φ; that is, the wave-function would not be single-valued. Thus, the acceptable wave-function solutions of Equation (1.22) are those where:

$$m = 0, \pm1, \pm2, \ldots \tag{1.36}$$

m is known as the magnetic quantum number and may be identified with the m_l employed in atomic spectra.

(b) The solution of the $R(r)$ and $\Theta(\theta)$ equations.

On substitution of $-m^2$ for $(1/\Phi)(\partial^2 \Phi/\partial\phi^2)$ into Equation (1.24) we obtain:

$$\frac{\sin^2\theta}{R} \cdot \frac{\partial}{\partial r}\left(r^2 \frac{\partial R}{\partial r}\right) - m^2 + \frac{\sin \theta}{\Theta} \cdot \frac{\partial}{\partial\theta}\left(\sin \theta \frac{\partial\Theta}{\partial\theta}\right)$$
$$+ r^2 \sin^2\theta \frac{8\pi^2 u}{h^2}\left(E + \frac{e^2}{r}\right) = 0 \tag{1.37}$$

On dividing through by $\sin^2\theta$ and on rearrangement of the resulting equation we obtain:

$$\frac{1}{R} \cdot \frac{\partial}{\partial r}\left(r^2 \frac{\partial R}{\partial r}\right) + \frac{8\pi^2 ur^2}{h^2}\left(E + \frac{e^2}{r}\right) = \frac{m^2}{\sin^2\theta} - \frac{1}{\Theta \sin \theta} \cdot \frac{\partial}{\partial\theta}\left(\sin \theta \frac{\partial\Theta}{\partial\theta}\right) \tag{1.38}$$

In Equation (1.38) θ occurs only on the right-hand side of the expression and r on the left. This permits each side of the equation to be equated to the same constant β. When this is done, and after multiplication of the r terms by R/r^2 and the θ terms by Θ, we obtain:

$$\frac{1}{\sin \theta} \cdot \frac{\partial}{\partial\theta}\left(\sin \theta \frac{\partial\Theta}{\partial\theta}\right) - \frac{m^2 \Theta}{\sin^2\theta} + \beta\Theta = 0 \tag{1.39}$$

$$\frac{1}{r} \cdot \frac{\partial}{\partial r}\left(r^2 \frac{\partial R}{\partial r}\right) - \frac{\beta R}{r^2} + \frac{8\pi^2 u}{h^2}\left(E + \frac{e^2}{r}\right)R = 0 \tag{1.40}$$

By solution of Equation (1.39) the azimuthal quantum number is obtained while from the solution of Equation (1.40) the principal quantum number results.

The solution of these equations is too detailed to be considered here. From the solution of the equation involving Θ only, however, the following results emerge:

(i)
$$\beta = l(l+1) \tag{1.41}$$

where l is a new quantum number;

(ii) l may take the values:

$$l = 0, 1, 2, 3, \ldots \tag{1.42}$$

(iii)
$$l \geqslant m \tag{1.43}$$

This new quantum number l may be identified with the new azimuthal quantum number employed in atomic spectra. It is to be noted that l may take the value of zero; this is to be contrasted with the old azimuthal quantum number k where this value was not permissible.

From Equation (1.40) finite solutions exist only for:

$$E = \frac{-\mu_0^2 e^4 c^4 \mu}{8h^2 n^2} = \frac{-hcR_H}{n^2} \tag{1.44}$$

$\mu = m_e m_p/(m_e + m_p)$, and R_H is the Rydberg constant for H and μ_0 is the magnetic permeability of a vacuum and:

$$l \leqslant n - 1 \tag{1.45}$$

n is identified with the principal quantum number and may take the values:

$$n = 1, 2, 3, \ldots \tag{1.46}$$

while permitted values of the azimuthal quantum number are: $l = 0, 1, 2, \ldots$ $n - 1$. It is interesting to note that the same expression for E emerges from quantum-mechanics considerations as was derived from the Bohr theory.

Thus, the quantum mechanical treatment given here reveals that at least three quantum numbers are necessary to characterize the hydrogen atom. Magnetic quantum numbers are necessary to explain the spectrum of atoms in magnetic inductions. As we have seen so far, the spectrum of the hydrogen atom appears to be explicable in terms of changes of just the principal number. However, when the hydrogen lines are examined with a spectrograph of great resolving power, splitting of the lines may be detected. This is illustrated in Fig. 1.12 for the first two lines of the Balmer series where it will be observed that each line has split into two with a spacing of 0.14 Å and 0.08 Å respectively. Thus, the splitting of the spectral lines cannot be accounted for in terms of n alone, and, as we shall see later, the azimuthal quantum number l has to be taken into account. When, however, the spectra of the alkali metals are considered, they can be only partially explained by the quantum numbers n and l alone. A new quantum number j (the inner quantum number) has to be introduced. This new

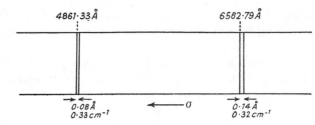

Fig. 1.12 First two doublets of the Balmer series

degree of freedom is necessary to take into account the spin of the electron where s is the electron spin quantum number. j is a quantum number compounded of l and s, and for the hydrogen atom:

$$j = l \pm s \tag{1.47}$$

The quantum numbers n, l, and j, and their appropriate selection rules can account for the spectra of the alkali metals. When a suitable magnetic induction is employed, a magnetic quantum number $\dot{m}_j$ has to be introduced where m_j may take the values:

$$j, j - 1, j - 2, \ldots, -j \tag{1.48}$$

On the basis of the four quantum numbers n, l, s, and m_j, it is possible to account for the spectra of the hydrogen atom and the alkali metals both in the presence and absence of a magnetic field. These are aspects which we shall explore later.

For the moment all that remains is to stress that the quantum-mechanical treatment given here has yielded three quantum numbers n, l, and m_l, although four quantum numbers are necessary to explain the spectra of, say, the alkali metals. However, this must not be regarded as a deficiency in quantum mechanics, but as an inadequacy in the way we have formulated our quantum-mechanical problem. In fact, our solution of the Schrödinger equation neglected to take into account the fact that the electron spins about its axis. When this is done there is no divergence between spectral observation and wave-mechanical theory.

The splitting of the Balmer lines may be explained by taking into account the quantum numbers n and l and the energy level diagram (Fig. 1.13). The letters s, p, d, and f are identified with the azimuthal quantum number (l) values of 0, 1, 2, and 3 respectively.

Further quantum mechanical treatment shows that the change in the azimuthal quantum number is limited to ± 1, i.e.

$$\Delta l = \pm 1 \tag{1.49}$$

and thus transitions are limited to those between adjacent columns in Fig. 1.13. In addition, l is permitted to take the values $0, 1, 2, 3, \ldots, (n - 1)$.

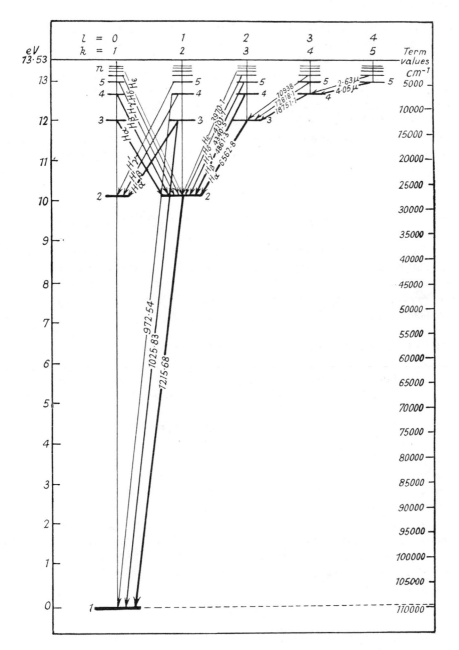

Fig. 1.13 Transitions between the energy levels in the hydrogen atom where the quantum numbers n, l, and k are taken into account. The right-hand subscripts $\alpha, \beta, \gamma, \ldots$ indicate respectively the first, second, third, $\ldots$ members of the Balmer series (After Grotrian [1.3]. Courtesy of Springer-Verlag, Heidelberg).

If these facts are employed in conjunction with the energy level diagram (Fig. 1.13), splitting of the Balmer lines is to be expected if the resolution of the spectrograph is adequate to detect it. In addition, this energy level diagram may be employed to account for the total electronic spectrum of the hydrogen atom. For example, the Lyman series involves transitions to the level where $n = 1$ and $l = 0$ from the levels where $l = 1$ and $n = 2, 3, 4, 5 \ldots$. Thus, in this case, no splitting of the spectral lines would be expected. In the diagram the value of the azimuthal quantum number (k) employed by the Bohr–Sommerfeld model is given. This is related to the aximuthal quantum number (l) from quantum mechanics by the equation:

$$l = k - 1 \tag{1.50}$$

In the treatment so far we have not needed to invoke a third quantum number to explain the observed spectrum of the hydrogen atom. However, this is no longer the case when we consider the spectra of the alkali metals even when the resolution is of a similar order to that employed for the hydrogen atom. In Section 1.4 we shall see that even in the absence of an external field it is necessary to employ three quantum numbers to account for the spectra of the alkali metals.

1.3   SPECTRA OF HYDROGEN-LIKE IONS

The spectra of the ions He^+, Li^{2+}, and Be^{3+} are in many ways similar to the spectrum of the hydrogen atom. The excitation energy required to produce the spectra of these ions is obviously much greater than that needed for the hydrogen atom. Emission spectra may be produced in electric discharges or arcs, whereas those of ions require the more extreme conditions of a spark spectra. The spectrum of the atom itself is indicated by placing the Roman numeral I after the symbol for the element; for example, the spectrum of the calcium atom could be referred to as Ca I. Similarly, the spectra of first and second ionized states of calcium would be referred to as Ca II and Ca III, respectively. If no Roman numeral is placed after the symbol for the element, it may be assumed that the spectrum referred to is that of the neutral element.

The general formula for the wavenumber of any line in the spectrum of the hydrogen atom or of lines in the spectra of hydrogen-like atoms (i.e. ions possessing one electron) is:

$$\sigma = RZ^2 \, (1/n_2^2 - 1/n_2^2) \tag{1.51}$$

The value of Z depends on the extent of ionization. For example, for He^+, Li^{2+}, and Be^{3+} Z would take the values $2, 3$, and 4 respectively. On substitution for Z and also for Rydberg's constant R (R varies slightly owing to the term containing μ, the reduced mass) the overall result is to displace the spectra of the hydrogen-like ions to shorter wavelengths. Equation (1.51) shows that if $n_1 = \infty$, then $\sigma_\infty = RZ^2/n_2^2$, and hence the wavenumber of the series limit σ

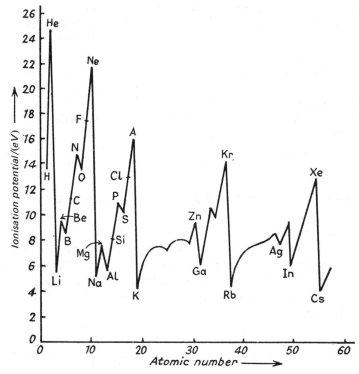

Fig. 1.14 Dependence of the ionization potential on the atomic number.

is dependent on Z^2, as is also the ionization potential. The ionization potential is the minimum amount of energy which must be supplied to an electron in the $n = 1$ level to remove it completely from the influence of the nucleus. Where $n_2 = 1$ and $n_1 = \infty$:

$$hc\sigma_\infty = hcRZ^2 \tag{1.52}$$

Thus, the ionization potential of He^+ is four times the value for the hydrogen atom, whilst that of Be^{3+} is sixteen times greater than this value.

The removal of an electron liberates unquantized kinetic energy, and it is observed experimentally that at a definite wavenumber σ_∞ for a given atom, the line structure gives way to a continuous region of electromagnetic radiation, the *continuum*, which cannot be resolved by any degree of dispersion. If a transition takes place from the ground state $n_2 = 1$ to an ionized state $n_1 = \infty$, the emission electron is removed from the ground state to a position of complete separation, where the attractive force between the nucleus and electron is insignificant. The lowest wavenumber σ_∞ at which the continum first begins gives a direct measurement of the energy required for this removal of the electron, and the energy $hc\sigma_\infty$ is termed the *ionization potential*. In Fig. 1.14 the ionization potential of the neutral atom is plotted against the atomic number of

23

each particular element. It will be noted that the alkali metals lie at the minima positions and the noble gases at the maxima. Between an alkali metal (e.g. Li) and the next higher noble gas (Ne) lie the ionization potentials of the elements whose atomic numbers fall in between these two. The position of the noble gases at the maxima illustrates the difficulty of removing an electron from a closed shell, while the position of the alkali metals at the minima indicates that the single external valence electron is easily ionized. This is in accordance with the highly electrovalent character of the alkali metals. The halogens, however, have almost closed shells, and it is much more difficult to remove an electron; in fact their chemical tendency is to gain an electron. In general, the position of an element in Fig. 1.14 may be reconciled with its chemical behaviour.

1.4   SPECTRA OF THE ALKALI METAL VAPOURS

The absorption spectra of the alkali metal vapours are in many ways similar to the absorption spectrum of the hydrogen atom. Each of these elements provides a number of term series, where each series is the difference between a fixed and a variable term and where the resulting differences give the wave-numbers of a series of lines. In the case of the lithium atom, four series of lines are obtained which are as follows:

$$\text{Principal series} \quad \sigma = 2s - np$$

$$\text{Sharp series} \quad \sigma = 2p - ns$$

$$\text{Diffuse series} \quad \sigma = 2p - nd$$

$$\text{Fundamental series} \; \sigma = 3d - nf$$

where, for example, lines in the Principal series are obtained from transitions from the p-levels to the 2s-level where the letter or number preceding s, p, d, or f indicates the value of the principal quantum number.

When the spectra of the alkali metals are examined in detail, it is found that in the principal and sharp series for a particular element each line is really a doublet, i.e. two lines. In fact, the sodium D-line, which is a member of the principal series, is actually two lines at wavelengths of 5896 and 5890 Å. This splitting may be traced back to the splitting of energy levels. To detect splitting of the lithium lines it is necessary to employ a spectrograph of high resolving power. Generally as one ascends the alkali metal series from Li, Na, K, Rb to Cs, the separation of the lines in the doublet increases. This is illustrated in the diagrammatic spectrogram (Fig. 1.15), which shows the fine structure in the principal series of the alkali metals. Table 1.4 gives the actual separation (in Å and cm^{-1} units) of the lines which constitute the first doublet of the principal series for the alkali metals (these are the lines on the extreme right in Fig. 1.15)

For all the alkali metals the two quantum numbers n and l, are inadequate to

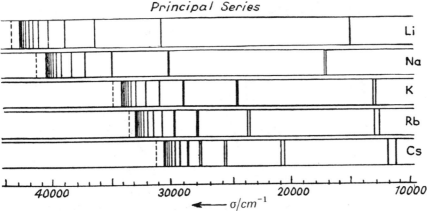

Fig. 1.15 Fine structures in the principal series of the alkali metals. The splitting of the lines in the doublets of sodium and potassium has been exaggerated. (Reproduced, by permission, from Introduction to Atomic Spectra by H. E. White [1.5]. Copyright 1934 McGraw-Hill Book Co. Inc., New York).

explain the spectra. For lithium, however, two quantum numbers almost suffice, although additional energy levels have to be considered to account for the 6 Å splitting of the sodium D-line, and similar doublets. The large splitting of the K, Rb, and Cs doublets clearly indicates that the spectra of these metals cannot be satisfactorily explained in terms of two quantum numbers. As we have seen, this led Sommerfeld to introduce a new quantum number j called the *inner quantum*

Table 1.4 The separation of the lines in the first doublet of the principal series for the alkali metals

Metal	Atomic No.	$\Delta\sigma$ (cm^{-1})	$\Delta\lambda$ (Å)
Lithium	3	0.34	0.15
Sodium	11	17	6
Potassium	19	58	34
Rubidium	37	238	147
Caesium	55	554	422

number; it will be seen that for monovalent elements j governs the total orbital angular momentum of the valence electron. Just as with l, a selection rule is supplied by quantum mechanics which limits the changes in j to 0 and ± 1. Thus, the selection rules are $\Delta l = \pm 1$ and $\Delta j = 0, \pm 1$; Δn may be any integral amount. The total angular momentum of a single valence electron is made up of two parts, the orbital angular momentum due to the electron describing an orbit and an additional angular momentum due to the spinning of the electron about an axis through its centre of mass. The value of the spin was first considered to be

25

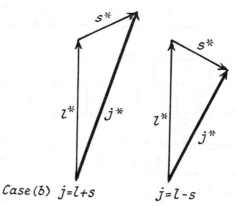

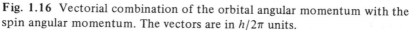

Fig. 1.16 Vectorial combination of the orbital angular momentum with the spin angular momentum. The vectors are in $h/2\pi$ units.

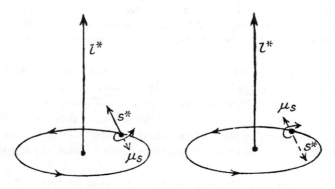

Fig. 1.17 Directions of the spin and orbital vectors with respect to the electron orbits on the s^* and l^* model. The magnetic moment μ_s due to the spinning electron is represented, and this acts in the opposite direction to the spin angular momentum.

half a quantum unit of angular momentum, i.e. $sh/2\pi = \frac{1}{2}\,h/2\pi$, where s is the electron spin quantum number. However, quantum mechanics has since shown that the exact expression for the angular momentum due to electron spin is $\sqrt{[s(s+1)]}\,h/2\pi$.†

In order to obtain the total angular momentum of the valence electron it is necessary to combine l and s vectorially (see Fig. 1.16). The procedure is that $\sqrt{[s(s+1)]}\,h/2\pi$ is vectorially added to $\sqrt{[l(l+1)]}\,h/2\pi$ and the resultant vector has the magnitude of $\sqrt{[j(j+1)]}\,h/2\pi$ where the 'j' under the square root sign is equal to $l \pm s$. For example, let $l = 1$ and $s = \frac{1}{2}$ (i.e. for a p-electron). This leads to:

† The shorthand symbols l^*, s^*, and j^* will frequently be written for $\sqrt{[l(l+1)]}$, $\sqrt{[s(s+1)]}$, $\sqrt{[j(j+1)]}$ respectively.

$$j^* \frac{h}{2\pi} = \frac{\sqrt{15}}{2} \cdot \frac{h}{2\pi} \text{ and } = \frac{\sqrt{3}}{2} \cdot \frac{h}{2\pi}$$

while the angle between the vectors:

$$l^* \frac{h}{2\pi} = \sqrt{2}\frac{h}{2\pi} \text{ and } s^* \frac{h}{2\pi} = \frac{\sqrt{3}}{2} \cdot \frac{h}{2\pi}$$

has to be such that the resultant $j^* h/2\pi$ can take its two values.

The two cases for combination of l^* and s^* are given in Fig. 1.17. To decide the direction of the vectors the following rule has to be applied. If in a loosely clenched righthand, with the thumb uppermost, the first finger is pointed in the direction of rotation, then the thumb indicates the direction of the vector representing that motion. Also, if the first finger is pointed in the direction of the current, the thumb indicates the direction of the magnetic field associated with that flow of current. If this rule is applied to: (i) the motion of the electron in the orbit, it indicates that the vector governing the orbital motion of the electron around the ellipse is perpendicular to the paper and pointing at the reader in both cases; (ii) the electron spinning on its axis in the two ways (i.e. spinning as the arrows actually indicate in Fig. 1.17 in the same or opposite sense to the motion of the electron in its orbit), then in one case it supports l and results in the value $l + s$, while in the other it opposes l giving the value $l - s$. The symbol μ_s employed in this figure is the magnetic moment resulting from the spinning electron, and this acts in the opposite direction to $s^* h/2\pi$ (see later).

There are then two values of j, i.e. $l + \frac{1}{2}$ and $l - \frac{1}{2}$, for each value of l. Each energy level (with the exception of $l = 0$) is now split into two levels with different energies; this enables the splitting of the lines in the spectra of the alkali metals to be accounted for when the selection rules $\Delta l = \pm 1$ and $\Delta j = 0, \pm 1$ are taken into account. The value that j takes for each l value is given in Table 1.5. It has been pointed out already that when $l = 0, 1, 2, 3$, the corresponding electrons have been termed s, p, d, and f electrons respectively. To characterize an energy state of an atom the capital letters S, P, D, and F are used, and in the case of a monovalent element they coincide with the small letters s, p, d, and f. This will be dealt with later in greater detail; S, P, D, and F will be used for a monovalent element whenever it is necessary to characterize the energy state of such an atom. Such symbols often have other information attached; for example, the corresponding j value is added as a right-hand subscript. Thus, when l is 1, then j would take the values 3/2 or 1/2, and the term symbol would then be $P_{3/2}$ and $P_{1/2}$. In addition, the value of the principal quantum number is frequently inserted as a number in front of the term.

The S term ($l = 0$) is a singlet level, while P, D, F, etc. are doublets. For example, it may be observed in Fig. 1.18 that the S and P levels for the sodium atoms are singlets and doublets respectively.

On comparison of the energy level diagram for hydrogen and sodium for a

Table 1.5 Possible *j*-values for a given term

Term	*l*-value	*j*-values		
S	0	1/2		
P	1	1/2	3/2	
D	2		3/2	5/2
F	3			5/2	7/2

fixed value of n and a variable value of l there is an appreciable variation in the energy level values of Na, whereas this is not perceptible on the H energy level diagram.

In Fig. 1.18 the term symbol has been inserted for each set of energy levels. Each term symbol has a left-hand superscript of 2, which is the value of what is known as the *multiplicity* of the term. It will be seen later that this is equal to:

$$2 \times \text{resultant electron spin quantum number} + 1$$

In addition, in Fig. 1.18 the splitting of the $^2P_{3/2}$ and $^2P_{1/2}$ levels is barely perceptible on the scale employed for the diagrams. The latter is even more true for the $^2D_{5/2}$ and $^2D_{3/2}$ levels where no attempt has been made to indicate the splittings, as also for the $^2F_{7/2}$ and $^2F_{5/2}$ levels.

The value of the principal quantum number is inserted as a number at the side of each energy level in Fig. 1.18. The emission lines are indicated by the lines possessing arrow tips, and it will be noted that each of these transitions obeys the selection rules $\Delta l = \pm 1$, and $\Delta j = 0, \pm 1$. There is no selection rule governing changes in n.

The doublets of the sharp series for sodium may be represented by:

$$\sigma_1 = hS_{1/2} - 3P_{1/2}; \sigma_2 = nS_{1/2} - 3P_{3/2} \qquad (1.53)$$

where, for a given value of n, σ_1 and σ_2 are the wavenumbers of the two lines in the doublet. The wavenumber separation $\Delta\sigma$ for the two lines in each doublet is:

$$\Delta\sigma = \sigma_2 - \sigma_1 = 3P_{3/2} - 3P_{1/2} \qquad (1.54)$$

and is clearly constant for each member of the sharp series (Fig. 1.19). The doublets of the principal series are given by:

$$\sigma_1 = nP_{1/2} - 3S_{1/2}; \sigma_2 = nP_{3/2} - 3S_{1/2} \qquad (1.55)$$

and, since the spacing between the $P_{1/2}$ and $P_{3/2}$ energy levels varies with the value of n, the separation of the lines in different doublets is not the same. The lines of the diffuse and fundamental series are made up of three components, and the wavenumbers are:

(a) for the diffuse series:

$$\sigma_1 = nD_{3/2} - 3P_{3/2}; \sigma_2 = nD_{3/2} - 3P_{3/2}; \sigma_3 = nD_{5/2} - 3P_{3/2} \qquad (1.56)$$

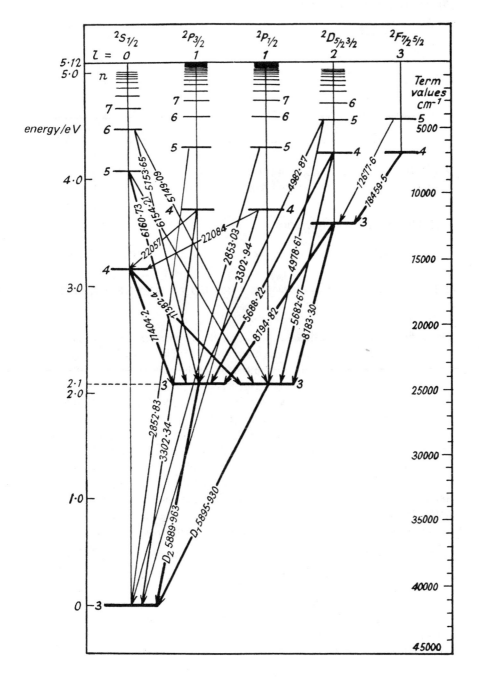

Fig. 1.18 Emission transitions of the sodium atom. The energy levels take into account the electron spin (After Grotrian [1.3]. Courtesy of Springer-Verlag, Heidelberg).

29

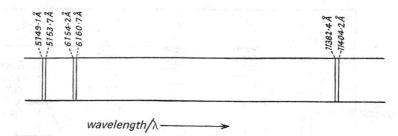

wavelength/λ ⟶

Fig. 1.19 First three doublets in the sharp series for sodium. The wavenumber separation of the lines in the doublets is approximately constant at 18 cm^{-1}

(b) for the fundamental series:

$$\sigma_1 = n\text{F}_{5/2} - 3\text{D}_{3/2}; \sigma_2 = n\text{F}_{5/2} - 3\text{D}_{5/2}; \sigma_3 = n\text{F}_{7/2} - 3\text{D}_{5/2} \quad (1.57)$$

1.5   ELEMENTS WITH MORE THAN ONE OUTER VALENCE ELECTRON

Having indicated the multiplet structure of the alkali metals, we can now consider elements which have more than one outer electron which may be excited and produce emission lines, e.g. He and the alkaline earths. The spectrum of an alkaline earth metal consists of a number of different line series. With a spectrograph of only moderate resolving power each of these lines in three of the series splits into triple lines (triplets) and in another series into six component lines. In the other series no line splitting can be observed (these lines are called singlets). In addition, the spectral lines of the elements He, Be, Mg, Zn, Cd, and Hg behave similarly, the degree of splitting increasing with atomic number. The most frequently used type of coupling is that proposed by Russell and Saunders, the so-called L, S coupling. The individual orbital momenta combine vectorially to give a resultant L which is integral. For atoms with two valence electrons with orbital angular momentum quantum numbers l_1 and l_2 respectively:

$$L = l_1 + l_2, \ l_1 + l_2 - 1, \ l_1 + l_2 - 2, \ldots, |l_1 - l_2| \qquad (1.58)$$

The resultant orbital angular momentum for each of these L values then becomes $\sqrt{[L(L+1)]}\,h/2\pi$. The electrons in the closed shells do not contribute towards the resultant L so only the valence electrons need be considered. If l_1 equals 2, and l_2 equals 1, then L has the values 3, 2, 1. The vector combination in $l_1^* h/2\pi$ and $l_2^* h/2\pi$ to give $L^* h/2\pi$ is illustrated diagrammatically in Fig. 1.20.

The changes in L are governed by a selection rule, $\Delta L = 0, \pm 1$; in addition, the selection rule $\Delta l = \pm 1$ still applies where the latter selection rule applies to the electron undergoing the quantum change. Different term symbols of the atom (for example S, P, D, and F) with more than one emission electron are characterized by different values of the resultant orbital angular momentum quantum number L of the electrons 0, 1, 2, 3, respectively, as are atoms with a single valence electron since then $L = l$. For an atom with more than one valence electron, the L-values and the corresponding term symbols can be readily evaluated. A few examples are given in Table 1.6.

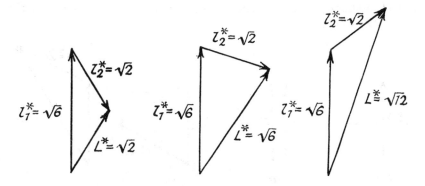

Fig. 1.20 Vector addition of $l_1^* h/2\pi$ and $l_2^* h/2\pi$ where $l_1 = 2$ and $l_2 = 1$. The vectors are in $h/2\pi$ units.

Table 1.6 Term symbols for some L-values for atoms with more than one valence electron[†]

Electrons	L-values	Corresponding term symbols
s + p	1	P
p + p	0, 1, 2	S, P, D
p + d	1, 2, 3	P, D, F
d + d	0, 1, 2, 3, 4	S, P, D, F, G

[†] The term symbols given here apply to non-equivalent electrons (see p. 83).

The spins of the emission electrons also combine vectorially to give a resultant electron spin quantum number, S, for the atom. In the case of two valence electrons the spins may be both parallel, i.e. ↑↑, or one may be parallel and the other antiparallel, i.e. ↓↑. The resultant S will then be either $+1/2 + 1/2 = 1$ or $+1/2 - 1/2 = 0$, and the resultant electron spin angular momenta would be $\sqrt{2}h/2\pi$ and 0 respectively. These points are illustrated in the vector diagrams in Fig. 1.21. When there are three unpaired electrons, S may take the values of 3/2 and 1/2 corresponding to the electron spin orientations of the types ↑↑↑ and ↓↑↑. Further, just as l and s combine to give j for a single electron, so the resultant orbital and spin angular momenta for atoms with more than one emission electron may be coupled by vectorial

31

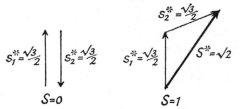

Fig. 1.21 Addition of the spin vectors $s^*h/2\pi$ and $s_2^*h/2\pi$ where $s_1 = s_2 = 1/2$. The vectors are in $h/2\pi$ units.

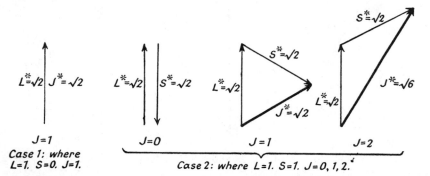

Fig. 1.22 Vector addition of $L^*h/2\pi$ and $S^*h/2\pi$: case (1) where $L = 1$, $S = 0$, and $J = 1$; case (2) where $L = 1$, $S = 1$, and $J = 0, 1, 2$. The vectors are in $h/2\pi$ units.

combination to give a series of J-values, according to:

$$J = L + S, \ L + S - 1, \ L + S - 2, \ldots, |L - S| \qquad (1.59)$$

Thus, there will be a total of $(2S + 1)$ values of J for every L value when L is greater than S, and $(2L + 1)$ values of J when S is greater than L. The total resultant angular momentum of the atom (excluding nuclear spin) is then given to be $\sqrt{[J(J + 1)]}\,h/2\pi$.

In this type of L,S coupling (Russell–Saunders coupling) it is implicit that the interaction between the spin momenta of the electrons is large (as is also the interaction of the orbital momenta of the electrons with one another) compared with the interaction of the electron spin and orbital motion. To illustrate the coupling of $L^*h/2\pi$ and $S^*h/2\pi$ an example will be considered and the $J^*h/2\pi$ values interpreted in terms of vector diagrams. For an atom with two valence electrons the resultant spin values are 0 and 1, and when $L = 1$, the corresponding J values are:

case (i) $J = 1$, when $S = 0$

case (ii) $J = 0, 1$, and 2, when $S = 1$

The corresponding vector diagrams are given in Fig. 1.22. For both cases (i) and (ii) the number of possible J values is $(2S + 1)$; this gives the number of levels into which a term of given L may split. It should be noted that when $L = 0$ and $S = 1$, then the only J value is 1, even though the multiplicity is 3;[†] this is in harmony with spectroscopic observation, where all S-levels are singlet levels. Table 1.7 gives the multiplicity to be associated with a given resultant spin S.

Table 1.7 Relation of *multiplicity* to resultant spin

Resultant spin (S)	Multiplicity (2S + 1)
0	Singlets
1/2	Doublets
1	Triplets
3/2	Quartets

Elements in Group II of the periodic table which have two valence electrons consist of singlet and triplet energy states corresponding to the one value and three values of J, respectively, for each L value. Elements in Group III have three valence electrons. S, then, may have the values 1/2 and 3/2, so that if L is greater than S, for each L value J has the values $L + 1/2$ and $L - 1/2$, and $L + 3/2, L + 1/2, L - 1/2$, and $L - 3/2$ respectively. Hence, there will be two sets of energy levels, one of doublets and one of quartets. These facts are in agreement with spectral observation and are summed up by the *alternation law* which states: *the terms of atoms or ions with an even number of valence electrons have odd multiplicities, and atoms or ions with an odd number of electrons have even multiplicities*. Even and odd multiplicities, therefore, alternate in successive groups of the periodic table, a fact which was known before it could be explained. The feasible multiplicities for a particular group in the periodic table are summed up in Table 1.8.

Table 1.8 Relation between multiplicities and group in the periodic table

Group in periodic table or number of electrons in outer shell	Possible multiplicities
1	2
2	1, 3
3	2, 4
4	1, 3, 5
5	2, 4, 6
6	1, 3, 5, 7
7	2, 4, 6, 8
8	1, 3, 5, 7, 9

[†] In a suitable magnetic field, however, the threefold degeneracy of this triplet level is removed, and this ^3S level splits into three component levels.

It should be noted that s, l, and j are employed for the quantum numbers of individual electrons, whereas S, L, and J are used to characterize the terms for atoms with one or more valence electrons.

It is now possible to appreciate the symbols which are the accepted method of identifying the various terms. As explained above, the letters S, P, D, F, G, H, ... are used when the value of L is 0, 1, 2, 3, 4, 5, ... respectively.[†] The letter is preceded by a superscript representing the multiplicity of the term and followed by a subscript giving the J value. In order to identify a particular term exactly, the value of the principal quantum number precedes the symbol.

In (L, S), i.e. Russell-Saunders, coupling, it was seen that there was such a strong interaction between the orbital angular momenta $l_1^* h/2\pi$, $l_2^* h/2\pi$, ... of the individual electrons, and also the spin orbital angular momenta $s_1^* h/2\pi$, $s_2^* h/2\pi$, ..., that they formed the resultants $L^* h/2\pi$ and $S^* h/2\pi$ respectively. For heavy atoms, however, the interaction between $l_1^* h/2\pi$ and the corresponding $s_1^* h/2\pi$ is stronger than the interaction amongst the orbital angular momenta for each other. The consequence is that $l_1^* h/2\pi$ and $s_1^* h/2\pi$ are coupled to give a resultant $j_1^* h/2\pi$, that is the $l^* h/2\pi$ and $s^* h/2\pi$ are coupled for each electron individually to form a resultant $j^* h/2\pi$ which is the total angular momentum for the particular electron. These $j_1^* h/2\pi$, $j_2^* h/2\pi$, ... then couple with one another to give a final resultant total angular momentum $J^* h/2\pi$ for the atom. This type of coupling is known as (j, j). A large number of energy levels may be classified in terms of (L, S) or (j, j) coupling or even some coupling scheme which falls between these extreme types.

To illustrate (j, j) coupling the electron configuration pd will now be considered where the electron spin quantum number has the value 1/2. For:

(i) the p electron, $l_1 = 1$ and $s_1 = 1/2$, and hence $j_1 = 3/2$ or $1/2$;

(ii) the d electron, $l_2 = 2$ and $s_2 = 1/2$, $j_2 = 5/2$ or $3/2$

The possible ways of combining these four j values to form the resultant J are:

(i) $j_1 = 1/2$ and $j_2 = 3/2$, $J = 1$ and 2

(ii) $j_1 = 1/2$ and $j_2 = 5/2$, $J = 2$ and 3

(iii) $j_1 = 3/2$ and $j_2 = 3/2$, $J = 0, 1, 2,$ and 3

(iv) $j_1 = 3/2$ and $j_2 = 5/2$, $J = 1, 2, 3,$ and 4

It should be noted that this (j, j) coupling scheme leads to the same number of terms as would have been the case for (L, S) coupling, and, in addition, it also gives the same J values.

For (j, j) coupling L and S remain undefined, and the selection rules $\Delta S = 0$

[†] The symbols S, P, D and F may be related to the sharp, principal, diffuse, and fundamental series, respectively. The letters G, H, ... then follow on in alphabetical order, omitting J.

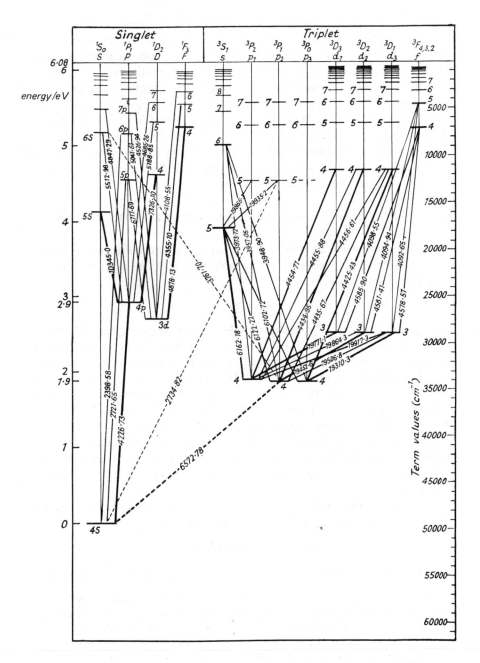

Fig. 1.23 Emission transitions of the calcium atom. (After Grotrian [1.3].
Courtesy of Springer-Verlag, Heidelberg).

and $\Delta L = 0, \pm 1$ no longer apply; the selection rules are $\Delta J = 0, \pm 1$ and $\Delta j = 0, \pm 1$.

Figure 1.23 is the energy level diagram for Ca where examples of the permitted transitions are indicated by the full lines connecting the different energy levels. For transitions where two electrons take part the selection rules are $\Delta l_1 = \pm 1$ and $\Delta l_2 = 0, \pm 2$, but if only one electron is involved, then the value of l may change by unity. For L,S coupling the selection rules are $\Delta J = 0, \pm 1$ and $\Delta L = 0, \pm 1$. In addition, transitions between the levels where $J' = 0$ and $J'' = 0$ are forbidden. Five series are observed for transitions between the singlet levels, and the terms for these are listed in the singlet part of the energy level diagram. It will be noted that no splitting of the spectral lines would be expected for transitions between the singlet levels. This contrasts strongly with the transitions between the triplet levels.

1.6   FORBIDDEN TRANSITIONS AND SELECTION RULES

Generally, in order that an atom may interact with electromagnetic radiation, it should possess a dipole moment. The variable electric dipole moment of an atom is a vector quantity M defined by:

$$M = \sum_i e_i r_i$$

where r_i is the length of the vector from the origin to the position of the ith particle which bears the charge e_i. It is by means of this variable electric dipole moment that the electric vector (of magnitude E) of an electromagnetic wave may, in general, interact with the atom. The interaction energy is $M \times E$, and if this value is inserted into the wave equation, then the probability of a transition between two states n and m may be evaluated. If from this calculation it emerges that the transition probability is zero, the transition is said to be forbidden as regards an electric dipole type of transition.[†] If in practice the transition took place, the spectral line might be classed as a forbidden line. This is certainly true as regards its classification with respect to an electric dipole interaction.

It is a most difficult procedure to evaluate the transition probability. In the wave-mechanical approach, selection rules such as:

$$\Delta L = 0, \pm 1 \text{ and } \Delta J = 0, \pm 1$$

emerge for electric dipole radiation, and these have been frequently employed

---

[†] If the transition moment R^{nm} differs from zero for a transition between the two states n and m, a certain probability exists for absorption or emission of radiation between these states. If, however, $R^{nm} = 0$, the transition is regarded as forbidden to the electric dipole mechanism. The permanent electric dipole moment may be evaluated by letting $n = m$ in the R^{nm} equation, and for an atom this, of course, turns out to be zero, whereas for a molecule this may either be positive or zero.

to decide whether a line is forbidden or permitted. Thus, a transition which disobeyed one of these selection rules might be classed as forbidden to electric dipole radiation. However, the majority of selection rules are not rigorous and, in fact, may be true only to a first approximation. For example, the selection rule $\Delta S = 0$ only rigorously holds for light atoms. As the atomic number increases the coupling gradually changes from the (L, S) type in light atoms to that of the (j, j) for heavier atoms. For example, carbon and silicon have practically pure (L, S) coupling, whereas the heavier members of group IVB, germanium, tin, and lead, tend more and more to the (j, j) type as the atomic number increases. Pure (j, j) coupling, however, is relatively rare but is approached by some of the electronic states of some of the heavier elements such as lead and thallium. In the (j, j) type of coupling the quantum number corresponding to the total electron spin is no longer clearly defined, and thus it would be wrong to employ the selection rule $\Delta S = 0$. Furthermore, if the concept of forbidden lines is based on the definition that only transitions which obey the selection rules for electric dipole radiation are the permitted ones, this would automatically class transitions involving magnetic dipole or an electric quadrupole (see later) as forbidden transitions, even though they obey the selection rule for that type of radiation.

In general, ordinary spectral lines arise from the interaction of an electric dipole oscillation with the electric vector of an electromagnetic wave, and this, in fact, is the only type of transition which we have considered so far. However, radiation can also result from magnetic dipole and electric quadrupole interaction. We shall now attempt to attach meaning to the term quadrupole moment.

In the system of charges

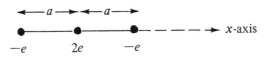

the sum of the products of the distances and charges (i.e. $\Sigma_i e_i x_i$ where the sum of the products lies along the x-axis where the charges are located) would equal zero. $\Sigma_i e_i x_i$ is the electric dipole moment for the system. However, although such an assembly of charges has zero electric dipole moment, it produces and electric field outside the system and is attributed to what is known as the *quadrupole moment*. This field is weaker than that of the electric dipole, since the potentials of the electric quadrupole and dipole fall off respectively as $1/r^3$ and $1/r^2$. The quadrupole may be regarded as coming from the whole

assembly of charges, while its action is governed by the quadrupole moment.[†]
In the case considered the quadrupole moment is $\Sigma_i e_i x_i^2$ where this expression,
unlike $e_i x_i$, is not zero.

The example chosen to illustrate the meaning of the term quadrupole
moment is not a very suitable one as regards its similarity to an atom itself. In
fact, there is no exact pictorial way of illustrating its meaning, and the most
satisfactory approach is a mathematical one. For this a review on quadrupole
moments by Buckingham [1.4] should be consulted.

If, for a particular transition, the probability of its occurrence by means of
its electric dipole moment turns out to be zero, as indicated by wave-mechanical
calculation, it is still feasible that the transition may take place by a magnetic
dipole or a quadrupole mechanism. However, the rate of emission or absorption
of energy by magnetic dipole or quadrupole interaction is usually negligibly
small compared with that of the electric dipole process. In fact, it can be shown
by calculation that magnetic dipole and electric quadrupole transition
probabilities are only 10^{-5} and 10^{-8} respectively of the electric dipole transition
probability. Thus, in general, magnetic dipole or quadrupole interaction leads
to spectral lines of low intensity, and it is for this reason that the lines are
described as forbidden. Hence, transitions which cannot occur by means of an
electric dipole mechanism are, in general, classed as forbidden.

It is only in exceptional cases that the spectroscopist is concerned with such
forbidden lines. Further consideration of spectral lines which occur by magnetic
dipole or a quadrupole moment mechanism is given in the chapter on
astrochemistry (Vol. 3).

By means of wave mechanics the selection rules for electric dipole, electric
quadrupole, and magnetic dipole radiation for atoms can be evaluated, and some
of these are listed in Table 1.9.

Table 1.9 Selection rules for electric dipole, electric
quadrupole, and magnetic dipole radiation of atoms

	Electric dipole	*Electric quadrupole*	*Magnetic dipole*
ΔJ	0, ±1	0, ±1, ±2	0, ±1
ΔL	0, ±1	0, ±1, ±2	0, ±1
ΔS	0	0	0

Electric and magnetic fields may result in a normal selection rule being
disobeyed. For example, in the presence of an electric field, spectral lines
due to electric dipole radiation may sometimes by observed which disobey the
selection rules $\Delta L = 0, \pm 1$ and $\Delta l = \pm 1$, where the latter selection rule applies
to the electron undergoing the quantum change.

[†] This quadrupole moment results from a system of charges. In Chapter 3 a
further type of quadrupole is considered which arises when the nuclear spin
quantum number is greater than 1/2, giving rise to what is known as a nuclear
quadrupole moment.

1.7 SPACE QUANTIZATION

So far our treatment of the motion of the electron in a hydrogen-like atom has been confined in the main to two dimensions, and these dimensions have been specified by reference to the quantum numbers, n and l. In the more general sense the motion of the electron is three-dimensional, and hence an additional quantum number is necessary to describe each energy state. This new quantum number does not in any way govern the size or shape of the orbit but determines its orientation with respect to some axis in space. In order to establish a fixed axis imagine the atom placed in a uniform magnetic induction field B^*. As a result of the field the orbital plane precesses about the field axis, and this precession of the orbital plane is analogous to the precession of a gyroscope in the gravitational field of the earth. The quantum conditions, which must be

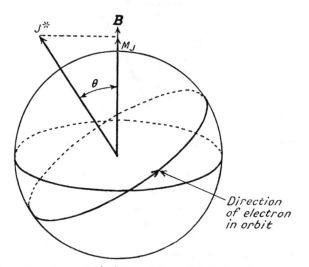

Fig. 1.24 The orientation of $J^*h/2\pi$ and the orbital plane in a magnetic field. The vectors are in $h/2\pi$ units (Reproduced, by permission, from Introduction to Atomic Spectra by H. E. White [1.5]. Copyright 1934 McGraw-Hill Book Co. Inc., New York)

introduced, imply that the orbit is space quantized, that is the orbital plane may take up only certain discrete orientations to the applied field. It can be shown that the angle θ (i.e. the angle between the field direction and the vector governing the orbital motion which is normal to the orbital plane) is governed by the values of the quantum numbers M_J and J, where M_J is termed the magnetic quantum number. These points are illustrated in Fig. 1.24 for one orientation of the orbital plane. Since the orbital plane precesses around B, then as $J^*h/2\pi$ is always perpendicular to this plane it follows that $J^*h/2\pi$ also precesses around B and translates a cone.

* Please see **Page vi** for an explanation of the use in this book of the symbol B

1.7.1   Magnetic moment and space quantization of angular momentum

When an emitting atom is brought into a magnetic field, each spectral line is split into a number of components. This splitting can be traced back to a splitting of energy levels in a magnetic field.

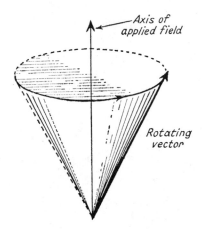

Fig. 1.25 Vector precessing in an applied field.

When a rotating vector is subjected to a torque, precession of the vector results around the direction of the applied torque. The case of the precession of a vector in the applied field is illustrated in Fig. 1.25.

A similar situation is encountered when an electron orbit is exposed to a magnetic field; the orbit has associated with it a magnetic moment owing to the motion of the electron. According to classical electrodynamics a current i conducted in a closed circuit of area A may be treated as a small magnet of magnetic moment $\mu = IA$, placed in the centre and standing perpendicular to the plane of the circuit. Since the rotation of an electron in its orbit is equivalent to the rotation of a negative point charge circumscribing the nucleus $\omega/2\pi$ times per second, the current $I = -(\omega/2\pi)e$. For a circular orbit radius r, the magnetic moment then becomes:

$$\mu = \left(-\frac{\omega e}{2\pi}\right)\pi r^2$$

that is

$$\mu = -\frac{\omega e r^2}{2}$$

The angular momentum (p) due to the rotation of an electron of mass m is:

$$p = mr^2\omega$$

whence
$$\mu = -\frac{e}{2m} \cdot p \qquad (1.60$$

For an atom it can be deduced from quantum mechanics that:
$$p = \sqrt{[J(J+1)]}\, h/2\pi$$
and hence
$$\mu = -\frac{e}{2m} \cdot \frac{h}{2\pi} \cdot \sqrt{[J(J+1)]} \qquad (1.61)$$

Thus, the quantization of the angular momentum p implies the quantization of the magnetic moment. When an atom is placed in a magnetic field, the total angular momentum vector $J^*h/2\pi$ is subject to a torque resulting from interaction between the atomic magnet and the external field B, and $J^*h/2\pi$ is made to precess about the direction of the field. The frequency of the precession is given by:
$$\nu_L = -\frac{B}{2\pi} \cdot \frac{\mu}{p} = \frac{Be}{4\pi m} \qquad (1.62)$$

ν_L is known as the Larmor frequency. It will be observed that the ν_L value is independent of the angle between $J^*h/2\pi$ and B. The quantum conditions imposed on this precessional motion are that the projection of the angular momentum $J^*h/2\pi$ on the field direction B will take only those values given by $M_Jh/2\pi$, where M_J takes the consecutive values differing from one another by unity from $M_J = +J$ to $M_J = -J$. Thus, M_J will be integral when J is integral, and half-integral when J is half-integral. From the relation
$$M_J = J, J-1, J-2 \ldots -J \qquad (1.63)$$
it can be seen that there are $(2J+1)$ values of M_J for a given value of J.

Fig. 1.26 illustrates the precession of the total angular momentum about the direction of the magnetic field; it also shows the two extreme components $M_Jh/2\pi$ of $J^*h/2\pi$ in the direction of B.

The possible orientations of $J^*h/2\pi$ to the direction of the magnetic field B are given in Fig. 1.27 for $J = \frac{1}{2}$ and also for $J = 1$, and these orientations are related to the splitting of the energy levels in the magnetic field.

If the rotation of the total angular momentum vector about the magnetic field is as in Fig. 1.28(a), then the electric current associated with this movement of negative charge may (by the normal physical convention) be regarded as flowing in the opposite direction in the electron orbit. If the convention explained on p. 27 is applied to this, then it follows that the magnetic vector associated with the flow of current will act in the opposite direction to the angular momentum vector, and thus the state of affairs is that shown in Fig. 1.28(c). Now the component of the magnetic moment in the direction of the field is $-\mu_J \cos\theta$, and the interaction energy is $\mu_J B \cos\theta$. If E_0 is the energy of the system in the field-free case, then when the field is applied the potential

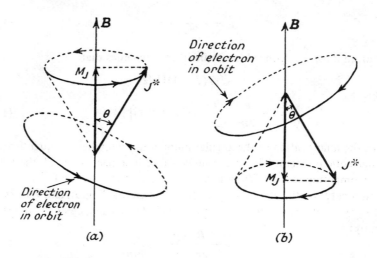

Fig. 1.26 Precession of the total angular momentum about an applied magnetic field (a) For one extreme orientation of $J^*h/2\pi$ and corresponding to $M_J = +J$. (b) For the other extreme orientation where $M_J = -J$. The vectors are in $h/2\pi$ units.

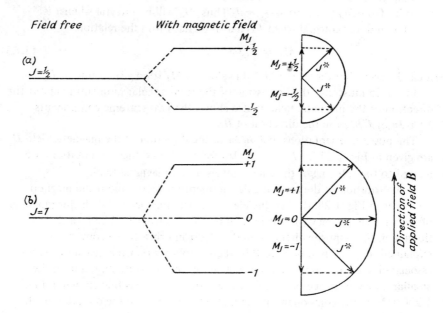

Fig. 1.27 Space quantization of $J^*h/2\pi$ and the splitting of the energy levels in a magnetic field. (a) For the case where $J = \frac{1}{2}$ and (b) where $J = 1$. The vectors are in $h/2\pi$ units.

energy of the system is given by:

$$E_{MJ} = E_0 + \mu_J B \cos\theta = E_0 - \mu_H B \qquad (1.64)$$

where μ_H is the magnetic moment component in the direction of the field.

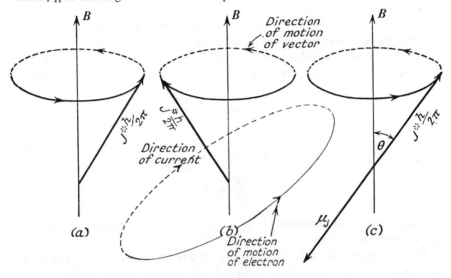

Fig. 1.28(a) Precession of the total angular momentum $J^*h/2\pi$ about an applied magnetic field B. (b) Direction of flow of electric current. (c) Precession of $J^*h/2\pi$ and the magnetic moment μ_J in a magnetic field B.

Since there are $(2J + 1)$ values of θ, that is $(2J + 1)$ orientations of the total electron angular momentum, it follows there will be $(2J + 1)$ different energy value in the presence of a suitable magnetic field. The amount of splitting of energy level $(E_{MJ} - E_0)$, as Equation (1.64) indicates, is directly proportional to the magnetic field strength B. If B is equal to zero, then $E_{MJ} = E_0$, but the space quantization remains, and the $(2J + 1)$ orientations of the electron orbits have the same energy, that is they are described as being degenerate.

So far no definite proof has been given for the space quantization of angular momentum. To a certain extent space quantization is almost implicit in the spectra studies of the Zeeman effect, but the most direct and convincing experimental proof comes from the Stern–Gerlach experiment.

1.7.2 The Stern–Gerlach experiment

The space quantization of angular momentum can be very clearly demonstrated by an experiment proposed by Stern (1921) and carried out in collaboration with Gerlach in 1922. The apparatus is illustrated in Fig. 1.29(a). A beam of atoms is produced in a heated oven, and the atoms are projected into a high vacuum and passed through a vertical collimating slit to yield a narrow atomic beam. This beam travels between the pole pieces of an electromagnet [see Fig. 1.29(b)]. The inhomogeneous magnetic field has a magnetic gradient which is fairly constant over the region through which the beam passes. The beam is detected on the condensation target AB. Initially a beam of silver atoms was employed.

Since the atoms constituting the beam have a magnetic moment, they may be regarded as extremely small magnets. In the presence of the inhomogeneous magnetic field, with a gradient $\partial B/\partial z$ acting along the z-axis [see Fig. 1.29(a)] perpendicular to the beam, the atomic magnets experience a force F_z given by:

$$F_z = \mu_J \cos \theta \, \frac{\partial B}{\partial z}$$

where $\mu_J \cos \theta$ is the component of the magnetic moment of the atoms in the direction of the field. When the magnetic field is uniform:

$$\frac{\partial B}{\partial z} = 0$$

The displacement force relative to the z-axis on the atomic magnets is then zero, and the beam undergoes no deflexion.

In a homogeneous magnetic field the atomic beam would move in a straight line, whereas in a non-homogeneous field F_z has a definite value, and the resultant force acts on the magnets causing a deflexion of the beam. The elementary magnet is then caused to deviate from its rectilinear path, the amount of deflexion depending on the non-homogeneity of the field. In fact, the non-homogeneity must be such that the field differs considerably over atomic dimensions. The inhomogeneity was produced by having one pole piece of the electromagnet as a knife-edge and the opposite pole with a groove [see Fig. 1.29(b)]. The field strength at the knife-edge is very much greater than elsewhere since the lines of force crowd together at this thin edge. The deflexion of each atomic magnet in the non-homogeneous field is dependent on the magnitude of μ_J and the orientation θ. Thus, for a particular atom magnet the beam will split into the number of values taken by $\cos \theta$.

When there is no magnetic field the silver beam forms a narrow line on the condensation target, but in the presence of a field the single line is split into two components. The lines observed were not parallel [see Fig. 1.29(c)]. The shape of the lines may be accounted for by the fact that the field gradient $\partial B/\partial z$ is constant only over the centre of the beam.

According to classical theory a broadening of the beam should have resulted since in a magnetic field all possible orientations of the magnetic moment ought to be present. The fact that the silver beam has split into two in the presence of the non-homogeneous magnetic field is definite proof of a very limited number of orientations. In fact, on the quantum theory the beam should be split into $(2J + 1)$ discrete beams, and this would correspond to $(2J + 1)$

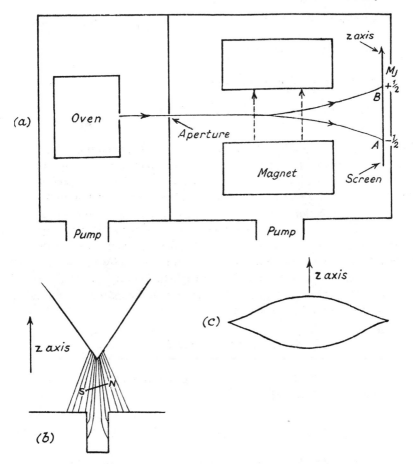

Fig. 1.29 Stern-Gerlach experiment for an atom which has $J = \frac{1}{2}$
(a) Diagrammatic representation when looking down on the apparatus. (b) The lines of force between the pole pieces and one orientation of the magnetic dipole. (c) Splitting pattern of the atomic beam as detected on the screen. Vertical slits were employed.

orientations of the magnetic moment with respect to the field. For silver then:

$$(2J + 1) = 2$$

hence,

$$J = \frac{1}{2}$$

that is $S = \frac{1}{2}$ and $L = 0$. Thus, in this case the experimental results are in full agreement with theory. It should be noted that this is definite experimental proof that l can and indeed must be able to take zero as one of its values.

1.8 ZEEMAN EFFECTS

1.8.1 The normal Zeeman effect

The magnetic moment of an atom[†] is given by

$$\mu = -\frac{e}{2m} \cdot \frac{h}{2\pi} \cdot J^* \qquad (1.65)$$

and results from the rotation of the electron, which has been treated as a negative point charge. The quantity $eh/2\pi m$ presents the smallest unit of magnetic moment and is called the *Bohr magneton* μ_B. If a homogeneous magnetic field of the order of 3 to 30 T is applied, precessional motion takes place about the direction of the magnetic field. The component M_J of J^* in the field direction must be either integral or half-integral. If the energy of an atom in the absence of a magnetic field is E_0, the total energy E_{MJ} in the presence of an external magnetic field strength B is given by $E_{MJ} = E_0 - \mu_J B \cos \theta$, or the change in energy due to the magnetic field is $E_{MJ} - E_0 = -\mu_J B \cos \theta$, where θ is the angle between the direction of the magnetic field and μ_J. Since $\cos \theta = M_J/J^*$, then on substitution for $\cos \theta$:

$$E_{MJ} - E_0 = -\mu_J B \frac{M_J}{J^*}$$

For an atom where $S = 0$, $\mu_J = \mu$ (see p. 52), and for such an atom, on substitution from Equation (1.65):

$$E_{MJ} - E_0 = +\frac{e}{2m} \cdot \frac{h}{2\pi} \cdot B \cdot M_J = \mu_B B M_J \qquad (1.66)$$

Another useful $E_{MJ} - E_0$ equation is obtained by substitution of the Larmor frequency of precession (ν_L) for $eB/4\pi m$ in Equation (1.66):

$$E_{MJ} - E_0 = h\nu_L M_J \qquad (1.67)$$

and this is the equation showing the dependence of the interaction energy on the velocity of precession.[‡]

Equation (1.67) can be used to account for the splitting of the spectral lines of elements in the presence of a suitable magnetic field. It will be seen later that these considerations of the normal Zeeman effect apply only to elements where the resultant spin quantum number is zero. From Equation (1.67) for a transition between the electronic energy states E' and E'':

$$\Delta E = (E_0' + h\nu_L M_J') - (E_0'' + h\nu_L M_J'') \qquad (1.68)$$

[†] This does not take account of the fact that a nucleus may also have a magnetic moment.

[‡] Equations of the type (1.66) and (1.67) have to be modified when $S \neq 0$ (see later).

$$\Delta E = (E_0' - E_0'') + h\nu_L M_J' - M_J'')$$

$$\nu = \nu_0 + \nu_L \Delta M_J \qquad (1.69)$$

where ν_0 is the frequency of the line in the absence of a magnetic field, and:

$$\nu_0 = \frac{E_0'' - E_0''}{h}$$

The selection rule for changes in M_J may be shown theoretically to be $\Delta M_J = 0, \pm 1,$[†] when the observations are made perpendicular to the applied magnetic field. It therefore follows from Equation (1.69) that when:

$$\Delta M_J = 1 \qquad \nu_1 = \nu_0 + \nu_L \qquad (1.70)$$

$$\Delta M_J = 0 \qquad \nu_2 = \nu_0 \qquad (1.71)$$

$$\Delta M_J = -1 \qquad \nu_3 = \nu_0 - \nu_L \qquad (1.72)$$

Thus, irrespective of the particular M_J' and M_J'' values concerned, only three lines could result when the selection rule is obeyed. This splitting of a spectral line into three in the presence of a suitable perpendicular magnetic field is known as the *normal Zeeman effect*. The lines are thus symmetrical about the centre line (ν_0) and a a frequency separation from it of ν_L. It can be predicted on the basis of classical mechanics that for perpendicular observations the line corresponding to $\Delta M_J = 0$ is polarized with the electric vector in a plane parallel to the magnetic field, and is referred to as a π *line*, while the two lines corresponding to $\Delta M_J = +1$ and -1 are plane polarized with the electric vector perpendicular to the field and called σ *lines*.[‡] For observations made parallel to the magnetic field only two lines result (since the $\Delta M_J = 0$ transition is not detected), and these correspond to $\Delta M_J = \pm 1$, where the light is circularly polarized and made up of σ components.

It follows from formula (1.69) that the lines belonging to any one transition governed by a particular ΔM_J value will all have the same energy, and hence only one line will result. This is illustrated in Fig. 1.30, for a perpendicular observation, where the three observed lines correspond to the three ΔM_J values.

1.8.2 The anomalous Zeeman effect

For all atoms where S is greater than zero, in the presence of a suitable magnetic field each spectral line is split into more than three components, and the *anomalous Zeeman effect* is obtained. There is, however, no divergence between experiment and theory. The additional lines are explained by the fact

[†] With the proviso that the combination $M_J'' = 0 \leftarrow M_J' = 0$ is forbidden when $\Delta J = 0$.
[‡] For an explanation of the terms σ and π White [1.5] should be consulted.

that the spacing of the upper energy levels for a particular L value differs from those of the lower levels which, of course, have a different L value. The amount by which each level is split depends on the values of $L, S,$ and J. Thus formula (1.66) has to be modified and actually becomes $E_{MJ} = E_0 + M_J g_J \mu_B B$, where

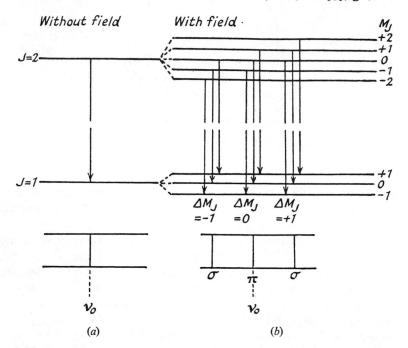

Fig. 1.30 Transitions and spectral lines resulting: (a) in the absence of a magnetic field, and (b) on the application of a magnetic field where the normal Zeeman effect results.

g_J is called the Landé g_J-factor whose value can be calculated when values of $L, S,$ and J are known. This may be illustrated by a consideration of one of the sodium D-lines. For the $^2P_{1/2}$ level of Na, $g_J = \frac{2}{3}$, while for the $^2S_{1/2}$ level $g_J = 2$. The relevant data for these two levels in the presence of a magnetic field are given in Table 1.10, where for a perpendicular observation the selection rule $\Delta M_J = 0, \pm 1$ still applies. The transitions obtained are shown in Fig. 1.31 where, as a result of the different magnitudes of splitting of the energy levels in the upper and lower states, four lines of different frequencies result.

No explanation has yet been given of (a) Why the energy equation was modified from $E_{MJ} + M_J \mu_B B$ to $E_{MJ} = E_0 + M_J g_J \mu_B B$; (b) the meaning of the g_J-factor;[†] (c) how the g_J-factor can be calculated from theory; and (d) why

[†] The g-factor is of fundamental importance in the understanding of electron spin resonance spectra and because of this has been allotted special consideration here.

Table 1.10 Interaction energy taking into account the g_J-factor for one of the sodium D-lines

Term	M_J values	g_J	$M_J g_J$	$E_{M_J} - E_0$ interaction energy
$^2P_{1/2}$	$+\frac{1}{2}, -\frac{1}{2}$	$\frac{2}{3}$	$+\frac{1}{3}, -\frac{1}{3}$	$+\frac{1}{3}\mu_B B, -\frac{1}{3}\mu_B B$
$^2S_{1/2}$	$+\frac{1}{2}, -\frac{1}{2}$	2	$+1, -1$	$+\mu_B B, \mu_B B$

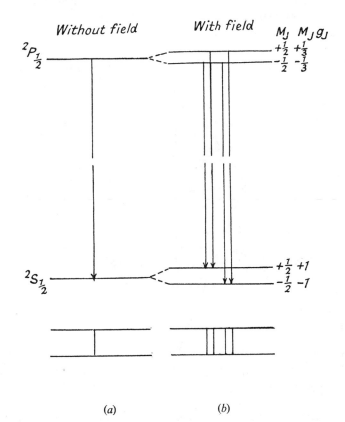

Fig. 1.31 Transitions and spectral lines for the $3^2P_{1/2} - 3^2S_{1/2}$ transition of the sodium atom, (a) in the absence of a magnetic field, and (b) on the application of a suitable magnetic field which produces the anomalous Zeeman pattern of lines.

when S is 0 and g_J is 1 the normal Zeeman effect is observed. These considerations and explanations will be our next objective, and the approach will be made by viewing the total magnetic moment of the atom, although in

49

the following case it will be assumed that the nucleus has zero magnetic moment.[†]

1.8.3   The magnetic moment of the atom and 'g'-factor

So far in considering the Zeeman effect the fact has been neglected that in addition to the electron rotating in an orbit it is also spinning about its own axis, and therefore will possess inherently a magnetic moment associated with the spin. Provided that the resultant spin is zero (as it is, for example in the case of the singlets of Zn and Cd) the theory given is in agreement with experiment regarding the number of observed Zeeman lines. When S is not zero the theory is totally inadequate as is shown experimentally, since the Zeeman pattern for a perpendicular observation is no longer generally composed of only three lines.

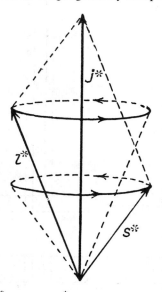

Fig. 1.32 Precession of $l^*h/2\pi$ and $s^*h/2\pi$ about their resultant $j^*h/2\pi$. The vectors are in $h/2\pi$ units.

When a vector quantity, such as an angular momentum, is subjected to a torque (whether from a magnetic or gravitational field) the result is that the vector precesses around the field direction. In an atom there are many vector quantities involving angular momenta and magnetic moments. These vector quantities exert torques on one another, and thus even in the absence of an external field they are caused to precess. An example of this occurs in the interaction of the orbit and spin vectors. Owing to the rotation of the electron

[†] In the chapter on nuclear magnetic resonance the re-orientation of the nuclear magnetic moment will be considered.

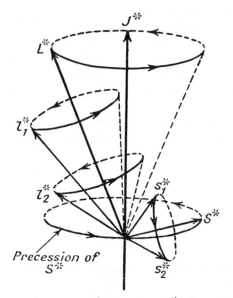

Fig. 1.33 Precession of $l_1^* h/2\pi$ and $l_2^* h/2\pi$ about $L^* h/2\pi$ and its precession about $J^* h/2\pi$. Similarly the precession of $s_1^* h/2\pi$ and $s_2^* h/2\pi$ about $S^* h/2\pi$ and its precession about $J^* h/2\pi$. All the vectors are in $h/2\pi$ units.

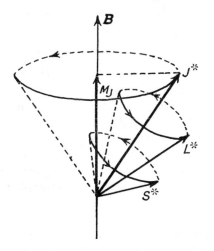

Fig. 1.34 Precession of $J^* h/2\pi$ about an applied magnetic field **B**. The vectors are in $h/2\pi$ units.

in its orbit and the spin about its axis, two magnetic fields are produced and the orbital $l^* h/2\pi$ and spin $s^* h/2\pi$ angular momenta are caused to precess. From Fig. 1.32 it can be seen that the motion possible for the vectors, if they are to give a constant resultant, is for $l^* h/2\pi$ and $s^* h/2\pi$ to precess around the same axis. This axis is that of the total angular momentum ($j^* h/2\pi$). A similar state of affairs exists for more than one electron: the $l^*h/2\pi$ values and $s^*h/2\pi$ values couple giving $L^*h/2\pi$ and $S^*h/2\pi$, respectively; $L^*h/2\pi$ and $S^*h/2\pi$ then precess around their resultant $J^*h/2\pi$ as shown in Fig. 1.33.

In a weak magnetic field, the magnetic moment associated with the total angular momentum $J^*h/2\pi$ causes the atom to precess like a top around the direction of the field. The precession of $L^*h/2\pi$ and $S^*h/2\pi$ around $J^*h/2\pi$ and the precession of $J^*h/2\pi$ about B are illustrated in Fig. 1.34, and the direction along which the effective magnetic moment acts will now be considered as it is essential to know this in order to derive a formula for g_J in terms of L^*, S^*, and J^*.

It has already been indicated that the projection of $J^*h/2\pi$ on B may take only the values:

$$+J, J-1, J-2, \ldots, -J$$

in multiples of $h/2\pi$ units. The resultant orbital angular momentum of an atom is given by:

$$P_L = \sqrt{[L(L+1)]}\,h/2\pi \tag{1.73}$$

and if its magnetic moment is μ_L then according to classical theory (see p. 54)

$$\frac{\mu_L}{P_L} = -\frac{e}{2m} = \frac{\mu_L}{L^*h/2\pi} \tag{1.74}$$

For the resultant electronic spin angular momentum several experiments show that the corresponding ratio is:[†]

$$\frac{\mu_S}{P_S} = -2\frac{e}{2m} = \frac{\mu_S}{S^*h/2\pi} \tag{1.75}$$

where p_S is the resultant spin angular momentum $\sqrt{[S(S+1)]}\,h/2\pi$ and μ_S is the corresponding magnetic moment. Thus the ratio between the magnetic moment and the resultant angular momentum vector for the electron spin is twice that for the orbital motion. The importance of this fact is that when the vector sum is made of μ_S and μ_L (as in Fig. 1.35) the resultant magnetic moment μ_{LS} for the atom cannot lie in direct line with the total angular momentum vector $J^*h/2\pi$. This may be seen in Fig. 1.35 by completing the magnetic moment parallelogram for μ_L and μ_S, where the resultant magnetic moment μ_{LS} for the atom is inclined by the angle θ to the axis of $J^*h/2\pi$. μ_{LS} could lie only along the direction of $J^*h/2\pi$:

(i) If μ_L were equal to μ_S. However this can never be the case.

(ii) If $\mu_S = 0$. Under these circumstances $S^* = 0$, and $L^* = J^*$, and in the presence of a magnetic field μ_{LS} and J^* would be in the same straight line. This, of course, is the state of affairs in the normal Zeeman effect, that is the case where $g_J = 1$.

Since the total angular momentum vector is invariant, $L^*h/2\pi, S^*h/2\pi, \mu_L,$

[†] The fact that the ratio of magnetic moment to angular momentum for the spinning electron is twice that for orbital motion has also been derived from quantum mechanics.

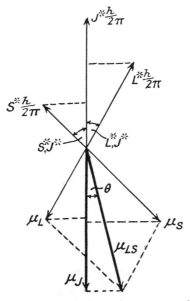

Fig. 1.35 Direction of the vectors $J^*h/2\pi$, $L^*h/2\pi$, and $S^*h/2\pi$, and the magnetic moments μ_{LS}, μ_L, and μ_S, where μ_{LS} is the magnetic moment of the atom (really μ) and is obtained by the vectorial combination of $\hat{\mu}_L$ and μ_S.

μ_S, μ_{LS} precess around $J^*h/2\pi$. Owing to this precession, only the component of μ_{LS} parallel to $J^*h/2\pi$ contributes to the magnetic moment of the atom. Thus, if μ_{LS} is resolved into one component parallel to J^* and another perpendicular to J^*, the perpendicular component will average to zero as a result of the rotation about J^*. The following may now be obtained:

   (i) A formula for the emitted frequency in an anomalous Zeeman transition in terms of M_J and g_J values for the upper and lower levels.

   (ii) A formula for g_J in terms of L^*, S^*, and J^*. In the derivation of this formula it must be assumed that only the component of μ_{LS} parallel to the field contributes to the magnetic moment.

1.8.4   Emitted frequencies in anomalous Zeeman transitions

When an atom is placed in a weak magnetic field the change in energy is given by:

$$E_{MJ} - E_0 = -\mu_J B \frac{M_J}{J^*} \qquad (1.76)$$

where μ_J is the component of the magnetic moment along the direction $J^*h/2\pi$ (see p. 46). Each energy level characterized by a particular J value is split into $(2J + 1)$ terms with a separation between consecutive levels which is now equal to $\mu_J B/J^*$. Let:

$$g_J \mu = \mu_J \qquad (1.77)$$

53

where μ is the magnetic moment of the atom and g_J is a constant. g_J is known as the Landé g-factor which takes into account that, in general, the magnetic moment of the atom is not in the same line as J^*.

Since

$$\mu = -\frac{e}{2m} \cdot \frac{h}{2\pi} \cdot J^* \tag{1.78}$$

then

$$\mu_J = -\frac{e}{2m} \cdot \frac{h}{2\pi} \cdot J^* \cdot g_J \tag{1.79}$$

and on substitution into Equation (1.76) for μ_J:

$$E_{MJ} = E_0 + \frac{e}{2m} \cdot \frac{h}{2\pi} \cdot M_J g_J B = E_0 + h\nu_L M_J g_J \tag{1.80}$$

where ν_L is the Larmor frequency (see p. 41).

For a transition:

$$\Delta E = (E_0' + h\nu_L M_J' g_J') - (E_0'' + h\nu_L M_J'' g_J'') \tag{1.81}$$

$$\Delta E = (E_0' - E_0'') + h\nu_L (M_J' g_J' - M_J'' g_J'') \tag{1.82}$$

$$\nu = \nu_0 + \nu_L (M_J' g_J' - M_J'' g_J'') \tag{1.83}$$

The separation of the lines in a Zeeman pattern then depends on g_J for both terms. A formula will now be derived for g_J in terms of L^*, S^*, and J^*.

1.8.5   The Landé g-formula

It follows from Equations (1.74) and (1.75) that:

$$\mu_L = -L^* \cdot \frac{h}{2\pi} \cdot \frac{e}{2m} \tag{1.84}$$

and

$$\mu_S = -2S^* \cdot \frac{h}{2\pi} \cdot \frac{e}{2m} \tag{1.85}$$

By reference to Fig. 1.35 the components of μ_L and μ_S along J^* are given by:

$$\text{Component of } \mu_L = -L^* \cdot \frac{h}{2\pi} \cdot \frac{e}{2m} \cdot \cos (L^*, J^*) \tag{1.86}$$

$$\text{Component of } \mu_S = -2S^* \cdot \frac{h}{2\pi} \cdot \frac{e}{2m} \cdot \cos (S^*, J^*) \tag{1.87}$$

On addition of Equations (1.86) and (1.87):

$$\mu_L + \mu_S = -\${L^*, \cos (L^*, J^*) + 2S^* \cos (S^*, J^*)\}\frac{h}{2\pi} \cdot \frac{e}{2m} \tag{1.88}$$

where

$$\mu_L + \mu_S = \mu_J \tag{1.89}$$

From Equations (1.89) and (1.88) it follows that:

$$\mu_J = -\{L^* \cos (L^*, J^*) + 2S^* \cos (S^*, J^*)\}\frac{h}{2\pi} \cdot \frac{e}{2m} \tag{1.90}$$

and on substitution for μ_J from Equation (1.79):

$$J^* g_J = L^* \cos(L^*, J^*) + 2S^* \cos(S^*, J^*) \tag{1.91}$$

On application of the cosine law to the S^*, L^*, and J^* vectors in Fig. 1.35:

$$S^{*2} = L^{*2} + J^{*2} - 2L^* J^* \cos(L^*, J^*) \tag{1.92}$$

$$L^{*2} = S^{*2} + J^{*2} - 2S^* J^* \cos(S^*, J^*) \tag{1.93}$$

that is

$$\cos(L^*, J^*) = \frac{L^{*2} + J^{*2} - S^{*2}}{2L^* J^*} \tag{1.94}$$

and

$$\cos(S^*, J^*) = \frac{S^{*2} + J^{*2} - L^{*2}}{2S^* J^*} \tag{1.95}$$

On substitution for $\cos(L^*, J^*)$ and $\cos(S^*, J^*)$ into Equation (1.91):

$$g_J = \frac{L^*}{J^*}\left\{\frac{L^{*2} + J^{*2} - S^{*2}}{2L^* J^*}\right\} + \frac{2S^*}{J^*}\left\{\frac{S^{*2} + J^{*2} - L^{*2}}{2S^* J^*}\right\} \tag{1.96}$$

$$g_J = \left\{\frac{L^{*2} + J^{*2} - S^{*2}}{2J^{*2}}\right\} + 2\left\{\frac{S^{*2} + J^{*2} - L^{*2}}{2J^{*2}}\right\} \tag{1.97}$$

$$g_J = \frac{3J^{*2} + S^{*2} - L^{*2}}{2J^{*2}} = 1 + \frac{J^{*2} + S^{*2} - L^{*2}}{2J^{*2}} \tag{1.98}$$

Since

$$J^* = \sqrt{[J(J+1)]}, L^* = \sqrt{[L(L+1)]}, \text{ and } S^* = \sqrt{[S(S+1)]} \tag{1.99}$$

$$g_J = 1 + \frac{J(J+1) + S(S+1) - L(L+1)}{2J(J+1)} \tag{1.100}$$

The Landé g_J-factor formula for an atom with a single valence electron is exactly the same form as Equation (1.100), except that L is replaced by l, S by s, and J by j in Equation (1.100) which then becomes:

$$g = 1 + \left\{\frac{j(j+1) + s(s+1) - l(l+1)}{2j(j+1)}\right\} \tag{1.101}$$

It may be observed from Equation (1.100) that when $J = L$ (i.e. $S = 0$) the value of g_J is unity, and the normal Zeeman effect would result, because the energy equation would then be $E_{M_J} = E_0 + \mu_B M_J B$. This is the case dealt with on p. 46).

The g_J-factor model accounts in full for the observed Zeeman patterns and gives an exceptionally close agreement with experiment. Determinations by the electron spin resonance method have confirmed by measurement that, for certain free radicals with one unpaired electron, the g_J-factor is almost exactly 2; this is the value to be expected from Equation (1.100) when $L = 0$.[†]

[†] In the case when $L = 0$ and $S \neq 0$ the symbol g_S is employed for the g-factor.

In the derivation of the g_J formula the quantum mechanical value for angular momentum was assumed to be of the form $\sqrt{[X(X+1)]}\,h/2\pi$ where X is J, L, or S. If, however, $Jh/2\pi$, $Lh/2\pi$, and $Sh/2\pi$ had been used for the magnitude of the vectors, a different formula would have resulted, giving different values of g_J. Since, however, Equation (1.100) gives values in agreement with experiment, this is excellent support for the quantum mechanical values of orbital angular momenta, that is angular momentum is of the form $\sqrt{[X(X+1)]}\,h/2\pi$ and not the classical value of $Xh/2\pi$, where $X = J$, L, or S.

The g_J-factor value depends on the values of J, S, and L. Different levels will therefore have different values of g_J. As a consequence of this the number of line components obtained in a suitable magnetic field will depend on the number of term components, that is on $(2J + 1)$, for each level rather than solely on the selection rule $\Delta M_J = 0, \pm 1$, as is the case in the normal Zeeman effect. To illustrate the suitability of the equation developed for g_J, we shall now examine the splitting pattern in the anomalous Zeeman effect for the D-lines of sodium where the data are given in Fig. 1.36. The case considered is that for the spectrum taken perpendicular to the applied magnetic field where the selection rule is $\Delta M_J = 0, \pm 1$. The σ and π polarization of the lines is indicated in Fig. 1.36.

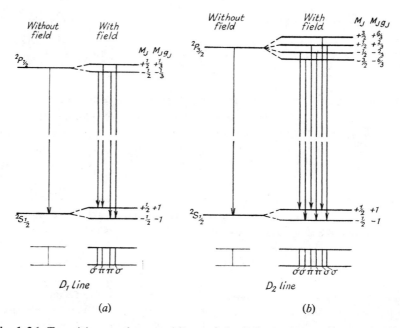

Fig. 1.36 Transitions and spectral lines of the D-lines of the sodium atom (a) in the absence and (b) in the presence of a suitable magnetic field. In the latter case, the anomalous Zeeman pattern of lines is produced.

Anomalous Zeeman effect of the sodium D-lines.

D₁ *line*		*Sodium lines*	D₂ *line*	
$3^2P_{1/2} - 3^2S_{1/2}$		*Transitions*	$3^2P_{3/2} - 3^2S_{1/2}$	
Upper level	*Lower level*	*Levels*	*Upper level*	*Lower level*
$S = \frac{1}{2}$	$S = \frac{1}{2}$	$(2S + 1) = 2$	$S = \frac{1}{2}$	$S = \frac{1}{2}$
$L = 1$	$L = 0$	L values	$L = 1$	$L = 0$
$J = \frac{1}{2}$	$J = \frac{1}{2}$	J values	$J = \frac{3}{2}$	$J = \frac{1}{2}$
$g_J = \frac{2}{3}$	$g_J = 2$	g_J factors	$g_J = \frac{4}{5}$	$g_J = 2$
2 values	2 values	$(2J + 1)$ values	4 values of M_J	2 values
of M_J	of M_J	of M_J		of M_J
$+\frac{1}{2}, -\frac{1}{2}$	$+\frac{1}{2}, -\frac{1}{2}$	M_J values	$+\frac{3}{2}, +\frac{1}{2}, -\frac{1}{2}, -\frac{3}{2}$	$+\frac{1}{2}, -\frac{1}{2}$
$+\frac{1}{3}, -\frac{1}{3}$	$+1, -1$	$g_J M_J$ values	$+\frac{6}{3}, +\frac{2}{3}, -\frac{2}{3}, -\frac{6}{3}$	$+1, -1$

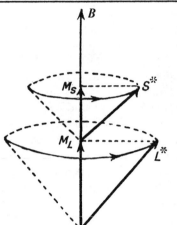

Fig. 1.37 Precession of the uncoupled $L^*h/2\pi$ and $S^*h/2\pi$ vectors in a sufficiently strong magnetic field. The vectors are in $h/2\pi$ units.

It can be confirmed with the help of formula (1.100) that the g_J value for the state $3^2P_{1/2}$ is $\frac{2}{3}$, and $\frac{4}{3}$ for the $3^2P_{3/2}$ state, and that four lines would be obtained for the $3^2P_{1/2} - 3^2S_{1/2}$ transition; for the $3^2P_{3/2} - 3^2S_{1/2}$ six lines would be obtained (see Fig. 1.36). It should be noted that what was originally termed the sodium D-line on p. 24 and later was shown to be really two lines, now becomes ten lines in the presence of a suitable magnetic field when the spectrum is taken perpendicular to the applied field.

1.9   THE PASCHEN—BACK EFFECT

The precession of $L^*h/2\pi$ and $S^*h/2\pi$ around their resultant $J^*h/2\pi$ vector has been dealt with, and it was seen how multiplet structure arises from transitions between the different sub levels of different electronic states. The greater the interaction of $L^*h/2\pi$ and $S^*h/2\pi$, the faster will be the precession of $L^*h/2\pi$ and $S^*h/2\pi$ around $J^*h/2\pi$, and the greater the multiplet splitting. In fact, multiplet fine structure separations are a measure of the interaction between $L^*h/2\pi$ and $S^*h/2\pi$ and the frequency of precession of $L^*h/2\pi$ and $S^*h/2\pi$ around $J^*h/2\pi$. A similar state of affairs exists when an atom is placed in a magnetic field; the velocity of precession of $J^*h/2\pi$ around the field direction depends on the strength of coupling of its magnetic moment with the field, and this is dependent on the magnitude of the field strength B. The greater value of B the greater will be the velocity of precession of the $J^*h/2\pi$ vector. This follows immediately from Equation (1.62) which is $B\mu_B/h =$ the velocity of precession (ν_L), where $B = eh/4\pi m$. Eventually, if B is continually increased, the velocity of precession of $J^*h/2\pi$ around B becomes of the same order as that of $L^*h/2\pi$ and $S^*h/2\pi$ around $J^*h/2\pi$. Under these conditions $L^*h/2\pi$ and $S^*h/2\pi$ are partially uncoupled, the motion of $L^*h/2\pi$ and $S^*h/2\pi$ becoming complex. As B is still further increased $L^*h/2\pi$ and $S^*h/2\pi$ become quantized separately and precess more or less independently around B. This is the Paschen-Back effect,[†] an is recognized by a complete change in the spectrum of the Zeeman pattern in the lines belonging to one multiplet. In the case of the sodium atom the uncoupling of $L^*h/2\pi$ and $S^*h/2\pi$ does not occur for values of B up to about 0.1 T. When, however, B is about ~ 1–10 T, then the frequency of precession of $L^*h/2\pi$ and $S^*h/2\pi$ around $J^*h/2\pi$ is of the same order as the precession of $J^*h/2\pi$ around the field, and the coupling of $L^*h/2\pi$ and $S^*h/2\pi$ begins to break down. If B is still further increased the coupling between orbital and spin angular momenta completely breaks down and we obtain the Paschen-Back effect, where $L^*h/2\pi$ and $S^*h/2\pi$ are quantized independently. Fig. 1.37 represents diagrammatically the vector model for the base where B is so strong that $L^*h/2\pi$ and $S^*h/2\pi$ precess independently around B.

The quantum conditions are that the projection of $L^*h/2\pi$ along the direction of B is equal to $M_L h/2\pi$ and that of $S^*h/2\pi$ along B equals $M_S h/2\pi$. The permissible values of M_L are:

$$L, L-1, L-2, \ldots, -L$$

while those of M_S is:

$$S, S-1, S-2, \ldots, -S$$

The selection rule for M_L when the spectrum is taken perpendicular to the applied field is:

$$\Delta M_L = 0, \pm 1$$

and that for M_S is:

$$\Delta M_S = 0$$

This theory will now be applied to the Paschen-Back effect for the case of the sodium D-lines, that is for the transitions $3^2 P_{3/2} - 3^2 S_{1/2}$ and $3^2 P_{1/2} - 3^2 S_{1/2}$.

[†] This type of effect is one of the important cases studied by the paramagnetic resonance method for crystalline solids containing unpaired electrons.

The values of M_L and M_S for these terms are:

^2P	$L = 1$	$M_L = +1;$	$0;$	$-1;$
	$S = \frac{1}{2}$	$M_S = +\frac{1}{2}; -\frac{1}{2}$	$+\frac{1}{2}; -\frac{1}{2};$	$+\frac{1}{2}; -\frac{1}{2};$
^2S	$L = 0$	$M_L = 0$		
	$S = \frac{1}{2}$	$M_S = +\frac{1}{2}; -\frac{1}{2};$		

It is very instructive to follow the sodium doublet through the cases when there is no field, a weak perpendicular field, and a strong perpendicular field, as is done in Fig. 1.38. It will be noted that when the field is sufficiently strong for the Paschen–Back effect to apply, five lines result from the sodium doublet, although six transitions take place. The spacing of the energy levels is such that two of the lines coincide.

The total interaction energy with the magnitude field in the Paschen–Back effect involves the following contributions:

(i) The energy of interaction of the spin magnetic moment (μ_S) is:

$$E_S = -\mu_S B \cos(\mu_S, B) \qquad (1.102)$$

(ii) The energy of interaction of the orbital magnetic moment (μ_L) is:

$$E_L = -\mu_L B \cos(\mu_L, B) \qquad (1.103)$$

(iii) Energy due to a small amount of remaining interaction between μ_S and μ_L:

$$\Delta E_{LS} = hcAM_L M_S \qquad (1.104)$$

where A is a coupling constant between the orbit and electron spin momenta. The total energy (E) is then given by:

$$E = E_0 + E_L + E_S + \Delta E_{LS} \qquad (1.105)$$

where $E_0 + \Delta E_{LS}$ gives all the forms of energy which the atom has in the absence of a magnetic field. On substitution of Equations (1.102) to (1.104) into (1.105):

$$E = E_0 - \mu_S B \cos(\mu_S, B) - \mu_L B \cos(\mu_L, B) + hcAM_L M_S \quad (1.106)$$

Now since

$$\mu_L = -\frac{e}{2m} \cdot L^* \frac{h}{2\pi} \quad \text{and} \quad \mu_S = -2\frac{e}{2m} \cdot S^* \cdot \frac{h}{2\pi} \quad (1.107 \text{ and } 1.108)$$

and

$$\cos(\mu_S, B) = \frac{M_S}{S^*} \quad \text{and} \quad \cos(\mu_L, B) = \frac{M_L}{L^*} \qquad (1.109 \text{ and } 1.110)$$

on substitution into Equation (1.106):

$$E = E_0 + 2\frac{e \cdot h}{2m \cdot 2\pi} BM_S + \frac{e \cdot h}{2m \cdot 2\pi} BM_L + hcAM_L M_S \quad (1.111)$$

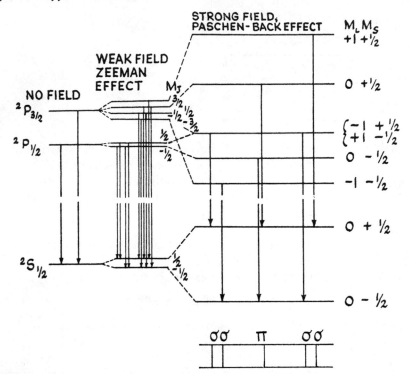

Fig. 1.38 Splitting of the $^2S_{1/2}$, $^2P_{1/2}$, and $^2P_{3/2}$ energy levels and the spectral lines (e.g. the D_1 and D_2 lines of the sodium atom) in the presence of a weak and a strong magnetic field. The spacing of the energy levels is such that the

$$(M'_L, M'_S) \rightarrow (M''_L, M''_S)$$

$$(0, \tfrac{1}{2}) \rightarrow (0, \tfrac{1}{2})$$

$$(0, -\tfrac{1}{2}) \rightarrow (0, -\tfrac{1}{2})$$

transitions result in a line of the same frequency. (*Courtesy of Dr. W. Jevons [1.6] and the Council of The Physical Society, London*)

Moreover, since:

$$\frac{h}{4\pi m} = \mu_{\mathbf{B}}$$

where $\mu_{\mathbf{B}}$ is the *Bohr magneton*, Equation (1.111) becomes:

$$E = E_0 + \mu_{\mathbf{B}}B(2M_S + M_L) + hcAM_LM_S \tag{1.112}$$

1.10 THE STARK EFFECT

1.10.1 Normal Stark effect

When an electric field is applied to molecules studied by the microwave technique, it can lead to most accurate determinations of electric dipole moments; for this reason the effect of an electric field is allotted special consideration here.

The splitting of a spectral line in a magnetic field was discovered by Zeeman in 1896, but it was not until 1913 that Stark demonstrated the splitting of the Balmer lines of the hydrogen atom in an electric field. The reason for this delay was the difficulty of producing the necessary very high potential gradient of the order of 10^7 Vm^{-1} along a discharge tube. A type of discharge tube employed in this work is shown diagrammatically in Fig. 1.39, where the distance between

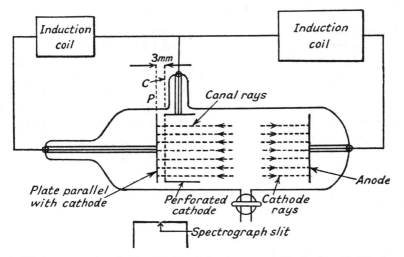

Fig. 1.39 An apparatus for the study of the transverse Stark effect indicating the position at which the slit of the spectrograph is located relative to the discharge tube.

the perforated cathode and the plate is 3 mm or less. The gas pressure in the discharge tube has a value such that the Crookes dark space is several centimetres long.[†] A very high potential can be maintained between the plate and the cathode without a discharge taking place since the mean free path of the ions is greater than the distance between the plate and the cathode.

An electric field does not interact with the magnetic moment of the atom. The action is of electric polarization, and the electric field displaces the centre of the negative charge and consequently the centres of negative and positive

[†] This is the region in which the energy is less than that required to produce ionization.

61

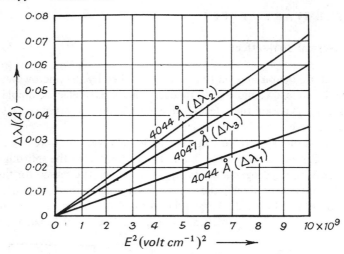

Fig. 1.40 Quadratic Stark effect for the potassium doublet $5^2P_{3/2,\ 1/2} - 4^2S_{1/2}$ at the wavelengths of 4044 and 4047 Å respectively. (*After Grotrian and Ramsauer [1.7]. Courtesy of S. Hirzel Verlag, Leipzig*).

charge no longer coincide. The extent of separation depends on the field strength E. The result is an electric dipole moment which can interact with the electric field. Owing to the electric torque to which the angular momentum $J^*h/2\pi$ is subjected, $J^*h/2\pi$ precesses about the direction of the field in such a way that the component of $J^*h/2\pi$ along the direction of the electric field is constant.

The stronger the field the greater the precessional velocity of $J^*h/2\pi$ around E. The energy shift for certain cases is given by $\Delta E = \mu_F E$, where μ_F is the induced electric dipole moment in the direction of the field. Since, however, the magnitude of μ_F itself is proportional to E, the splitting of the energy levels increases quadratically with E. This effect can be shown by plotting the wavelength shift against the square of the field strength, and if $\Delta\lambda\ \alpha E^2$, a straight line should be obtained. For example, in the case of the potassium doublet this may be observed for the transitions $5^2P_{3/2} - 4^2S_{1/2}$ and $5^2P_{1/2} - 4^2S_{1/2}$ at the wavelengths 4044 Å and 4047 Å, respectively (see Fig. 1.40).

As can be seen from the diagrammatic spectrogram in Fig. 1.41, the original K doublet is split into three components in the electric field[†] (see later). In accordance with what was indicated above, the plot of $\Delta\lambda$ against E^2 results in a straight line for all the three spectral lines.

The important difference between the Zeeman and Stark effects is that each pair of levels $M_J = +J$ and $M_J = -J$ arising from a given level has exactly the same energy in an electric field but different energies in a magnetic field. The

[†] These three potassium lines may be readily accounted for in terms of an energy level diagram, similar to that given for sodium in Fig. 1.42.

explanation is as follows. An electron may rotate either clockwise or anticlockwise in its orbit. On application of the rule on page 27 the component of the magnetic moment in the direction of the field ranges from $-\mu_J \cos \theta$ to $\mu_J \cos \theta$ and the corresponding interaction energies range from

$$E_{MJ} - E_0 = \mu_J B \cos \theta \quad \text{to} \quad E_{MJ} - E_0 = -\mu_J B \cos \theta$$

Thus, since there are $(2J + 1)$ values of $\cos \theta$, then there are $(2J + 1)$ values of the interaction energies in a magnetic field. The mechanism of interaction of the electric field with the atom is purely electrostatic. The dipole moment remains unaltered on reversing the direction of rotation (i.e. on passing from $+ M_J$ to $- M_J$), and (since $\Delta E = \mu_F E$) it follows that the energy will be the same for M_J components which are numerically the same but differ in sign.

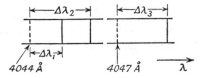

Fig. 1.41 Relation of the electric field component lines (full lines) to the field free components (broken lines) for the quadratic Stark effect of the potassium doublet $5^2 P_{3/2, 1/2} - 4^2 S_{1/2}$.

Although there are $(2J + 1)$ orientations of the orbit to the field, because of this degeneracy in an electric field there are only $(J + \frac{1}{2})$ different energy levels when J is half-integral, or $(J + 1)$ levels if J is integral. For example, the level $J = \frac{5}{2}$, instead of having six components, as it would have in a magnetic field, has only three components in an electric field.

The selection rules for M_J are identical with those for the Zeeman case. When the observation of the spectrum is made perpendicular to the applied field $\Delta M_J = 0, \pm 1$. The transitions $\Delta M_J = \pm 1$ give rise to lines which are polarized with the electric vector perpendicular to the field, while for $\Delta M_J = 0$ the line is polarized with the electric vector parallel to the applied field. For parallel observations $\Delta M_J = \pm 1$, and the lines are unpolarized.

As an example of the Stark effect the sodium D-lines may be considered; the transitions involved are:

$$3^2 P_{1/2} - 3^2 S_{1/2} \quad \text{and} \quad 3^2 P_{3/2} - 3^2 S_{1/2}$$

When $J = \frac{3}{2}, M_J = \pm \frac{3}{2}, \pm \frac{1}{2}$, and when $J = \frac{1}{2}, M_J = \pm \frac{1}{2}$, and three lines would result in the presence of the electric field, when the observation is made perpendicular to the field. This is indicated in Fig. 1.42.

The Stark effect is frequently more complex than outlined here. In fact, the difference in energy between the field-free value and the value in the electric field is given by:

$$\Delta T = AE + BE^2 + CE^2 + \ldots \tag{1.113}$$

63

This equation applies to the hydrogen-like atom in an electric field. In the case of the lower energy states of the hydrogen atom, if the values A, B, and C are inserted, it may be shown that for field strengths of $<10^7$ Vm^{-1} only the term AE is significant, and this dependence of ΔT on the first power of E is known as a *first-order Stark effect*. Dependence of ΔT on E^2 would be known as a *second-order Stark effect,* and this is the case which was considered in the potassium doublet example.

1.10.2   Stark effect in a strong electric field

If the electric field strength is increased, eventually the interaction energy between the electron and the field exceeds that between the orbit and spin motions; the coupling between $L^*h/2\pi$ and $S^*h/2\pi$ is then broken down, $L^*h/2\pi$ and $S^*h/2\pi$ becoming independently quantized. Because of the induced electric dipole, the orbital angular momentum is still capable of interacting with

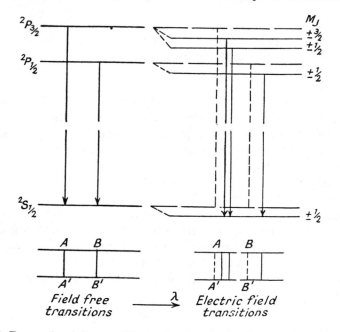

Fig. 1.42 Energy level diagram illustrating the Stark effect of the sodium doublet $3^2P_{3/2,\,1/2} - 3^2S_{1/2}$.

the applied electric field and is quantized with respect to it (as in Fig. 1.37, which is the case for a magnetic field). The orbital angular momentum $L^*h/2\pi$ then has the components $M_L h/2\pi$ the direction of the field. The electron spin, however, has no electric dipole associated with it and cannot interact with the electric field; its mechanism of coupling acts through its magnetic dipole moment and the magnetic field associated with the orbital motion of the electron. The magnetic field resulting from the precession of $L^*h/2\pi$ about E may be resolved into a component along E and another perpendicular where the latter averages to zero. Thus, the $S^*h/2\pi$ may be regarded as precessing about the direction of E and having the component $M_sh/2\pi$ along it.

1.11 WIDTH OF SPECTRAL LINES

If a spectral line is examined by means of a spectrograph, its width is dependent on the slit width employed. In general, the narrower the slit width then the less broad is the resulting spectral line on the photographic plate. Nevertheless, however narrow the slit width the sharpest spectral line has a finite breadth even for the best optical system. A typical spectral line has an intensity distribution of the type.

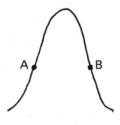

The breadth of the line is defined as the width in cm^{-1} between the two points (A and B) whose intensities are half that of the maximum intensity. For most of the hyperfine structure studies of spectral lines it is essential to reduce the line width as much as possible. In addition, the spectrograph has to be designed so that it is capable of resolving the lines once they have been reduced to this width. If, however, the detected spectral line is made up for example of two components, these cannot be resolved if the half-width of each of the doublet lines is of the order of, or exceeds the separation of, the two components. In such a case the line structure would not be observed however great the resolving power of the spectrograph. Three causes producing the breadth of a spectral line are: (i) Its 'natural' width; (ii) Doppler effect; (iii) External effects such as the intermolecular fields produced by ions and dipoles. Each of these causes will now be considered.

1.11.1 Natural width of a spectral line

From a quantum-mechanical approach it emerges that the energy levels of an atom are not a set of discrete values, and that each of the levels concerned in a particular transition $E_2 - E_1$ has a finite energy width ΔE_1 and ΔE_2. As a result of this an infinitely sharp line is not obtained but instead an intensity distribution with a half-width of the order of the sum of the two term widths.

From the Heisenberg Uncertainty Principle, the ΔE_1 and ΔE_2 values are related to the average times Δt_1 and Δt_2 the atom remains in the states E_1 and E_2 respectively by:

$$\Delta E \times \Delta t \approx h/2\pi$$

It thus follows that the greater the mean life of the atom in a particular state, the smaller will be the width of the observed line.

1.11.2   The Doppler effect

The frequency of emitted radiation depends on the velocity of the source relative to that of the observer. The observed frequency increases if the motion is towards the observer and decreases if the motion is in the opposite direction. Only for those atoms which have no velocity component in the direction of the observer will the frequency of the emitted light be equal to the natural frequency of the spectral line. If each of the atoms in a given gaseous sytem had roughly the same velocity due to the overall motion of the gas (e.g. in the atmosphere of a star), the width of the spectral line would be unaffected, but the line would be displaced to higher or lower frequencies according to the direction of motion. When, however, the centre of mass of the system is fixed, and the atoms are in random motion and have a Maxwellian distribution of velocities, the line is broadened but not displaced, that is the spectral line emitted by this gas consists of a range of frequencies symmetrically disposed about the natural frequency.

From the Doppler principle it follows that if a source of light of wavelength λ_0 is moving in the line-of-sight, with a velocity v relative to the observer, the apparent wavelength (λ) measured by the observer will be:

$$\lambda = \lambda_0 \, (1 + v/c)$$

where c is the velocity of light. Motion away from the observer produces a shift to longer wavelengths, and such motion is taken as being positive.

Since the range of velocities increases with temperature so must the range of frequencies comprising the spectral line increase. The temperature of the atoms may be calculated in terms of the broadening of the spectral line and the mass (m) of the atoms.

1.11.3   External effects

One of the major causes in the broadening of spectral lines is that due to the pressure of the gas. When two atoms collide, if the time the two atoms are in collision is large compared with the mean time between collisions, the atom which is emitting or absorbing energy will be subjected to a strong atomic field. This leads to a shift of the spectral lines to the red, and the effect is particularly noticeable at high pressures. Such pressure broadening of the lines is dependent on the density, temperature, and nature of the gas within which the absorbing or emitting atom is situated. The effect on the spectral line for two species A and B in collision, where A is the neutral atom which is emitting, is dependent on the nature of B and the type of force operating between them. If B is a noble gas atom, the interacting force would be of the van der Waals type. However, if B were an ion, Coulomb forces would be present, and in, for example, a high electric current density discharge, where ions are produced, such interaction could bring about appreciable broadening of the spectral line and could outweigh the other effects.

1.12   HYPERFINE STRUCTURE OF SPECTRAL LINES

The examination of individual multiplet components with a spectrograph of the highest possible resolving power sometimes shows them to be composed of a number of lines. This *hyperfine structure* is spread over a narrow wavelength range (e.g. a few tenths of an Ångström unit) and can only be explained by concluding that the atomic nucleus is responsible for this further splitting.

Many elements are composed of isotopes which, although having an identical arrangement of extranuclear electrons, differ from one another by virtue of their differing masses. Since the mass of the nucleus enters into Rydberg's constant through the reduced mass of the atom μ, different isotopes will have slightly different values of Rydberg's constant. This variation in R_∞ will be particularly marked for hydrogen and deuterium. In fact, from the known Rydberg's constants for the two isotopes it is possible from theory to consider one particular line of a certain isotope, and calculate the wavelength at which the corresponding line for the other isotope would appear. This will now be illustrated for the 4861.33 Å line of the hydrogen atom with respect to the corresponding deuterium line.

For this 4861.33 Å hydrogen atom line, the result is:

$$\frac{1}{\lambda_H} = R_H \left(\frac{1}{2^2} - \frac{1}{4^2} \right) \tag{1.114}$$

For the corresponding deuterium line

$$\frac{1}{\lambda_D} = R_D \left(\frac{1}{2^2} - \frac{1}{4^2} \right) \tag{1.115}$$

Thus

$$\lambda_D/\lambda_H = R_H/R_D \tag{1.116}$$

and

$$\lambda_D/\lambda_H - \lambda_H/\lambda_H = R_H/R_D - 1 \tag{1.117}$$

$$\frac{\Delta\lambda}{\lambda_H} = -\frac{R_D - R_H}{R_D} \tag{1.118}$$

or

$$\Delta\lambda = -\lambda_H \frac{R_D - R_H}{R_D} \tag{1.119}$$

where $\lambda_D - \lambda_H = \Delta\lambda$. When the value of R_D and R_H are inserted in Equation (1.119) for the 4861.33 Å line of H, the corresponding line of D can be calculated to be shifted by a $\Delta\lambda = -1.32$ Å.

Thus, the deuterium line lies close to the hydrogen line at a wavelength 4860.01 Å. This result and others were verified experimentally by Urey.

Many cases of isotopic shift are not as simple as that for the hydrogen and deuterium lines considered. In fact, for the rare earths and heavier elements the main contribution to isotopic shift is not due to a difference in mass but results from the deviation of the nuclear magnetic field from being purely Coulombic.

When it was shown that certain elements contained only one isotope but still had hyperfine structure (e.g. bismuth had no isotopes, yet the 4722 Å line contained six hyperfine components) an additional type of explanation had to be sought. Pauli in 1924 suggested that the effect might be accounted for in terms of the nucleus spinning, and that owing to this rotation of (positive) charge there would be an associated magnetic moment and a corresponding angular momentum. The nuclear angular momentum is $\sqrt{[I(I + 1)]}h/2\pi$, where I is known as the *nuclear spin quantum number* and may take integral or half-integral values. I may differ for different isotopes and has the value of $\frac{1}{2}$ for the hydrogen atom but for deuterium is unity. If the nucleus has no spin, (i.e. $I = 0$), or has $I = \frac{1}{2}$, the nuclear charge has a spherical distribution; however, when I is greater than $\frac{1}{2}$ the nuclear charge is distributed ellipsoidally.

The magnetic moment μ associated with nuclear spin is:

$$\mu = \frac{e}{2m_p} \, p_I \qquad (1.120)$$

where m_p is the mass of the proton and p_I is the nuclear angular momentum given by:

$$p_I = I^* \cdot h/2\pi \qquad (1.121)$$

Substituting Equation (1.121) into (1.120):

$$\mu = \frac{e}{2m_p} \cdot \frac{h}{2\pi} \cdot I^* \qquad (1.122)$$

where the quantity $e/2m_p \cdot h/2\pi$ is known as the *nuclear magneton* μ_N that is:

$$\mu_N = eh/4\pi m_p \qquad (1.123)$$

Since the mass of the proton is 1836 times the mass of an electron, the nuclear magneton is 1/1836 times the Bohr magneton. Experimental observations indicate that, as with an electron, the magnetic moment to be associated with a spinning nucleus is greater than that given by Equation (1.122). By analogy with the transient of an electron a nuclear g-factor (g_N) may be introduced (see p. 73).[†]
The resultant electron angular momentum $J^*h/2\pi$ couples with the

[†] To be strictly accurate the right-hand sides of formulae (1.120) and (1.122) have to be multiplied by the nuclear g-factor (i.e. g_N see later).

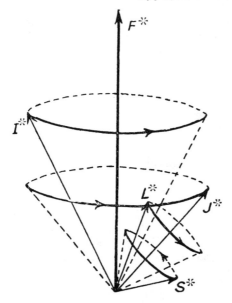

Fig. 1.43 Precession of $L^*h/2\pi$ and $S^*h/2\pi$ about their resultant $J^*h/2\pi$ and the precession of $J^*h/2\pi$ and $I^*h/2\pi$ about their resultant $F^*h/2\pi$. The vectors are in $h/2\pi$.

nuclear angular momentum $I^*h/2\pi$ in a way exactly similar to the coupling between $L^*h/2\pi$ and $S^*h/2\pi$. The coupling takes place through the magnetic moment associated with the spinning nucleus interacting with the resultant electron magnetic moment. Thus, I and J couple to give a resultant F where $F^*h/2\pi$ represents the total angular momentum of the atom for both the nucleus and the extranuclear electrons. Figure 1.43 illustrates how $L^*h/2\pi$ and $S^*h/2\pi$ couple to give $J^*h/2\pi$, and how $J^*h/2\pi$ and $I^*h/2\pi$ couple to give $F^*h/2\pi$.

The resultant quantum number F may take the values

$$F = J+I, J+I-1, J+I-2, \ldots, |J-I| \qquad (1.124)$$

There are then $(2J + 1)$ values of F for every J value when I is greater than J, and $(2J + 1)$ values of F for every J value when J is greater than I. The nuclear spin angular momentum may take up different orientations relative to the resultant electron angular momentum, and for each different orientation of the former with respect to the latter there is a different energy. For a given value of J the splitting of the energy levels owing to different values of F is characterized by the equation:

$$E_F = \frac{A}{2}\{F(F+1) - I(I+1) - J(J+1)\} \qquad (1.125)$$

where A is an interaction constant, and E_F is the interaction energy between the nuclear angular momentum and the electron angular momentum.

The selection rules for F in electron transitions between hyperfine levels are exactly similar to those for J in fine structure, that is $\Delta F = 0, \pm 1$ and the transition $F'' = 0 \leftarrow F' = 0$ is forbidden. As an example of the application of this selection rule the 4122 Å line of bismuth may be considered. This line is split into four hyperfine components. In the upper and lower electronic energy states $J = \frac{1}{2}$, and since $I = \frac{9}{2}$ (see p. 76) then F has the values 5 and 4.[†] This is illustrated in Fig. 1.44, and, in addition, the transitions are given and related to the four hyperfine lines.

1.12.1   Zeeman effect of hyperfine structure

In a magnetic field a space quantization of $F^*h/2\pi$ takes place exactly as for $J^*h/2\pi$. The precession of $J^*h/2\pi$ and $I^*h/2\pi$ about $F^*h/2\pi$ and the precession of $F^*h/2\pi$ about the magnetic field are illustrated in Fig. 1.45. The projection of $F^*h/2\pi$ in the direction of the field gives the value of $M_F h/2\pi$, there being $(2F + 1)$ values of M_F.

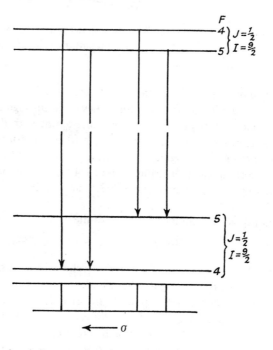

Fig. 1.44 Energy level diagram showing transitions giving rise to hyperfine lines for the 4122 Å line of bismuth. (*Reproduced, by permission from Atomic Spectra and Atomic Structure by G. Herzberg* [1.8]. *Copyright* 1944 *Dover Publications. Inc., New York*).

[†] It will be noted that in the higher electronic state the F terms are inverted. For a consideration of this p. 393 in White [1.5] should be consulted.

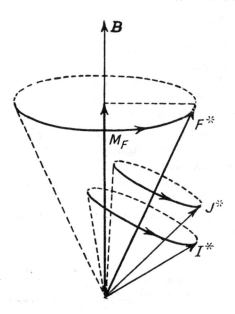

Fig. 1.45 Precession of $J^*h/2\pi$ and $I^*h/2\pi$ about their resultant $F^*h/2\pi$, and the precession of $F^*h/2\pi$ about a magnetic field. The vectors are in $h/2\pi$ units.

To illustrate the space quantization of $F^*h/2\pi$ in a very weak magnetic field consider the transition $7^2P_{3/2} - 7^2S_{1/2}$ in doubly ionized thallium. Under high resolution the line resulting from this transition is seen to be made up of three components, while in a suitable magnetic field the hyperfine lines each break up into a symmetrical pattern similar to the anomalous Zeeman pattern of fine structure in a suitable magnetic field. The energy level diagram for the $7^2S_{1/2}$ and $7^2P_{3/2}$ states of doubly ionized thallium is given in Fig. 1.46. For these states J' is equal to $\frac{3}{2}$ and $J'' = \frac{1}{2}$. The value of the nuclear spin quantum number for thallium is $\frac{1}{2}$. Hence the values of F in the upper state are 2 and 1, while in the lower state they are 1 and 0. The values of M_F are:

State	F values	M_F values
Upper	$\begin{cases}2\\1\end{cases}$	$\begin{aligned}&2, 1, 0, -1, -2\\&1, 0, -1\end{aligned}$
Lower	$\begin{cases}1\\0\end{cases}$	$\begin{aligned}&1, 0, -1\\&0\end{aligned}$

The selection rules for hyperfine structure are analogous to those for fine structure, and hence when observation is made perpendicular to the applied field, then:

$$\Delta M_F = \pm 1 \text{ for } \sigma \text{ components}$$

71

and

$$\Delta M_F = 0 \text{ for } \pi \text{ components}$$

The transitions responsible for the hyperfine structure components in the absence of a field are shown in the centre of Fig. 1.46; the splittings in a weak magnetic field are shown on the right of the figure. The theoretical number of lines in the presence of the field based on the selection rule $\Delta M_F = 0, \pm1$ would be twelve. To detect this Zeeman effect of the hyperfine structure components of a spectral line, and not to break down the coupling between the nuclear spin angular momentum and the total angular momentum of the electrons, would, in general, require a magnetic field strength of the order of 0.02 T; this would lead to a spacing of the energy between consecutive magnetic sub-levels of about 10^{-3} cm^{-1}. This leads to lines so close together in the resulting electronic transitions that it is not normally possible to separate them even with a spectrograph of high resolving power.

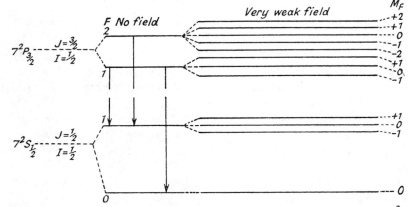

Fig. 1.46 Splitting of the F levels in the presence of a magnetic field for $7^2S_{1/2}$ and $7^2P_{3/2}$ states of Tl III. The transitions between the F levels in the absence of a magnetic field are indicated.

1.13   THE BACK–GOUDSMIT EFFECT

When the magnetic field strength is increased the rate of precession of $F^*h/2\pi$ around the field B increases. When at sufficiently high field strengths the velocity of precession of $F^*h/2\pi$ is of the same order as that of $J^*h/2\pi$ and $I^*h/2\pi$ around $F^*h/2\pi$, the coupling breaks down between $J^*h/2\pi$ and $I^*h/2\pi$, and these consequently become independently space quantized. Since the energy of interaction between $J^*h/2\pi$ and $I^*h/2\pi$ is small, a comparatively weak magnetic field (e.g. 0.3 T) will cause this type of effect which is known as the *Back–Goudsmit effect*. The components M_J and M_I in the direction of the field have the values: (a) $M_J = J, J - 1, J - 2, \ldots, -J$, that is there are $(2J + 1)$ values of the component M_J; (b) $M_I = I, I = 1, \ldots, -I$, i.e. $(2I + 1)$ values of

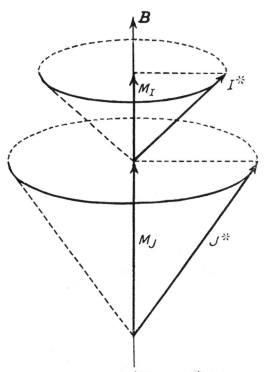

Fig. 1.47 Independent precession of $J^*h/2\pi$ and $I^*h/2\pi$ in a magnetic field of sufficient strength to produce this uncoupling (Back–Goudsmit effect). The vectors are in $h/2\pi$ units.

M_I. Fig. 1.47 is the vector diagram for $J^*h/2\pi$ and $I^*h/2\pi$ when they have uncoupled.

1.13.1 The 'g'-factors and the effects of various magnetic field strengths

Before considering an equation for the energy of the atom in a weak magnetic field such that only $J^*h/2\pi$ and $I^*h/2\pi$ uncouple, it is necessary to introduce another g-factor. So far we have mainly considered the Landé g_J-factor for the electron. However, other g-factors exist; for example, to relate the mechanical moment of the spinning nucleus to its magnetic moment μ a g_N-factor is required as given in the equation:

$$\mu_N = g_N \frac{e}{2m} \cdot \frac{h}{2\pi} \cdot \frac{1}{1836} \cdot \sqrt{[I(I+1)]} \qquad (1.126)$$

where

$$\frac{e}{2m} \cdot \frac{h}{2\pi} \cdot \frac{1}{1836} = \mu_N \qquad (1.127)$$

the nuclear magneton μ_N, e is the charge on the proton, m_e is the mass of the electron, and $1836m$ is the mass of the proton. The nuclear magneton is the unit of magnetic moment of the nucleus. Nuclear moments are often measured in nuclear magneton units, the magnitude of which is $5.050824 \times 10^{-27}\,JT^{-1}$ g_N is a non-dimensional constant known as the nuclear g-factor; both the magnitude and sign of g_N are dependent on the internal structure of the nucleus. If the vectors $I^*h/2\pi$ and $J^*h/2\pi$ couple together, this gives a resultant total angular momentum $F^*h/2\pi$ for the whole atom. In the relating of μ_F to F a g_F-factor is involved as is shown in the following equation:

$$\mu_F = -g_F\mu_B\sqrt{[F(F+1)]} \qquad (1.128)$$

In a similar way to that by which the value for the g_J-factor was derived in terms of J, L, and S, it is possible to obtain g_F in terms of g_J, g_I, F, J, and I. The basic equation is:

$$\mu_F = \mu_J \cos(J^*, F^*) - \mu \cos(I^*, F^*) \qquad (1.129)$$

This equation can be derived by inserting the magnetic vectors on to the angular momenta diagram in Fig. 1.43 then resolving μ_J and μ along the direction F^*.

On substitution in Equation (1.129) of the magnitude of μ_F, μ_J and μ from Equations (1.128), (1.79), and (1.126), and employing the cosine law:

$$g_F = g_J \left\{ \frac{F(F+1)+J(J+1)-I(I+1)}{2F(F+1)} \right\}$$

$$+ \frac{g_N}{1836} \left\{ \frac{F(F+1)+I(I+1)-J(J+1)}{2F(F+1)} \right\} \qquad (1.130)$$

As g_J and g_N are of the same magnitude, and the second term on the right-hand side is divided by 1836, it follows that this term may normally be neglected.

Having introduced g_I it is now possible to consider the energy equation when a Back–Goudsmit effect is taking place, that is in a magnetic field of sufficient strength that $J^*h/2\pi$ and $I^*h/2\pi$ have uncoupled. If E_0 is the energy value of the original electron level, then the difference in energy between a particular J level (E_H) and the original level in this magnetic field is given by:

$$E_B - E_0 = g\mu_B M_J B - g_N \frac{\mu_B}{1836} M_J B + B_I M_I M_J\dagger \qquad (1.131)$$

where B is an interaction constant. The second term on the right-hand side of Equation (1.131) takes into account the interaction of the nuclear angular momentum with the magnetic field, while the third term is the interaction energy between the orbital and nuclear angular momenta.

† The second term on the right-hand side of Equation (1.131) is small compared with the first and third and is often neglected.

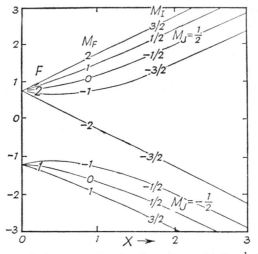

Fig. 1.48 Variation of the energy levels of an atom with $J = \frac{1}{2}, I = \frac{3}{2}$ as the magnetic field strength is increased, X is a non-dimensional field parameter. (*After Smith* [1.9]. *Courtesy of Methuen and Co. Ltd.*)

In Fig. 1.48 the effect of applying increasing strengths of very weak and weak magnetic fields on the energy levels of an atom where $I = \frac{3}{2}$ and $J = \frac{1}{2}$ may be observed. The plot is that of $E/\Delta E$ against X, where X is a non-dimensional field parameter, and:

$$X = (g_I - g_J) \frac{\mu_N}{\Delta E} B$$

where ΔE is the energy shift from the field free value. At X values around 1 the energy level values correspond with the space quantization of $F^* h/2\pi$, and there are $(2F + 1)$ values of M_F for each value of F. When X is in the region of a value of $\frac{7}{4}$, the splitting of the energy levels is that corresponding to the Back-Goudsmit effect.

Another way of illustrating the splitting of the energy levels in a very weak magnetic field is that employed in Fig. 1.49. The example chosen is that for $I = \frac{1}{2}$ for a $^2S_{1/2}$ term. The cases considered are:

(i) No interaction between $J^* h/2\pi$.

(ii) Interaction between $J^* h/2\pi$ and $I^* h/2\pi$.

(iii) The effect of a very weak magnetic field.

(iv) A weak magnetic field.

For this $^2S_{1/2}$ term where $I = \frac{1}{2}$ the F values are 0 and 1. When $F = 0$, then the corresponding M_F value would also be 0, but for $F = 1$, the M_F values are $0, +1$, and -1. Case (iv) above is that where the Back-Goudsmit effect operates, and it will be observed that the levels are symmetrically disposed with respect to the original level. The four energy levels ab, cd, ef, and gh represent the four values which E_H in Equation (1.131) may take when the appropriate

M_J and M_I values are inserted at this magnetic field strength B. Levels ab, cd, and ef result from the displacement of the former M_F levels of value $+1, 0, -1$ for the $F = 1$ value while level gh may be related to the $F = 0$ value.

The selection rules for the types of transitions involved in the Back–Goudsmit effect have a similar form to those employed for the fine structure in strong magnetic fields and are $\Delta M_I = 0$ and $\Delta M_J = 0, \pm 1$ where the $\Delta M_J = \pm 1$ leads

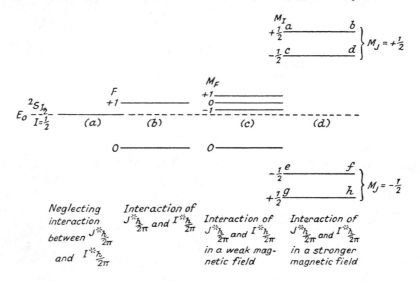

Fig. 1.49 Various cases of splitting of a $^2S_{1/2}$ energy level for an atom with $I = \frac{1}{2}$.

to σ components, while $\Delta M_J = 0$ gives π components. As may be discerned from Equation (1.131) and the fact that M_I has $(2I + 1)$ values, each term with a constant M_J is split into $(2I + 1)$ components, but since I is constant for a given nucleus, the number of components is constant for all terms. In fact the determination of the number of hyperfine lines resulting from the splitting of a spectral line in the presence of such a magnetic field provides an excellent method for the determination of I, as is illustrated in the following example of Back and Goudsmit.

For the bismuth 4722 Å line J in the upper state is $\frac{3}{2}$, and in the lower electronic state $\frac{1}{2}$ so that $M_J = \frac{3}{2}, +\frac{1}{2}, -\frac{1}{2}, -\frac{3}{2}$ and $+\frac{1}{2}, -\frac{1}{2}$, respectively. There are $(2I + 1)$ energy levels for each M_J value, and since M_I does not alter in going from one electronic state to another (i.e. $\Delta M_I = 0$), it follows that $(2I + 1)$ line components will be observed for each $M_J' - M_J''$ transition. Hence with a spectrograph of sufficient resolving power is merely necessary to count the line components for each such transition and equate the number to $(2I + 1)$. For example in the case of ^{209}Bi ten lines were observed; therefore $2I + 1 = 10$, and hence the nuclear spin quantum number of ^{209}Bi is $\frac{9}{2}$.

1.14 EFFECT OF VARIOUS MAGNETIC STRENGTHS ON NA D-LINES

Finally, to summarize the effect of the different strength of magnetic field on the Na D-lines:

(1) Jackson and Kuhn studied the Zeeman effect on the hyperfine structure of the 5890 Å Na D-line and photographed the lines in absorption. They found there was a hyperfine Zeeman effect in a magnetic field of 0.16 T, where the resulting spectrum could be analysed in terms of the M_F' and M_F'' values. At a field strength of about 0.3 T the line components altered and the Back–Goudsmit effect was observed where the external magnetic field had produced uncoupling of $J^*h/2\pi$ and $I^*h/2\pi$, each of which had become separately quantized and was precessing independently around B.

(2) At field strengths up to about 4 T for the Na D-lines the Zeeman effect is observed and transitions between the M_J' and the M_J'' values of different term values may be detected. Above $\sim$4 T the first sign of the Paschen–Back effect is observed,[†] and this is revealed by an asymmetrical distribution of the lines in the Zeeman pattern.

1.15 TYPE OF INFORMATION DERIVED FROM ATOMIC SPECTRA

1.15.1 Quantum numbers and the structure of the atom

It is now proposed to summarize how quantum numbers have emerged from spectral studies.

The spectrum of the hydrogen atom was explicable almost exactly by the difference of two terms. For a given series of spectral lines one of these terms was fixed and the other varied. Each of the spectral lines in such a series could be related to an integer which, when its value was inserted in the variable term, enabled the calculation of the wavenumber of a particular line. For example, in the Lyman series the wavenumbers were calculated from the formula:

$$\sigma = R_H - R_H/n^2$$

where $n = 2, 3, 4, 5, \ldots$, and R_H is the Rydberg constant. Each of these values

---

[†] The strength of the field required to produce these effects depends very much on the atom concerned, and in the case of the oxygen atom a field of about 9 T is necessary before a partial Paschen–Back effect is observed, while a field of 20 T is required to produce a complete Paschen–Back effect in the oxygen atom.

of n could be identified with a particular spectral line and in addition could be identified with a particular orbit of the electron. Furthermore, these values of n could be interpreted in terms of the Bohr theory to govern the angular momentum where the value of the angular momentum is $nh/2\pi$. An integer which fixes the angular momentum in this manner is known as a *quantum number*, n being, in fact, the principal quantum number. By means of the principal quantum number the spectrum of atomic hydrogen could be readily explained. However, when a spectrograph of extremely high resolving power is employed, the Balmer lines of hydrogen are found to be doublets, and the Bohr model becomes inadequate and the azimuthal quantum number (l) has to be introduced.

The spectra of the alkali metals can be partially explained by the quantum numbers n and l, as can also to some extent the spectra of helium and the alkaline earths. The reason that helium and the alkaline earths have more series than the alkali metals cannot be explained by using n and l alone. In addition, on only moderate resolving power being employed, certain lines in the principal series of Na, K, Rb, and Cs are seen to be doublets (see Fig. 1.15). These facts cannot be explained without the introduced, of a further degree of freedom, that is a new quantum number.

As we have seen, the quantum number j was introduced and was compounded of l and s; where $j = l \pm s$. The new degree of freedom really took into account the spin of the electron where s is the spin quantum number. With more than one valence electron the l's are coupled to yield L and the s's to give the resultant S, while the vector sum of L and S gives J. $J^*h/2\pi$ is the total angular momentum of the atom, $L^*h/2\pi$ the resultant orbital angular momentum, and $S^*h/2\pi$ the resultant electron spin angular momentum. For atoms with only one valence electron, j, l, and s are respectively equivalent to J, L, and S. With all the quantum numbers then available it was possible to explain the observed spectra of the alkali metals in the absence of a magnetic or an electric field.

So far all observed phenomena have been explained in terms of n and l, and also j and thus in two dimensions. On an atom being subjected to a magnetic field a third dimension may be introduced for the direction of the applied field is an arbitrary axis in space in terms of which the orientation of the two-dimensional plane of the orbit may be described. Further splitting of the spectral lines is detected in the presence of a magnetic field which cannot be accounted for solely by the quantum numbers n, l, and j.

From what has already been said it might be expected that an additional quantum number was introduced to account for the various possible orientations of the orbital plane with respect to the field. This is the magnetic quantum number (m_J) where $m_j h/2\pi$ is the value of the total electronic angular momentum in the direction of the field. For atoms with more than one valence electron m_j is replaced by M_J. Since all the integers or half-integers introduced so far have been subject to quantum conditions, it might be expected that not all orientations of the orbital plane with respect to the field are possible, but

only those for which m_j is an integral or half-integral multiple of $h/2\pi$ units.[†]
Thus, j must be space quantized. This space quantization is confirmed by the
Stern–Gerlach experiment. On the basis of four quantum numbers, n, l, s, and
m_j, it is possible to explain the Zeeman and anomalous Zeeman effects; even
when the nature of the external field is altered from magnetic to electric the four
quantum numbers are still adequate to account for the normal Stark effect
though M_L and M_S are required for strong electric fields. In addition, the
building up of the periodic table can also be explained. It is not until the
resolving power is increased almost to extremes that difficulties are encountered
by new factors, resulting once again in the splitting of the spectral lines. It is
difficult to visualize how this further splitting arises and what extra degree of
freedom must be specified in order to account for this hyperfine structure. To
resolve this problem the nucleus must be considered, and since the mass of the
nucleus may sometimes differ for the same element (isotopes), the multiplet
structure could result from this variation of mass. However, this explanation
goes only so far, since some element yield hyperfine lines but have no isotopes.
Thus, eventually, the only possibility which remains in the mechanical model of
the atom is to predict a nuclear spin governed by the nuclear spin quantum
number I which is subject to the usual quantum condition. The nuclear spin
angular momentum $I^*h/2\pi$ vectorially adds to $J^*h/2\pi$ to give the overall
angular momentum $F^*h/2\pi$ of the whole atom.

In conclusion, then, the introduction of a new quantum number becomes
necessary in order to explain the line splitting. The process has been one of
observation and initially of hypothesis where a new quantum number is added
when an extra degree of freedom is found.

1.15.2 Determination of J and g_J values
from the Zeeman pattern

If the magnetic splitting results in a completely resolved Zeeman pattern of a
spectral line, the J and g_J values of the energy levels concerned may be
determined from the spectrum. This experimental determination of g_J is
most important since its value indicates the type of coupling in the atom, and
it is found that in many atoms the L,S coupling for the electrons is often not
the ideal Russell–Saunders type. If the Zeeman pattern of a spectral line has
been completely resolved, then:

(a) If the lesser J value is J_1 its value is obtained by counting the number of

---

[†] Here the quantum-mechanical value is not $\sqrt{[m_j(m_j + 1)]}\,h/2\pi$ or
$\sqrt{[M_J(M_J + 1)]}\,h/2\pi$ but is $m_j h/2\pi$ or $M_J h/2\pi$. In the cases of j, l, s and F, J,
L, S, and I, however, the angular momentum takes the form
$\sqrt{[X(X + 1)]}\,h/2\pi$, where X is one of these quantum numbers.

π polarized components and equating them to $(2J_1 + 1)$.[†] As the π and σ components overlap it is then better to photograph the π polarized component separately.

(b) The g'_J and g''_J factors may be obtained. When an electronic transition takes place between two magnetic energy sub-levels, it follows from Equation (1.82) that the energy difference between one of the lines and that of the unsplit line is given by:

$$\Delta E = (M'_J \cdot g'_J - M''_J g''_J) \times h\nu_L \tag{1.132}$$

The corresponding frequency difference is given by:

$$\Delta\nu = (M'_J \cdot g'_J - M''_J g''_J)\nu_L \tag{1.133}$$

where $M'_J - M''_J = 0$ or ± 1. From the number of π components both J' and J'' can be determined, and thus the M'_J and M''_J values follow. Since $\Delta\bar{\nu}$ is readily determinable from the spectrum and the Larmor frequency can be calculated from the values of e, m, and B, then from Equation (1.133) the value of $M'_J \cdot g'_J - M''_J g''_J$ is obtained. Thus, to determine g'_J and g''_J another equation of the form of (1.133) is required. A number of these equations is obtained from the other line component, and the best values of the two unknowns g'_J and g''_J are assigned.

1.15.3   Building up of the periodic table of elements

By the introduction of five quantum numbers, n, l, s, m_j, and I, the atom can be characterized, while the valence electrons can be characterized by n, l, s, and m_j. From the quantum numbers n, l, j, and m_l, and a principle formulated by Pauli in 1925, called the *Pauli exclusion principle*, it is possible to build up the electronic configuration of the elements in harmony with the periodic table. Briefly, Pauli's principle states that two electrons in an atom may not possess identical sets of the four quantum numbers n, l, m_l and m_s, where $m_l = l$, $(l - 1), (l - 2), \ldots, -l$ and $m_s = +\frac{1}{2}$ or $-\frac{1}{2}$. This is the case of an atom in a sufficiently strong magnetic field such that the coupling between $l^*h/2\pi$ and $s^*h/2\pi$ is broken, and the $l^*h/2\pi$ and $s^*h/2\pi$ vectors are thus space quantized independently of each other in the direction of the field. Each horizontal line in

---

[†] When, however, $J_1 = J_2 = J$, since the $M_J = 0 \leftarrow M_J = 0$ transition is forbidden, the number of π components is only $2J$. Thus, by equating the number of π components to $2J$ the J-values for both the levels can be determined.

Table 1.11 gives the four quantum numbers which characterize an electron.[†] In general, the number of electrons with a given value of l which may exist with the same value of n is $2(2l + 1)$. The maximum number of electrons which may exist in an atom for a given value of n is the sum of the maximum number of electrons permitted in the $s(l = 0)$, $p(l = 1)$, $d(l = 2)$, and $f(l = 3)$ sub-shells. Thus, the total number of electrons is given by:

$$\sum_{l=0}^{l=n-1} 2(2l + 1) = 2n^2 \tag{1.134}$$

If, therefore, $n = 1$, the total number of electrons = 2

2	8
3	18
4	32

The building up of the periodic table follows the *Aufbauprinzip* of Bohr and Stoner. Electrons are added one after another to orbital positions about a nucleus and distribute themselves in accordance with the Pauli exclusion principle and according to the principle that the energy of the resulting system should be a minimum.

When n is equal to 1, the maximum number of electrons in this shell is 2.1^2. Since the values which l can take are:

$$n - 1, n - 2, \ldots 0$$

it follows that $l = 0$. Hence $m_l = 0$ and:

$$m_s = +\tfrac{1}{2} \text{ and } -\tfrac{1}{2}$$

For the case $n = 2$, the maximum number of electrons in this shell is 2.2^2. The possible values of l are 1 and 0. For the $l = 0$ value $m_l = 0$ and $m_s = +\tfrac{1}{2}$ and $-\tfrac{1}{2}$; this identifies the two 2s electrons in Table 1.11. When $l = 1$, m_l may take the values $+1, 0, -1$. For each of these m_l values there are two m_s values of $+\tfrac{1}{2}$ and $-\tfrac{1}{2}$. Thus, the six 2p electrons in Table 1.11 may be accounted for. In this manner the electronic configurations of the elements is built up and may be correlated with the periods in the periodic classification of the elements.

The possible states of an electron by the application of the Pauli exclusion principle having been determined, the term types of the ground and excited

[†] According to the more modern theory of quantum numbers it might be expected that the four quantum numbers, n, l, j, m_j, would be the appropriate ones to characterize the atom, rather than to select a most arbitrary case of an atom in a strong magnetic field. However, these give the same number of possible states for an electron as the $n, l, m_l m_s$ quantum numbers. The latter were chosen here because they are in such wide use, although the former are really more suitable in terms of the actual existence of the atom. Nevertheless, solely from the point of view of building up the periodic table either method leads to a satisfactory model.

Table 1.11 Build-up of the periodic table

n values	l values	m_l values	m_s values	Number of electrons in shell, and type
(a) 1	0	0	$+\frac{1}{2}$	2(1s)
1	0	0	$-\frac{1}{2}$	electrons
(b) 2	0	0	$+\frac{1}{2}$	2(2s)
2	0	0	$-\frac{1}{2}$	electrons
2	1	$+1$	$+\frac{1}{2}$	
2	1	$+1$	$-\frac{1}{2}$	
2	1	0	$+\frac{1}{2}$	6(2p)
2	1	0	$-\frac{1}{2}$	electrons
2	1	-1	$+\frac{1}{2}$	
2	1	-1	$-\frac{1}{2}$	
(c) 3	0	0	$+\frac{1}{2}$	2(3s)
3	0	0	$-\frac{1}{2}$	electrons
3	1	$+1$	$+\frac{1}{2}$	
3	1	$+1$	$-\frac{1}{2}$	
3	1	0	$+\frac{1}{2}$	6(3p)
3	1	0	$-\frac{1}{2}$	electrons
3	1	-1	$+\frac{1}{2}$	
3	1	-1	$-\frac{1}{2}$	
3	2	$+2$	$+\frac{1}{2}$	
3	2	$+2$	$-\frac{1}{2}$	
3	2	$+1$	$+\frac{1}{2}$	
3	2	$+1$	$-\frac{1}{2}$	
3	2	0	$+\frac{1}{2}$	10(3d)
3	2	0	$-\frac{1}{2}$	electrons
3	2	-1	$+\frac{1}{2}$	
3	2	-1	$-\frac{1}{2}$	
3	2	-2	$+\frac{1}{2}$	
3	2	-2	$-\frac{1}{2}$	

electronic states of an atom will now be considered. We shall examine the two cases of:

(a) Atoms with one external electron.

(b) Atoms with two or more outer electrons which are non-equivalent, that is, the electrons which have either different n or l values and for which Russell—Saunders coupling applies to the individual momentum vectors associated with the electrons.

In the case of a completed subshell, for which the value of the orbital angular momentum of the electrons is constant, and for the case of a completed shell [in which the electrons all have the same value of the principal quantum number (n)] the resultant orbital $(L^*h/2\pi)$ and spin $(S^*h/2\pi)$ angular momenta are zero. Consequently in computing possible term types of an atom it is necessary to consider only those electrons external to complete shells or subshells.

Case (a)

For a hydrogen-like atom the lowest (ground) state is where $n = 1, l = 0$. Since $S = s = \frac{1}{2}$ the multiplicity $(2S + 1) = 2$ and the value of $J = j = \frac{1}{2}$ [see Equation (1.59)]. The ground state of the hydrogen atom is thus a $^2S_{1/2}$ state. Excited states of the hydrogen atom correspond to other pairs of n and l values; these are according to the value of l of $0, 1, 2, \ldots$ $^2S, ^2P, ^2D, \ldots$ states respectively. For lithium and sodium the $(1s), (1s, 2s, 2p)$ shells respectively are full and need not be considered. The ground state of Li is a $^2S_{1/2}$ term as is also that of the Na atom.

Case (b)

In order to evaluate the terms of an atom with two outer electrons, the possible values of L, the resultant orbital angular momentum quantum number, must first be found. To illustrate the procedure the example of an atom with an outer p-electron and a d-electron will be considered. The possible L values obtained by vectorial combination of $l = 1$ and $l = 2$ are, according to Equation (1.58), 1, 2, and 3 so that these non-equivalent electrons form P, D, and F terms. The two spin orientations are ↑↑, ↑↓ so the resultant spin angular momentum quantum number $S = 1$ or 0, the multiplicities are 3 and 1 respectively, and there are six terms: $^1P, ^1D, ^1F, ^3P, ^3D, ^3F$. The terms corresponding to some other electron configurations are given in Table 1.12.

Table 1.12 Terms from two non-equivalent electrons

Electron configuration	Terms
s s	^1S ^3S
s p	^1P ^3P
s d	^1D ^3D
p p	^1S ^1P ^1D ^3S ^3P ^3D
d d	^1S ^1P ^1D ^1F ^1G ^3S ^3P ^3D ^3F ^3G

For three non-equivalent electrons the term types are evaluated by forming the L resultant between two of the electrons, as was done in the previous example, and then vectorially combining these L values with the l value of the remaining electron. A simple example is that of $s + p + d$ when $L = 1, 2$, and 3. The spin combinations are obtained by first combining s_1 and s_2. This leads to singlet (i.e. $S = 0$) and triplet terms $(S = 1)$. From the singlet term only a doublet term may arise on combination with s_3, whereas the triplet term leads to a doublet plus a quartet $(S = \frac{3}{2})$. Thus, the multiplicities are 2, 2, and 4 respectively. The

83

feasible terms are therefore: ^2P, ^2D, ^2F, ^2P, ^2D, ^2F, ^4P, ^4D, and ^4F. The terms for some other instances of three non-equivalent electrons are given in Table 1.13.

Table 1.13 Terms from three non-equivalent electrons

Electrons configuration	Terms
s s p	^2P ^2P ^4P
s s d	^2D ^2D ^4D
s p p	^2S ^2P ^2D ^2S ^2P ^2D ^4S ^4P ^4D

An important rule exists which helps in deciding the relative order of the different energy states for a particular electron configuration having the same n and L values. The rule is one by Hund which states: *Of the terms given by equivalent electrons (i.e. same n and L value) those with the greatest multiplicity lie deepest, and of these the lowest is that with the greatest L values.* Hund's rule has been found to apply to many electron configurations in atoms especially for those involving the ground state. For the nitrogen atom which has the electronic structure:

$$1s^2 2s^2 2p^3$$

there are three equivalent valence electrons which are to be associated with the optical spectrum. These $2p^3$ electrons are characterized by the terms ^4S, ^2P, and ^2D. On application of Hund's rule it is seen that the ^4S is the ground state, and the next above it will be the ^2D then the ^2P.

So far in this treatment of term types the right-hand subscript, that is, the J value has been omitted. The subscripts for singlet, doublet, and triplet terms will now be considered. For singlet terms we have:

$$^1S_0; \, ^1P_1; \, ^1D_2; \, ^1F_3$$

for doublet terms:

$$^2S_{1/2}; \, ^2P_{1/2}; \, ^2D_{3/2}; \, ^2F_{3/2}$$

$$^2P_{3/2}; \, ^2D_{5/2}; \, ^2F_{7/2}$$

while for triplet terms:

$$^3S_1; \, ^3P_0; \, ^3D_1; \, ^3F_2$$
$$^3P_1; \, ^3D_2; \, ^3F_3$$
$$^3P_2; \, ^3D_3; \, ^3F_4$$

The J values can be readily verified by vectorially combining the L and S values. In order to decide which member of the multiplet lies lowest a second rule is helpful. This rule states: *Multiplets formed from equivalent electrons are normal when less than half the shell is occupied but inverted when more than half the shell is filled.* A normal term is defined as one in which the multiplet with the smallest J values lies deepest on the energy level diagram, and an inverted term is one in which the largest J value lies lowest. For two equivalent p-electrons as in the carbon atom the electronic structure of which is:

$$1s^2 2s^2 2p^2$$

the terms are ^1S, ^1D, and ^3P. The ground state is ^3P, and of the $^3P_0$, $^3P_1$, and $^3P_2$ multiplets the lowest will be that with the smallest J value, that is the $^3P_0$, since the 2p subshell is less than half full. The oxygen atom, on the other hand, has the electronic structure:

$$1s^2 2s^2 2p^4$$

The terms are again ^1S, ^1D, and ^3P, the ground state being a ^3P term. However, since the 2p subshell is more than half full, the multiplets will be inverted and the ground state will be the $^3P_2$ term. Table 1.14 summarizes the ground electronic state for some of the commoner elements.

For a detailed treatment of the building up of the periodic table White [1.5] should be consulted, while a more elementary treatment is given by Kronig [1.10].

1.15.4 Inadequacy in Dirac quantum-mechanical theory

Since the nineteen-thirties great interest has centred on the small differences between the observed spectrum and that predicted by the quantum-mechanical approach of Dirac. The discrepancy was first noted in the fine structure of the 6563 Å line in the Balmer series of the hydrogen atom. It was not until 1945, however, that a microwave experiment by Lamb and Retherford on the hydrogen atom showed beyond doubt that some displacement of the energy levels (other than what would be expected in the Dirac approach) had occurred. Such displacements are termed Lamb shifts.

Table 1.14 Ground electronic state for some of the commoner elements

Group in periodic table	Elements		Ground electronic state
1	Hi, Li, Na, K, Rb, Cs.	Cu, Ag, Au.	$^2S_{1/2}$
2	Be, Mg, Ca, Sr, Ba.	Zn, Cd, Hg.	$^1S_0$
3	B, Al, Ga, In, Tl.		$^2P_{1/2}$
4	C, Si, Ge, Sn, Pb.		$^3P_0$
5	N, P, As, Sb, Bi.		$^4S_{3/3}$
6	O, S, Se, Te.		$^3P_2$
7	F, Cl, Br, I.		$^2P_{3/2}$
8	He, Ne, Ar, Kr, Xe.		$^1S_0$

Some of the important features in the Lamb–Retherford approach are listed below:

(1) They examined the possibility of stimulating transitions between fine structure levels of the same principal quantum number, although they did not seek to detect the transitions by following absorption but looked for changes in the atoms themselves.

(2) The transition from the $2^2S_{1/2}$ level to the ground state $1^2S_{1/2}$ is forbidden, and it was anticipated that the $2^2S_{1/2}$ state would have a long life. In fact, their method depends on this $2^2S_{1/2}$ state being metastable while the $2^2P_{3/2}$ and $2^2P_{1/2}$ levels are not.

(3) They produced atomic hydrogen by thermally dissociating H_2 in a tungsten oven at 2500 K, and the hydrogen atoms were then drawn into an evacuated atomic beam apparatus. The resulting beam of hydrogen atoms was then cross bombarded by an electron beam which had sufficient energy to excite the hydrogen atoms to the $2^2S_{1/2}$, $2^2P_{1/2}$, and $2^2P_{3/2}$ levels. The hydrogen atom beam traversed a homogeneous magnetic field produced by an external magnet, and the $2^2S_{1/2}$, $2^2P_{3/2}$, and $2^2P_{1/2}$ levels could be separated into their Zeeman components. By varying the strength of the magnetic field it then becomes possible to employ a fixed-frequency oscillator to induce transitions between the $2^2S_{1/2}$ level and one of the 2^2P levels.

(4) The atoms were detected by directing them on to a tungsten plate. This produced an ejection of electrons, which were then collected on an anode connected to a sensitive d.c. amplifier.

Transitions were induced from the Zeeman components of the metastable $2^2S_{1/2}$ state to one of the unstable 2^2P levels, and when they were in a 2^2P state, they decayed at once to the ground state. This resulted in a reduction in the metastable atom beam intensity. When this took place resonance was inferred, that is the frequency necessary to induce a transition between the Zeeman components of $2^2P - 2^2S_{1/2}$ at a particular magnetic field strength was determined.

(5) From a study of such transitions between the magnetic energy levels, their results indicated that a frequency separation of about 1000 MHz must be postulated between the $2^2S_{1/2}$ and $2^2P_{1/2}$ states at zero field strength in order to explain the experimental shifts.

One of the requirements of the Dirac theory was that for hydrogen-like atoms states of a particular n value having terms with the same j value should be degenerate. However, the Lamb-Retherford work indicated that the $2^2S_{1/2}$ level was not degenerate with the $2^2P_{1/2}$ but lay about 1000 MHz above it. Eventually Lamb was able to determine the $2^2P - 2^2S_{1/2}$ separations to within 0.1 MHz. Similar work on He^+ showed that the corresponding levels were separated by 14 000 MHz; again according to the Dirac theory these ought to have been degenerate.

The Lamb-Retherford type of experiments led to further theoretical development, and methods have evolved of theoretically treating the atom not as an isolated unit but one engaged in interaction with the electromagnetic field.[†] Such an approach has accounted not only for the Lamb shifts but has also predicted a small alteration in the value of the magnetic moment of the electron. The latter has been confirmed by recent measurements on the magnetic moments of free atoms.

In many ways the spectral studies of the hydrogen atom have been most rewarding in the testing of the theories of atomic structure. In fact, the inadequacies of successive theories have emerged through studies at increased resolving power.

1.15.5 Qualitative emission spectroscopy

When a chemical element is suitably excited, it emits a characteristic spectrum which is unique for that element. The spectrum is generally recorded photographically. Excitation can be accomplished by various means, the most powerful methods being the electric arc and spark; the d.c. arc has the advantage that more material is excited. The material under examination is placed in a hollow bored out of the negative electrode, usually made of graphite. has a fairly high electrical conductivity high current discharges can be readily maintained. The extremely high temperature developed is suitable even for excitation of the most refractory material.

The spectrum emitted by the excited elements may be quite simple and have comparatively few lines (e.g. for the alkali metals and alkaline earths), or may be very complex (e.g. for the rare earths and the transition elements). A spectrograph of good resolving power is necessary for a detailed examination of complex spectra. Theoretically, all elements can be detected spectrographically, but in reality the method is restricted to the metals and metalloids.

The dispersing elements of the spectroscope can be made of glass, but the range is then limited to the region in which glass is transparent, approximately 3400 Å to 10 000 Å. The range can be extended by the use of quartz optical components and by the use of diffraction gratings.

One advantage of qualitative emission spectroscopy is that it is not necessary to examine all the lines in the spectrum of an element, but only the so-called *persistent lines*. The persistent lines are those which are last to disappear as the concentration of the element is gradually decreased. The identification of the persistent (the *raies ultimes*) lines proves without doubt the presence of the corresponding element. The persistent lines are generally identified by comparison with a standard. For example, there is available commercially a

---

[†] Optical spectroscopy has now shown the displacement in the energy levels of $n = 1, 2,$ and 3 of deuterium compared with that to be expected from the Dirac approach. Very high resolving power was employed, and the Doppler broadening was minimized by working at the temperature of liquid hydrogen. The measured shifts were: $2^2 S_{1/2}$ 0.0369 cm^{-1} and $3^2 S_{1/2}$ 0.0083 cm^{-1} which were in satisfactory agreement with the values of 0.0353 and 0.0105 cm^{-1} which were calculated on the basis of this new theory.

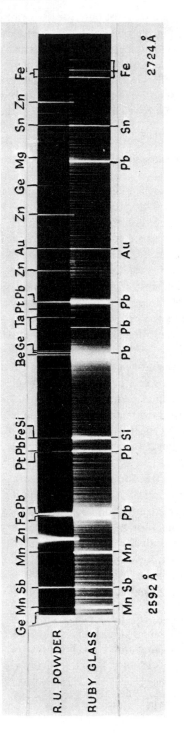

Fig. 1.50 Identification of elements present in ruby glass by means of the emission spectrum between 2592 Å and 2724 Å of R.U. powder. In these spectrograms the emission lines are white (Courtesy The Research Laboratories of the General Electric Co. Ltd., Wembley, England).

powder called the R.U. powder, which is prepared so that on excitation it emits the persistent lines of some fifty elements which can be used as a standard. The lines of unknown wavelength of the unidentified element may probably be matched up with corresponding lines in the standard. The identification of Mn, Pb, Sb, Au, Sn, and Fe in ruby glass by means of the *raies ultimes* spectra is illustrated in Fig. 1.50.

It is rarely necessary in chemical analysis to determine the wavelength of any lines in absolute values, though this can be done very accurately using an interferometer (as in the case of the standard Cd line which was the ultimate wavelength standard) or by the method of interpolating an unknown line between two lines the wavelengths of which are known [1.11]. The usual procedure is to compare the unknown spectrum with a standard spectrum. The operation is conveniently carried out on a comparator which brings both spectra into focus simultaneously, and line coincidences can be sought by racking both spectra backwards and forwards independently. With experience it is possible to recognize groups of lines immediately, and this considerably assists in the interpretation of the spectrogram.

Examples of the types of problems to which qualitative analysis can be applied are listed below, where the principle is to identify certain atomic lines:

(1) Rapid qualitative analysis, including an approximate quantitative estimation by the intensity of the various lines.

(2) Examination of materials which are subject to rigid specification, for example, aluminium and zinc alloys.

(3) Testing precipitates to ensure freedom from impurities.

(4) Analysis of fine chemicals, for example, A.R. quality reagents.

(5) Analysis of ores, refractory materials, glasses, slags, clay, and scrap material.

(6) Forensic analysis.

(7) Food, drugs, and other organic chemicals.

(8) Two of the important analytical problems which emerged from nuclear energy work were:

(i) The analysis of impurities in materials containing proportions of, for example, uranium and plutonium.

(ii) Isotopic abundance analysis.

On account of the complexity of the spectra straightforward spark methods are often not satisfactory in the analysis of impurities in, for example, uranium, plutonium, and rare earth metals, and steps have to be taken to reduce the background effects due to the spectrum of the uranium itself. In one Russian method the material is initially heated in a furnace and the impurities are driven off. These impurities are collected on an electrode which is employed in the arc or spark process. In this manner a higher concentration of the impurity can be investigated since the spectrum is no longer dominated by the many lines of the material present in excess.

Two elements with isotopes that are important in nuclear energy work are hydrogen and uranium. Although the estimation of the abundance ratio of hydrogen—deuterium mixtures has largely been made by means of molecular spectroscopy (see Vol. 3), use has also been made of the relative intensity of the emission lines of H and D to determine their relative concentrations.

Two two most important isotopes of uranium are ^{238}U and ^{235}U, the latter being present ot less than 2 per cent in natural uranium. Of these two isotopes only the ^{235}U undergoes fission with thermal neutrons and consequently the abundance ratio of these isotopes will alter as fission proceeds in a thermal reactor. The reactor performance will depend on the proportion of ^{235}U in the fuel elements and, in addition, the chain reaction will no longer be self-sustaining, should the ^{235}U content be insufficient. Thus, the relative proportions of the various isotopes is extremely important. Much of this work has been done by means of mass spectrometry, but determinations have been made using certain lines in the atomic spectrum of uranium. In Fig. 1.51 part of the spectra of the isotopes ^{238}U, ^{236}U, and ^{235}U is given together with an iron calibration spectrum below. In addition, a thorium spectrum has been photographed above for accurate wavelength measurement. This thorium spectrum was produced by the microwave source tube method already considered. The spectra were obtained by means of a plane grating spectrograph. The lack of coincidence of the lines due to the three isotopes will be noted: in particular for the 4244.4 Å lines the wavelength differences are:

$$^{238}U - ^{236}U = 0.18\,\text{Å}$$

$$^{236}U - ^{235}U = 0.07\,\text{Å}$$

In general, for isotopic analysis work very high resolution is necessary, and potentially the use of an interferometer of the Fabry—Perot type is attractive. For analytical uses interferometers have presented serious disadvantages. However, recently these have been overcome to some extent, and in particular the use of special coatings has increased the light-gatherine power.

1.15.6   Quantitative spectrochemical analysis [1.12a, b]

The intensity of an emitted spectral line of an element depends partly on the source of excitation and the concentration of the element. Provided that suitable means are available for measuring the intensity of the emitted radiation the concentration of the element can be determined. The accuracy of the method depends on:

(i) The range of standard samples available.

(ii) Maintaining the excitation conditions and the exposure time constant for all the spectra to be employed in the estimation.

Prior to 1925 quantitative analysis of substances from their atomic spectra was largely a matter of visually comparing line densities on the photographic plate of samples of unknown compositions with the densities of corresponding

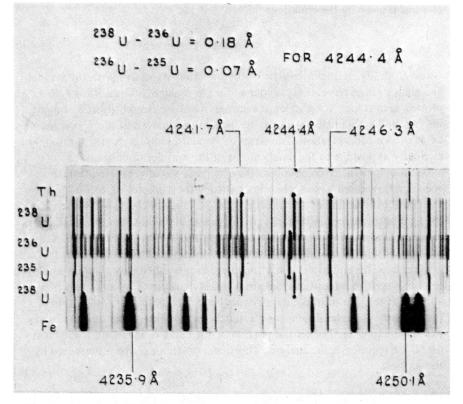

$$^{238}U - {}^{236}U = 0.18 \text{ Å}$$
$$^{236}U - {}^{235}U = 0.07 \text{ Å} \quad \text{FOR} \quad 4244.4 \text{ Å}$$

calibration spectrum lies below and a standard thorium spectrum above
the uranium spectra. (Courtesy of Dr. L. Bovey and the United Kingdom
Atomic Energy Authority, Harwell).

lines produced from a series of standard samples of known composition. In 1925,
however, Gerlach [1.12b] introduced the internal standard method which today
is the basis of practically all quantitative spectrographic techniques. The method
depends on the comparison of the intensity of a line of the element being
determined (analysis line) with that of an element present or added in fixed
concentration (standard line). In the examination of metals and alloys the
internal standard line is usually a weak line of the main constituent. If a suitable
internal standard is not present, a strong line of an element not originally present
in the sample may be employed. This element is added in a fixed small amount
to the sample under investigation. The selection of the internal standard line is
such that variations in the excitation conditions would affect it in a manner
identical to that experienced by the analysis line (such line pairs are said to be
homologous). In this way the intensity ratio of the analysis line to the standard
line would not be influenced by the excitation process but would depend on the
concentration of the element in question.

The relationship between the light intensity and the concentration (c) of an element is given by the equation:

$$\log \frac{I_2}{I_1} = n \log c + \log a \qquad (1.135)$$

where I_1 and I_2 are the intensities of the radiation associated with the standard and analysis lines respectively, and a and n are constants. Thus, if $\log I_2/I_1$ is plotted against $\log c$ for a series of samples of known composition, a straight line should be obtained which may be used in the analysis of samples containing an unknown concentration. The series of standard samples should approximate as closely as possible to the composition of the samples to be analysed.

Although intensity comparisons may be made visually, where greater accuracy is required a photoelectric densitometer is employed. Since, however, the density of the line on the photographic plate is not the same as intensity of the light absorbed or emitted, the photographic plate must be calibrated. As no exact mathematical expression exists relating the density of the image to the intensity of the light producing it, the relationship has to be expressed graphically. This emulsion calibration curve as it is called may then be employed to obtain relative intensities from given image densities. The curve which consists of the plot of image density (ordinate axis) against the logarithm of the intensity increases slowly at first, reaches a maximum, and then decreases with over-exposure. The central portion of the curve is linear and is the most useful for spectrographic analysis. This linear portion may be represented by the formula:

$$d = \gamma \log I - K \qquad (1.136)$$

where d is the density of the image, I the intensity of light falling on the plate, and K and γ are constants. γ, the slope of the line, is called the contrast. Since the contrast of a photographic emulsion varies with:

  (a) type of emulsion,
  (b) age of the emulsion,
  (c) development conditions,
  (d) wavelength of light,

For accurate spectrographic analysis a calibration curve is necessary for each photographic plate for each particular wavelength examined. For details of some of the methods available for obtaining emulsion calibration curves Strouts, Gilfillan, and Wilson [1.12c] should be consulted.

Direct-reading spectrographs, which employ for example photomultiplier tubes as the radiation detector, eliminate some of the disadvantages and inaccuracies of the photographic emulsion and are commercially available. Such instruments, however, are not at present suitable for the determination of constituents present in small concentration, and their limited flexibility makes them more suitable for routine analysis.

The chief sources of error in quantitative spectrochemical analysis are:

  (a) Inhomogeneity of the sample.
  (b) Instability of the excitation process.

(c) Determination of the line density.

(d) An unsuitable concentration of the element under investigation.

The arc, though more sensitive than the spark, is very erratic, and this variation may lead to an error in the determination of up to 50 per cent of the amount estimated. The spark, however, is more stable, and errors can be kept down to 2 per cent of the amount involved. Ideal concentrations range from about 0.005 to 2.00 per cent, but on the lower limit trouble is encountered in correcting for background intensity.

Examples of the type of quantitative analysis problems studied are:

(a) The determination of the amounts of impurities in aluminum, namely, Fe, Si, Ti, Cu, Mn, Mg, Zn, and Ni.

(b) Sodium in duralumin.

(c) Estimation of the amounts of impurities in Cu alloys, namely, Bi, Sn, Pb, Fe, Al, Si, Sb, As, Zn, P, Se, and Te.

In conclusion, the applications of spectrographic analysis are many and varied, and the advantages over traditional chemical analysis are speed and sensitivity. The sensitivity may well be of the general order of several parts per million.

1.15.7   Atomic spectra studies of some heavier elements

The spectra which arise from the rare earths (La to Lu) and heavier atoms including the actinide series (Ac to Cm) are very complex. For an atom such as sodium the number of lines recorded in conventional sources is of the order of fifty whereas in such heavier atoms thousand of lines may be obtained for each element. The complexity of the spectra from these elements may be attributed to the fact that a large number of the electronic states have a high multiplicity and large angular momenta.

To study these many-line spectra, high-dispersion instruments are desirable, and grating spectrographs sometimes in conjunction with interferometers are employed. Several large grating spectrographs now exist which under ideal conditions can potentially determine wavelengths to an accuracy of 1 part in 10^7.

Wavelength determination of the lines in a spectrum is usually carried out by superimposing the accuractely known wavelengths from another element(s) on the spectrogram. Both the relative positions of the lines of known and unknown wavelengths are then measured by means of a travelling microscope under which the photographic plate is mounted on a movable carriage. This carriage may be moved backwards or forwards by means of a screw coupled to a drum normally calibrated in divisions corresponding to $100\,\mu m$ of travel while an attached vernier permits estimation to $10\,\mu m$. For a grating spectrograph there is approximately a linear dependence of the positions of the lines on the photographic plate on the wavelength. This enables linear interpolation between standard reference lines to be made.

The accuracy of the standard lines is one of the governing factors in the ultimate accuracy of the determination. The iron arc which is a widely used standard is unsatisfactory for some investigations because of the scarcity of lines in some regions and the lack of exact reproducibility of the line wavelengths. Thorium has been tried as a better standard since this element has many more lines uniformly spaced.

In addition to high accuracy wavelength measurement and the necessity for a good standard wavelength source, a means of rapidly measuring these tens of thousands of lines is also desirable. The conventional method of linear interpolation is a lengthy and fatiguing process so automatic measuring methods have been developed. In such a method the position of the screw governing the travelling microscope is recorded automatically, and photoelectric devices have replaced the measuring microscope, hence reducing the strain in the visual examination of photographic plates.

For studies on the spectra emitted from rare earth and transuranic elements it is desirable to have a steady, intense, and long-lived source giving fine lines. For hyperfine structure or isotope shift studies the lines have to be as narrow as possible. One suitable source used by several spectroscopists in the United States and by Bovey and Wise in England [1.13a] consists of sealing the halide of the metal under examination into a quartz tube together with a rare gas at a low pressure. The metal atom is obtained in an excited electronic state by directing microwave radiation of 2450 MHz (12 cm wavelength) into the quartz tube. In this way a strong atomic emission spectrum of the metal results, and this approach has the advantage in the case of poisonous materials, e.g. plutonium, of being entirely sealed. The preparation of the source involves heating anhydrous aluminium iodide with the oxide of the rare earth or transuranic element at 773–873 K and subsequent sublimation *in vacuo* of the metal iodide into an evacuated quartz tube. Neon at low pressure is also added to the tube prior to sealing off. For relatively involatile halides, the source tube is heated in a furnace.

When a large number of spectral lines is obtained in a narrow wavelength range, it is desirable to increase the dispersion of the lines as much as possible and also to achieve the best accuracy of measurement of the relative positions of the lines. In the visible region at a wavelength of about 5000 Å (20 000 cm^{-1}) in order to determine the wavenumber to 0.01 cm^{-1}, it is essential to measure the wavelength to 0.002 Å. Such an accuracy in wavelength is tending to the limit which even the best grating instruments can achieve. It can be shown that the error $\Delta\sigma$ in the wavenumber σ is related to the error $\Delta\lambda$ in the wavelength λ by the relation $\Delta\sigma = \Delta\lambda/\lambda^2$. Thus, the advantage of working at wavelengths in the infrared rather than the visible region is the greater accuracy of wavenumber determination. With this in view atomic spectra studies of the rare earth and transuranic elements have recently been extended from the photographic region into the infrared between 1 and 3 μm. This development has been made possible by:

  (i) The introduction of sensitive detectors, for example, the lead sulphide photoconductive cell. This is a semiconductor device which when cooled to 233 K or less can detect radiation over the range 0.4 to 3.5 μm and allows fast scanning of the spectrum. Its response time is much quicker than that of the thermocouple which is in wide use for molecular infrared spectra studies.

  (ii) The availability of relatively cheap Merton NPL replica gratings suitable for infrared work.

  (iii) Improvements in sources for exciting the emission spectrum of the element (e.g. the microwave source tube).

  Employing the microwave source tube, Bovey and Steers [1.13b] have examined the atomic spectra of plutonium, americium, neptunium, lutetium,

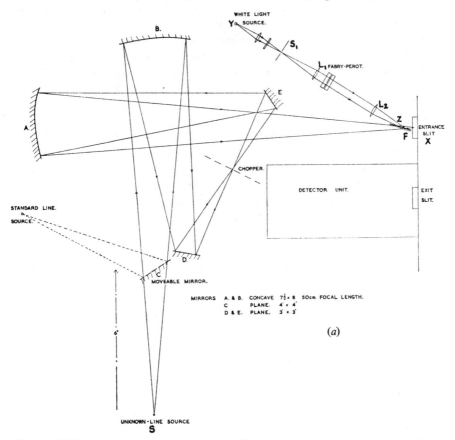

Fig. 1.52(a) Optical arrangement for the illumination of the entrance slit of a spectrometer employed for infrared studies on atomic spectra. The three sources of illumination of the slit are: (i) An unknown line source *S*, (ii) a standard line source, and (iii) light from a pair of Fabry–Perot plates. (*Courtesy of Dr. L. Bovey and the United Kingdom Atomic Energy Authority, Harwell*)

and uranium in the 1 to 3 μm region; their apparatus is illustrated in Fig. 1.52(a) and (b). In Fig. 1.52(a) the radiation from the element (labelled unknown line source S), which has been excited by the microwave radiation, undergoes reflection at the mirrors B and D and is interrupted by a chopper (see Vol. 2) rotating at 800 Hz. After further reflection at mirrors E and A the radiation enters the slit of a plane grating spectrometer. The movable mirror C enables a standard line source to be directed on to the entrance slit X of the spectrometer. As is schematically indicated in Fig. 1.52(b) light from this entrance slit is collimated by the concave mirror G on to the reflection grating, is then diffracted, and subsequently the spectrum is brought to a focus on the exit slit by means of the concave mirror H.[†] The radiation is then directed on to the detector units as illustrated in Fig. 1.52(a) and (b). The infrared radiation from the sample is detected by means of a lead sulphide photoconductive cell cooled with a mixture of solid CO_2 and acetone, the output of which is amplified by a 800 Hz amplifier and fed to a pen recorder.

Owing to the lack of standard wavelengths in the 1–3 μm region the method of wavelength measurement adopted was the use of interference fringes from a Fabry–Perot interferometer. This method had been previously developed by other workers in this field. The arrangement is shown along YZ in Fig. 1.52(a) and diagrammatically in Fig. 1.52(b). White light from a filament lamp is focused on to a slit S_1 which is situated at the focal point of the lens L_1. The parallel light from L_1 passes through a pair of Fabry–Perot interferometer plates and is focused on to a small section of the entrance slit of the spectrometer by lens L_2 and mirror F. The interferometer plates are parallel and have a high reflectivity and small absorption. This results in as much light being reflected as is transmitted. The interferometer may be considered as a pass band filter for wavenumbers (σ) given by:

$$\sigma = \text{integer}/2 \times n \times d \qquad (1.137)$$

where n is the refractive index of the medium between the plates and d is the distance between them.

As the grating is rotated to scan the spectrum a series of maxima and minima light fringes are produced, and these are detected by means of a photomultiplier the output of which, after amplification, is used to drive a second pen on the recorder. Provided that the Fabry–Perot plates are maintained at a constant temperature and the air removed from between them, the fringes would occur at equal wavenumber intervals given by the above expression. By superimposing several standard lines on the fringe trace, the absolute wavelength of each fringe can be readily evaluated, and these are then used as accurate wavelength standards which have the advantage of being equally spaced. Lines from the spectrum of the noble gases are used as standards. Those in the infrared region

---

[†] This arrangement of concave mirrors and plane grating is known as the Fastie–Ebert mounting.

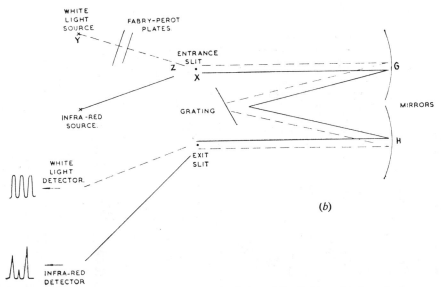

Fig. 1.52(b) Diagrammatic arrangement of the white light and infrared sources and their respective detectors in relation to the entrance and exit slits of the grating spectrometer employed in Fig. 1.52(a). (*Courtesy of Dr. L. Bovey and the United Kingdom Atomic Energy Authority, Harwell*)

are usually few and weak, and therefore higher orders of the stronger lines from the visible region were employed; their wavelengths were accurately known because they were determined by interferometer means.

Fig. 1.53 shows an actual record in the region 28 690 to 28 125 Å taken on the spectrometer already described. Lines due to Ne and Am can be observed and also the calibration fringes (the faint lines) which have a constant spacing. If the wavelengths of the neon lines are assumed, the wavelengths of the Am lines may be readily evaluated by merely taking into account that the fringe spacing is $0.63 \, \text{cm}^{-1}$.

1.16   SOME MORE RECENT STUDIES

The more recent trends in atomic spectra work have been more in the province of the physicist than the chemist. It is proposed to pick out a few studies and give appropriate references to enable the student to gain a broader insight into some of the more recent developments.

1.16.1   Spectral line width

Many recent studies have been made on the Doppler width, natural width, and the influence of pressure broadening on line width (see p. 65), and these are of fundamental interest. Kuhn [1.14] has given an excellent account of the factors influencing the width and shape of spectral lines.

97

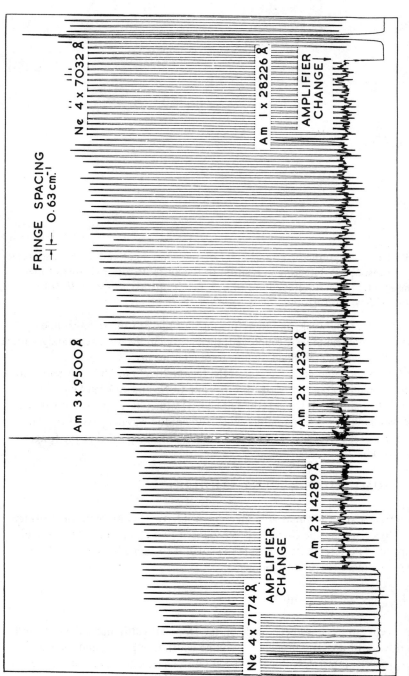

Fig. 1.53 Pen recorder trace of spectra with calibration fringes superimposed. (*Courtesy of Dr. L. Bovey and the United Kingdom Atomic Energy Authority*)

Hindmarsh, Petford, and Smith [1.15] have studied the effect of collision broadening and spectral shifts on noble gases at pressures of less than one atmosphere. From applying a classical theory of 'foreign gas' broadening, and from measurements of collision broadening and shift, they evaluated the constants C_6 and C_{12} of the Lennard-Jones potential·(V) where:

$$V = \frac{C_{12}}{r^{12}} - \frac{C_6}{r^6} \tag{1.138}$$

for the energy of interaction between two atoms separated by a distance r. The resulting values of C_6 were of the expected order of magnitude, and where measurements on one spectral line broadened by a number of noble gases are available, these are found to be approximately proportional to the polarizability of the noble gas.

Vaughan and Smith [1.16] examined the collision broadening and shift due to low pressure of He, Ne, Ar, and Kr on two emission lines of krypton at temperatures of 80 and 295 K. Different perturbing gases were found to produce widely different temperature dependence of line shift and line broadening. The results were interpreted in terms of an impact theory of collision broadening, assuming the Lennard-Jones (equation) interaction between radiating and perturbing atoms. Semi-quantitative agreement of theory and experiment was found in the case of the helium perturbers; the results for neon perturbers are described with striking accuracy.

A useful review article on the broadening and shift of spectral lines is one by Ch'en and Takeo [1.17].

Interesting work has been carried out on the strength of a spectral line and its relation to various concepts such as the matrix elements of the dipole moment. Comparisons have been made between measured line strengths and those computed from theory; such studies have important applications in the field of astrophysics. This type of work has been reviewed by Foster [1.18].

1.16.2 Translational spectra of compressed gases

Infrared and far-infrared studies of highly compressed simple gases have demonstrated that such gases have a translational spectrum. Thus, in a compressed mixture of helium and argon at room temperature a translational spectrum is obtained (see Fig. 1.54).

The absorption of compressed gases is quoted in the form of an absorption coefficient (A_σ) for each particular wavenumber where:

$$A_\sigma = \frac{1}{\rho^2 l} \ln \frac{I_0}{I} \tag{1.139}$$

where I_0 is the intensity of radiation transmitted by an empty absorption cell at wavenumber σ, and I is the intensity of radiation transmitted by the cell filled

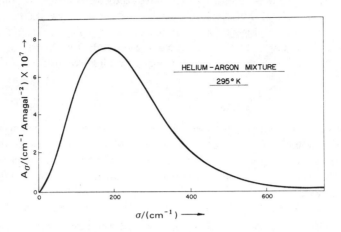

Fig. 1.54 The translational spectrum of a helium–argon mixture at room temperature at 295 K. A_σ is the absorption coefficient and is defined by Equation 1.139). (After Bosomworth and Gush [1.19]).

with gas of density ρ at the path length (l) of the cell which is given in metres. Various densities of gases are employed.

1.16.3   Astrochemistry

In general, atomic spectra have provided and should still provide an abundance of information on the nature of the Universe.

The sun has a large number of absorption and emission lines; in fact, it provides its own continuous radiation which some of the elements in the sun absorb. The latter may be detected on the earth by their characteristic absorption frequencies. Over half the elements known on the earth have been detected in the sun's absorption spectrum. Other elements may also be present since some elements may well not absorb radiation sufficiently strongly to be detected. The high-temperature regions of the sun have been studied by the hydrogen atom line at 1215 Å. This most intense line appreciably influences the constituents of our upper atmosphere.

The main differences in the spectral characteristics of a star are now believed to be attributable to the different temperatures of stars, and many stars are now thought to have similar composition. The temperatures of stars range from about 3000 to 80 000 K and are used in their classification. The sun is classified as a G-type star and has a characteristic temperature of 5500 K although its temperature variation from the photosphere to corona regions is from 4600 K to over a million degrees. Some of its spectral features are metallic lines, H lines, Ca I, Ca II, Fe I lines, and CN, C_2, and CH bands.

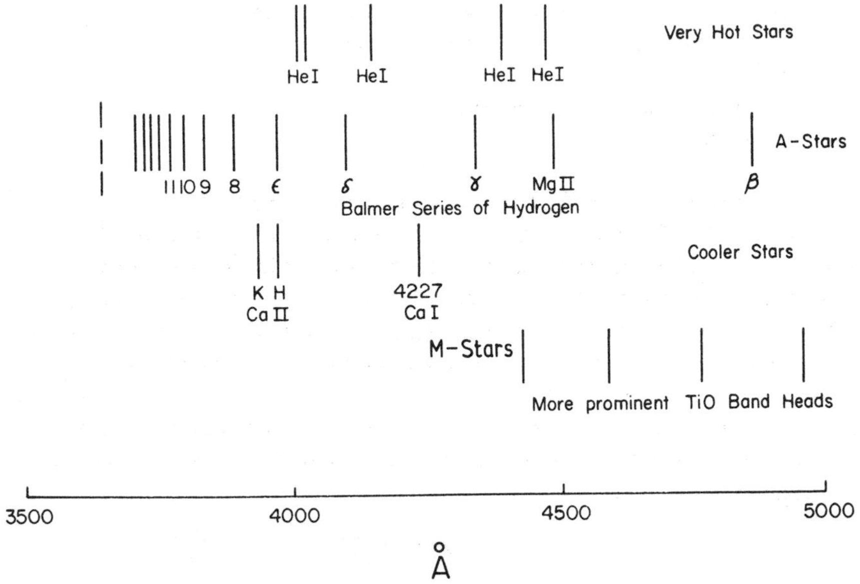

Fig. 1.55 Absorption features of some different types of stars between 3500 and 5000 Å. (After Evans [1.21]).

Some of the prominent features found in different types of stars are represented in Fig. 1.55. Sirius is an example of an A-type star, is white, and has a temperature of about 10 000 K. Some others of its spectral features, in addition to the Balmer lines, are Fe, Na, and Ca II lines. Betelgeuse is an M-type of star, is red in appearance, has a temperature of about 3000 K, and, in addition to the prominent TiO bands, has a Ca I line at 4226 Å and Ca II and Fe I lines as well as bands from a variety of diatomic species.

Only a few features on the application of atomic spectra to astrochemistry have been indicated here. A much fuller account is given in Vol. 3 in the chapter on astrochemistry.

1.16.4   The negative hydrogen ion

A number of interesting theoretical and experimental studies has been made on the simplest negative ion. One of the important reactions in the sun is:

$$H + e \; (+ \text{ kinetic energy}) \rightarrow H^-$$

This results in a continuous emission spectrum and accounts for a considerable fraction of the light from the sun. Theoretically it has been shown that H^- has only a 1S quantum state, and therefore the spectrum of the ion would be expected to consist only of a continuum, and in fact, the continuous absorption radiation by H^- has recently been studied experimentally. Theoretical studies have been made on the ground state (1^1S) of the H$^-$, and it has been found [1.22, 1.23] that the term value is $6083.0958 \, cm^{-1}$.

1.16.5   Optical pumping and double resonance

Optical pumping is a procedure for bringing about changes in the population distribution of atoms and ions amongst their energy states as a result of the absorption of radiation. The resulting excited atom or ion subsequently spontaneously emits radiation but need not necessarily return to its initial electronic state, and as a consequence of this the assembly of atoms or ions is transferred from an equilibrium to a non-equilibrium state. This process is termed optical pumping. Subsequent collision of the excited atoms or ions and a relaxation process may occur, which tend to restore the equilibrium state.

The process of optical pumping as first conceived as a two-step process is represented in Fig. 1.56.

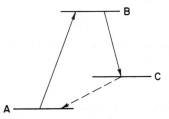

Fig. 1.56 Optical pumping scheme (After Cohen-Tannoudji and Kastler [1.24])

The first step from A to B in Fig. 1.56 involves an induced absorption of polarized light while the second step from B to C involves spontaneous emission of light. Thus, the population ratio of levels A and C is changed; that is C is increased relative to A. From C to A the relaxation occurs which reverts the atom or ion to its initial state. The increase in population of C relative to A is dependent on the intensity of the applied 'pumping' light and on the velocity of the relaxation processes tending to restore thermal equilibrium amongst the A and C state populations. If the relaxation processes are insufficiently slow, the optical pumping will be ineffective.

In Chapter 3 of Vol. 2 the case of the maser (microwave amplification stimulated emitted radiation) is considered, and this (as also does the laser) involves the non-equilibrium distribution of energy which is the first step in the action of a maser.

In Fig. 1.56 either (a) the C state could be an electronic level distinct from the A state, or (b) A and C may be two magnetic or hyperfine sublevels of the same electronic state which is most probably the ground state or a metastable long-living state.

One method of studying the excited states is termed the *double resonance* method and this may involve (a) an optical resonance fluorescence in an atomic vapour, and also (b) simultaneous exposure of the atomic vapour to a constant magnetic field and a radio-frequency field.

Hence, the name double resonance is given to this phenomenon since the optical and radio-frequency transitions occur simultaneously. A masterly account of the double resonance method and optical pumping technique has been given by Kastler [1.25].

The first study of double resonance was made on the resonance line of Hg which involves the transition $6^3P \rightarrow 6^1S$ (at 2537 Å). By the use of polarized optical resonance radiation of 2537 Å, selective excitation of the Zeeman sublevels of the excited state is brought out. This is achieved experimentally by directing plane polarized light from a low-pressure mercury lamp through a silica tube containing mercury vapour at low pressure, and resonance fluorescence results. The light from the mercury lamp is polarized with the electric vector in the z-direction while its direction of travel is the x-direction (see Fig. 1.57). The fluorescent light emerging in the z-direction from the silica tube is detected by a photomultiplier. A Helmholtz coil and a radio-frequency coil produce a constant magnetic field in the z-direction and an oscillatory field in the y-direction.

If one of the isotopes of mercury is employed which has a nuclear spin quantum number of zero, then for the experimental arrangement as given in Fig. 1.57 only π-light will be absorbed, and as a consequence of this only the Zeeman level $M_J = 0$ will be populated in the excited state which leads to the emission of only π-light. Since the photomultiplier lies along the z-direction, no resonance fluorescence is detected. If a constant magnetic field in the z-direction and an oscillatory field in the y-direction are applied to the Hg vapour on the quartz tube, then, when the applied radio-frequency becomes

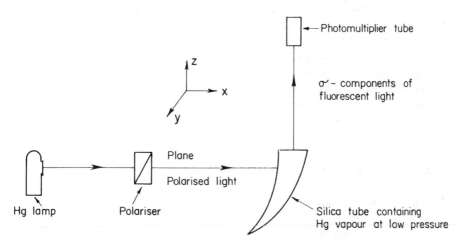

Fig. 1.57 The component features in the double resonance experiment (The Helmholtz coil and the radio-frequency coil are not represented).

equal to the Larmor frequency, transitions take place from the $M_J = 0$ level to the $M_J = \pm 1$, and light is emitted in the z-direction as a result of the σ-components of the Zeeman effect. Thus, when the radio-frequency resonance occurs, the population in the Zeeman levels is altered, and this may be detected by the σ-components of the Zeeman effect. Hence, the suitable radio-frequency which brings about these transitions gives the spacing of the Zeeman levels in the 6^3P_1 (excited) state of the Hg atom. The transitions may be represented as in Fig. 1.58 where the optical π-transitions are indicated by vertical lines and σ-transitions by slanting lines.

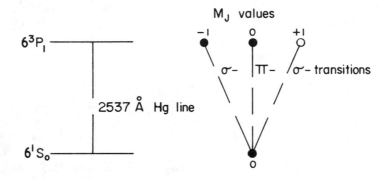

Fig. 1.58 Double resonance in the Hg line $6^3P_1 - 6^1S_0$ for an isotopic series which has a nuclear spin quantum number of zero.

In general, there can be several variations of the double resonance experiment involving different directions of the beams and fields, and involving different orientations of polarizers. All have in common, though, that the radio-frequency resonance alters the population of the Zeeman levels. However, the detection may be accomplished by a change of either polarization or intensity of the fluorescent light.

Double resonance and optical pumping studies have been used to:
 (i) yield nuclear spin quantum numbers and magnetic moments;
 (ii) gain information on multiple quantum transitions;
 (iii) investigate excited states of atoms;
 (iv) give details on nuclear hyperfine interactions and hyperfine splitting constants; this is achieved by studying each of the resonance frequencies necessary for the constant magnetic field.

One important application of optical pumping techniques is in solid-state lasers to obtain population inversion.

1.16.6   Chemical analysis

A steady advance has been made in the past fifteen years in the variety of commercial instruments available for the identification of elements and estimation of their concentration. It is proposed to consider briefly two procedures which have found widespread use.

(A) Atomic absorption spectro-photometry

Varian Instruments have produced a number of atomic absorption spectro-photometers, the most recent of which is the sophisticated AA6. The instrument described here however is the Atomspek Mk3 manufactured in England by Rank Hilger.

In the Mk3 the sample is burnt in (a) nitrous oxide—acetylene, or (b) air—acetylene, or (c) air—propane flame, or (d) argon—hydrogen flame and an absorption spectrum of the elements produced by burning the specimen in which the flame is taken by directing continuous radiation through the flame. The continuous radiation is produced from a hollow cathode lamp. There is a full range of burners in order to determine the elements at optimum sensitivity. 67 elements may be determined on the Atomspek Mk3 which instrument functions in the 1930—8530 Å range and has a wavelength accuracy better than 5 Å below 5000 Å. Resonance lines are used for the identification of e.g. Al (3093 Å), Pb (2170 Å), K (7665 Å), and Cu (3248 Å). An example of the

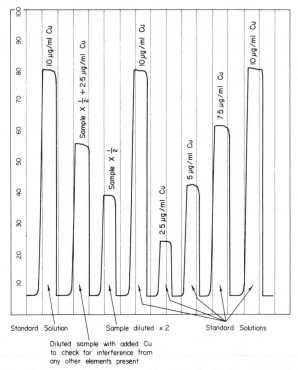

Fig. 1.59 Determination of copper in a liquid sample (Courtesy of Rank Hilger).

105

procedure employed for the determination of copper contamination in a liquid sample is given in Fig. 1.59 where the trace of linear absorbance against time is given for a number of standard solutions which have a linear dependence on the copper concentration. From this calibration the concentration of copper in the sample can be estimated. A comparison may also be made with (a) the sample diluted twice + 2.5 µg/mlCu and (b) the sum of the spectrum of the sample diluted twice and that of 2.5 µg/ml Cu. Since (a) and (b) agree within the limits of analysis, it may be concluded that there is no interference from any other elements present.

A wide variety of applications is found for atomic absorption spectrophotometry, and some of those for the Atomspek are in ferrous and non-ferrous metallurgy, medicine and biology, mining and geochemistry, the analysis of glass, water, organic compounds, and in pollution control. Thus, it may be appreciated that atomic absorption spectrophotometry is employed by a wide variety of scientists and for many different purposes.

A number of books [1.29–1.31] is now available on Atomic Absorption Spectrophotometry.

(B) X-ray fluorescence spectroscopy

Figure 1.60 shows the main features and layout of an X-ray spectrometer. The procedure is to direct an X-ray beam from an X-ray tube on to a sample. If the X-ray beam has adequate energy, an electron is ejected from one of the inner shells of an atom. An electron falls back from the next higher level, and the atom itself emits X-ray radiation. Each atom thus emits its own characteristic X-rays since the electronic change is associated with discrete energy changes between the various inner electron shells of the atom. In addition, since the intensity of this radiation is a function of the number of atoms excited, it may be used in quantitative analysis.

In Fig. 1.60 the X-ray fluorescent radiation from the sample is directed through a collimator (Soller slits) which consists of a number of parallel metallic slits. The resulting parallel beam of radiation is directed on to an analysing crystal which functions according to Bragg's law:

$$n\lambda = 2d \sin\theta \qquad (1.140)$$

where n, the order of diffraction, is usually 1, d is the lattice spacing in the crystal, and 2θ the angular rotation of the detector; hence, the range of wavelength which can be analysed is determined by both d and 2θ. Since the sensitivity of the detectors is not uniform over the whole spectrum, a combination of scintillation and proportional counters is employed. The intensity measureed by the detector may be visually displayed or recorded on a print-out system. A Rank Hilger Fluovac may be employed for X-ray fluorescence work. This instrument employs a reference scintillation counter and associated electronics to monitor the primary X-ray beam by counting on a fixed copper

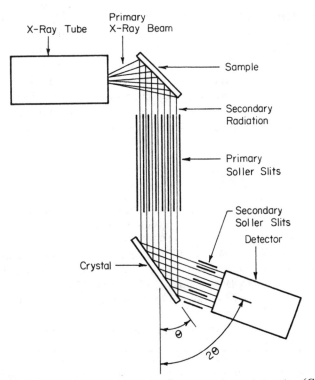

Fig. 1.60 Schematic diagram of an X-ray fluorescent spectrometer (Courtesy of Rank Hilger).

sample placed in a small portion of the beam. The practice is to take analytical counts over the time necessary to collect a pre-set number of reference counts. Counts are taken at the elemental peak for a series of standard samples and the count rate is plotted against the concentration to yield a straight line which may then be employed to estimate the concentration of an unknown from the counts obtained.

The X-ray fluorescent method is an attractive method of analysis for the major components of a sample and can yield a high degree of accuracy. The method has found a breadth of application in, for example, the analysis of metals, slags, refractories, and waste ducts. Haycock [1.32] has described the use of the Fluorovac in the determination of barium, zinc, and calcium in lubricating oils and additives. He outlines some of the problems created by the inter-elemental effects and how these may be overcome by the use of internal standard techniques.

1.17 BOOKS ON ATOMIC SPECTRA

For a wider treatment of atomic spectra and its quantum-mechanical aspects

White [1.5], Herzberg [1.8], and Condon and Shortley [1.33] should be consulted. Three specialized monographs are available: one on Molecular Beams by Smith [1.9], another on Fine Structure in Line Spectra and Nuclear Spin by Tolanksy and one on The Spectrum of Atomic Hydrogen [1.35]. For a comprehensive account a recently revised book by Kuhn [1.14] is recommended, and one by Hindmarsh [1.36] should be consulted. The account given here has been deliberately biased with a view to providing the necessary background for the microwave, nuclear magnetic resonance, and electron spin resonance techniques.

1.18   REFERENCES

1.1   Pearse, R.W.B. and Gaydon, A.G., *The Identification of Molecular Spectra*, Chapman and Hall, London (1950).

1.2   Kuhn, H. *Z. Phys.* **76**, 782 (1932).

1.3   Grotrian, W. *Graphische Darstellung der Spectrum von Atom und Ionen mit ein, zwei, und drei Valenzelektronen,* Springer Verlag, Berlin (1928).

1.4   Buckingham, A.D. *Quart. Rev.*, **13**, 183 (1959).

1.5   White, H.E. *An Introduction to Atomic Spectra*, McGraw-Hill, New York and London (1934).

1.6   Jevons, W. *Report on Band Spectra of Diatomic Molecules*, The Physical Society, London (1932).

1.7   Grotrian, W. and Ramsauer, G. *Phys. Z.*, **28**, 846 (1927).

1.8   Herzberg, G. *Atomic Spectra and Atomic Structure*, Dover Publications, New York (1944).

1.9   Smith, K.F., *Molecular Beams*, Methuen, London (1955).

1.10   Kronig, R. De L. *The Optical Basis of the Theory of Valency*, Cambridge University Press, London (1935).

1.11   Sawyer, R.A. *Experimental Spectroscopy*, Chapter 9, Chapman and Hall, London (1944).

1.12   a. Twyman, F. *Metal Spectroscopy*, Griffen, London (1951).
        b. Gerlach, W, and Schweitzer, E. *Foundations and Methods of Chemical Analysis by the Emission Spectrum*, Hilger, London (1931).
        c. Strouts, C.R.N., Gilfillan, J.H. and Wilson, H.N. *Analytical Chemistry*, Vol. II, Oxford University Press, Oxford (1955).

1.13   a. Bovey, L., and Wise, H. *Recent Developments in Light Sources Excited by Microwaves*, H.M.S.O., London (1959).
        b. Bovey, L. and Steers, E.B. *The Optical Spectra of Some Rare Earth and Transuranic Elements in the 1−3 micron Region*. H.M.S.O., London (1959).

1.14   Kuhn, H.G. *Atomic Spectra*, Longmans, London (1969).

1.15   Hindmarsh, W.R. Petford, A.D. and Smith, G. *Proc. Roy. Soc.*, **A297**, 296, (1967).

1.16   Vaughan, J.M., Smith, G., *Phys. Rev.*, **166**, 17 (1968).

1.17   Chen, S.Y. and Takeo, M. *Rev. Mod. Phys.*, **29**, 20 (1957).

1.18   Foster, E.W., *Rep. Progr. Phys.*, **27**, 469 (1964).

1.19   Bosomworth, D.R. and Gush, H.P., *Canad. J. Phys.*, **43**, 751 (1965).

1.20   Ewen, H.I. and Purcell, E.M. *Nature* **168**, 356 (1951).
1.21   Evans, D.S. *Observations in Modern Astronomy*, The English
        Universities Press. Ltd., London (1968).
1.22   Pekeris, C.L. *Phys. Rev.* **126**, 1470 (1962).
1.23   Midtdal, J., *Phys. Rev.* **138A**, 1010 (1965).
1.24   Cohen-Tannoudji, C. and Kastler, A. *Progr. In Optics,* **5**, 1 (1966).
1.25   Kastler, A. *Physics To-day* **20**, 34 (1967).
1.26   Popescu, D., Popescu, I. and Richter, J. *Z. Physik,* **226**, 160 (1969).
1.27   Eliseev, V.V., Sholin, G.V., *Opt. Spectrosc.*, **30**, 292 (1971).
1.28   Niemax, K. *Physics Letters*, **38A**, 141 (1972).
1.29   Munoz, J.R. *Atomic Spectroscopy and Analysis by Flame Photometry*,
        Elsevier, New York (1968).
1.30   Rubeska, I. and Moldan B., *Atomic Absorption Spectrophotometry*,
        Iliffe, London (1967).
1.31   Reynolds, R.J., Aldous, K. *Atomic Absorption Spectroscopy*, Iliffe,
        London (1970).
1.32   Haycock, R.E. *Journal of the Institute of Petroleum*, **50**, 123 (1964).
1.33   Condon, E.U. and Shortley, G.H. *The Theory of Atomic Spectra*,
        Cambridge University Press, London (1935).
1.34   Tolansky, S. *Fine Structure in Line Spectra and Nuclear Spin*, Methuen,
        London (1948).
1.35   Series, G.W. *The Spectrum of Atomic Hydrogen*, Oxford University
        Press, Oxford (1957).
1.36   Hindmarsh, W.R. *Atomic Spectra*, Pergamon, Oxford (1967).

2 Nuclear magnetic resonance (n.m.r.) spectroscopy

2.1 GENERAL THEORY OF HIGH RESOLUTION N.M.R. SPECTROSCOPY

2.1.1 Introduction

Nuclear magnetic resonance spectroscopy (n.m.r.) studies the behaviour of certain atomic nuclei, namely those which have magnetic moments arising from their possessing 'spin', in the presence of applied magnetic fields. The applied field is responsible for the setting up of nuclear energy levels between which transitions may be caused to occur by absorption of suitable electromagnetic radiation; because the pattern of energy levels is a property both of the nuclei in a molecule and of their electronic environment and relationship to each other, the experiment is of enormous value to chemists both as a theoretical and a structural tool. It is fortunate that some of the most suitable nuclei for study are among the most important and widespread in chemistry, e.g. ^1H, ^{13}C, ^{31}P, ^{19}F, and it is true to say that the development of n.m.r. spectroscopy since the first successful experiments in 1945 [2.1] has been of paramount importance to the chemist. Useful descriptions of the n.m.r. experiment are available both in classical and quantum-mechanical terms; it is necessary to consider both, since some aspects of the technique are best dealt with from one point of view while others are best considered in the alternative way.

2.1.2 Quantum mechanical description

The particles which make up an atomic nucleus, neutrons and protons, possess a property which is described as *spin angular momentum*. Nuclear structure is complex since the sub-particles have orbital as well as spin motions which combine together in various ways to give a resultant spin angular momentum for the nucleus [2.2] which may or may not be zero.

All nuclei with odd mass number possess 'spin', and the spin angular momentum vector is measured in units of $h/2\pi = \hbar$ (h = Planck's constant) and given the symbol $I\hbar$. This is quantized, i.e., only certain values may occur, and it is necessary to define a quantum mechanical operator $\bar{I}$ corresponding to this property of spin. When $\bar{I}$ operates upon an eigenfunction (a nuclear spin wavefunction) ψ_n it generates eigenvalues I according to the equation:

$$\bar{I}\psi_N = [I(I+1)]^{\frac{1}{2}}\,\psi_N \qquad (2.1)$$

I is the nuclear spin quantum number, and for nuclei of odd mass number it is an odd integral multiple of $\frac{1}{2}$ i.e. $I = n/2$ where n is an odd integer.

Nuclei having even mass number may have even nuclear charge, in which case $I = 0$, or odd nuclear charge, in which case I is an integer (1, 2, 3, etc.). In chemistry, however, n.m.r. studies are almost exclusively with nuclei of spin $I = \frac{1}{2}$, and all future discussion in this chapter will be concerned with such nuclei, e.g. ^{1}H, ^{19}F, ^{13}C, ^{31}P, ^{15}N, ^{29}Si.

If a nucleus has spin, it behaves as a spinning finite spatial distribution of charge, and a magnetic moment μ_N arises which is proportional to the magnitude of the spin, i.e.

$$\mu_N = \gamma_N I\hbar \qquad (2.2)$$

The proportionality constant γ_N is a fundamental nuclear property known as the gyromagnetic ratio; it may be expressed in terms of a dimensionless constant g_N (the nuclear g-factor) and the nuclear magneton $\mu_N = e\hbar/2M$ (e and M are the charge and mass of the proton), and Equation (2.2) may be written as:

$$\mu = g_N \mu_N I \qquad (2.3)$$

Consider that an isolated nucleus of spin $\frac{1}{2}$ is placed in a steady magnetic induction B_0^*. There is an interaction between the magnetic moment μ_N and the field, and the resulting energy is expressed quantum-mechanically as the eigenvalues of a Hamiltonian operator:

$$\bar{\mathcal{H}} = -\mu B_0 \qquad (2.4)$$

If it is defined that B_0 is in the $+z$ direction of a system of Cartesian coordinates with their origin at the nucleus, it is necessary to consider the component of the nuclear magnetic moment vector along this direction. Quantum theory requires that the z component of the nuclear spin vector can only take up one of a set of discrete values, m_z, which are eigenvalues of the equation:

$$\bar{I}_z \psi_N = m_z \psi_N \qquad (2.5)$$

$\bar{I}_z$ is actually one of a set of spin angular momentum operators $\bar{I}_x, \bar{I}_y, \bar{I}_z$, and $\bar{I}^2$ which are analogous to the more familiar operators for electron spin angular momentum $\bar{S}_x, \bar{S}_y, \bar{S}_z$, and $\bar{S}^2$. m_z may only have the values $+I$, $(I-1), \ldots, -1$, and for a spin $\frac{1}{2}$ nucleus may therefore be $\pm\frac{1}{2}$.

* Please see Page vi for an explanation of the use in this book of the symbol B

111

Equation (2.4) may now be written as:

$$\bar{\mathcal{H}} = -g_N\mu_N B_0 \bar{I}_z \qquad (2.6)$$

The energy levels of the system are the eigenvalues of the Schrödinger equation in which the operator $\bar{\mathcal{H}}$, defined by Equation (2.6), operates upon the nuclear spin wavefunction ψ_N:

$$-g_N\mu_N B_0 \bar{I}_z \psi_N = E_N \psi_N \qquad (2.7)$$

Since Equation (2.5) defines the eigenvalues of $\bar{I}_z$, Equation (2.7) becomes simply:

$$\begin{aligned}
-g_N\mu_N B_0 \bar{I}_z \psi_N &= -g_N\mu_N B_0 m_z \psi_N \\
&= -g_N\mu_N B_0 (\pm \tfrac{1}{2}) \psi_N \qquad (2.8)
\end{aligned}$$

Two energy levels are produced, one corresponding to $m_z = +\frac{1}{2}$ (conventionally, spin α, corresponding to a lowering of energy) and the other to $m_z = -\frac{1}{2}$ (conventionally, state β, corresponding to an increase in energy.) In classical terms, the α state corresponds to alignment of the nuclear moment parallel to the field, and state β corresponds to alignment antiparallel to the field. The situation is shown diagrammatically in Fig. 2.1.

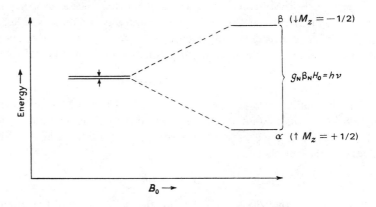

Fig. 2.1 The energy of nuclear (spin $\frac{1}{2}$) levels in a magnetic induction B_0.

The energy difference between the two levels is $g_N\mu_N B_0$, which corresponds to a frequency separation ν given by the Planck relationship:

$$\nu = \frac{g_N\mu_N B_0}{h} \qquad (2.9)$$

This is the so-called *resonance condition*.

In a sample containing many nuclei, some will have spin α and some spin β; there will be a Boltzmann distribution between the two levels such that:

$$N_\beta/N_\alpha = \exp\left(-g_N\mu_N B_0/kT\right) \tag{2.10}$$

where N_β and N_α are the numbers of nuclei in the β and α states respectively.

For magnetic inductions realizable in the laboratory, this corresponds to there being fractionally more nuclei in the lower level (α) than in the upper (β), and the system of nuclei is magnetically polarized; for example at $B_0 = 1$ tesla, the excess population in the ground state is a few nuclei per million (for ^{1}H it is about 7 per million). The possibility therefore exists that nuclei may be transferred between the levels by an appropriate oscillating electromagnetic field of frequency ν, *with net absorption of energy*; this is the basis of the n.m.r. experiment.

The resonance condition is defined by Equation (2.9) which suggests that the experimental observation of nuclear resonance absorption may be carried out either by fixing ν and varying B_0 until resonance is observed, or by fixing B_0 and varying ν. In practice both methods have found widespread application, and nowadays a third technique is of importance in which a pulse of radiation near to the frequency ν is applied to the system of nuclei to excite the resonance very rapidly; the decay of magnetic polarization is observed as a function of time. This technique will be dealt with later.

Because different nuclei have different nuclear g-factors, the resonance condition (2.9) varies from nucleus to nucleus; for a constant B_0 of 2.35 tesla, the values of ν for four important nuclei are shown in Table 2.1.

Table 2.1 The resonance condition for several important nuclei

Nucleus	Frequency ν/(MHz) *for a constant magnetic induction of 2.35 tesla*
^{1}H	100.04
^{19}F	94.13
^{13}C	25.15
^{31}P	40.48

At resonance, the probability of the radiation of frequency ν inducing a transition in the sense $\alpha \rightarrow \beta$ (absorption) is exactly equal to the probability of its inducing a transition in the sense $\beta \rightarrow \alpha$ (emission). Because there are more nuclei in the lower energy state, a net absorption will occur. A little thought, however, will reveal that the excess population in the lower level will gradually diminish with time, while the population of the upper level will increase until they become equal; there will then be no net absorption and the signal, of a spectrometer designed to detect resonance, will disappear! This phenomenon is known as *saturation*, and if it were not for the fact that there exist means whereby the spin system exchanges energy with its surroundings and tends to restore the Boltzmann equilibrium (in opposition to the effects of net absorption of the applied radiofrequency) the experiment would be of little practical value;

these restoring processes are known as *relaxation,* and they effectively provide a continuous supply of (excess) nuclei in the lower level for excitation. It should not be imagined that the relaxation processes are of spontaneous emission, since because the energy levels are so close together the probability for spontaneous emission is vanishingly small. They are, in fact, transitions induced by the thermal noise spectrum of the medium surrounding the nuclear spin system (i.e. the liquid structure or solid lattice in which there are many degrees of freedom other than those concerned with spin); this has components at suitable radio-frequencies, and induces transistions in the same way as does the applied frequency ν, but with a net tendency to restore the Boltzmann distribution inst-ead of destroying it. In fact, it is precisely these relaxation processes which allow the setting-up of the Boltzmann distribution, when B_0 is applied to the system, in the first place. It is convenient to consider two different relaxation mechanisms, each of which is effectively a first-order rate process characterized by its own time constant; *spin–lattice* relaxation, of time constant T_1, occurs because there is exchange of energy between the spin states and the surrounding medium, and *spin–spin* relaxation, of time constant T_2, occurs with exchange of energy between different nuclear spins. It is useful to think of T_1 processes as affecting the lifetimes of population of spin energy levels while T_2 processes affect the relative energies of the spin levels rather than lifetimes; we shall return to relaxation processes later in the chapter.

2.13 Classical description

The quantum mechanical description of the basis of the n.m.r. experiment is necessary for our eventual understanding of spectral analysis (see Section 2.3.), but in order to appreciate some important modern experimental techniques (pulsed n.m.r.) it is helpful to think in classical terms.

If the spinning nucleus is regarded as being equivalent to a current circulating in a closed loop, it behaves like a magnetic dipole whose magnetic moment vector μ is given by:

$$\mu = I\pi r^2 \tag{2.11}$$

where i is the equivalent current and r is the radius of the loop. A charge of q/c coulombs rotating at $v/2\pi r$ revolutions per second is equivalent to a current:

$$I = 5qv/\pi rc \text{ amperes} \tag{2.12}$$

where c is the velocity of light and v is the velocity of the rotating charge.

Substituting Equation (2.12) into (2.11) gives:

$$\mu = 5qvr/c \tag{2.13}$$

If the mass of the nucleus is M, the spin imparts an angular momentum vector p given by:

$$p = Mvr \tag{2.14}$$

The magnetic moment and angular momentum vectors are collinear, and substitution of Equation (2.14) into (2.13) gives the relationship:

$$\mu = 5qp/Mc \qquad (2.15)$$

The ratio μ/p $(=5q/Mc)$ is the *gyromagnetic ratio* γ (mentioned earlier) which is a characteristic nuclear property.

If a spinning charged particle is placed in a magnetic field of strength B_0, with its magnetic moment vector μ inclined at an angle θ to the direction of this field (defined as the $+z$ direction) it will experience a torque L which tends to align it parallel to the field. The magnetic field causes the angular momentum to change, and the rate of change with time is equal to the torque exerted on the magnetic moment by the applied field:

$$dp/dt = L \qquad (2.16)$$

and from simple electromagnetic theory:

$$L = \mu \times B_0 \qquad (2.17)$$

Substituting Equations (2.15) and (2.17) into (2.16) gives:

$$\frac{dp}{dt} = \frac{5q}{Mc} p \times B_0 \qquad (2.18)$$

The torque causes p to precess about B_0 with an angular frequency ω_0 rad s^{-1} defined by:

$$dp/dt = d\omega_0 \qquad (2.19)$$

Substituting Equation (2.19) into (2.18) gives:

$$\omega_0 = \gamma B_0 \qquad (2.20)$$

This is the Larmor equation, which may be rewritten in linear frequency units ν_L Hz:

$$\nu_L = \gamma B_0/2\pi \qquad (2.21)$$

It is important to note that this precession frequency is independent of θ, the angle of inclination of p to B_0. The process is illustrated diagrammatically in Fig. 2.2.

If a small secondary magnetic field B_1 is applied at right angles to the main field B_0, i.e. in the xy plane, then at a particular point on the precessional path the nuclear dipole experiences a combination of B_0 and B_1 which tends to change the angle (but not the precession rate) by an amount $+\delta\theta$; at a point π radians further along the precessional path, however, the combination of B_0 and B_1 will tend to change θ by $-\delta\theta$. Integrating this over the whole precessional path indicates that if B_1 is fixed in the xy plane it cannot bring about any net change in the orientation (and thereby the magnetic energy) of

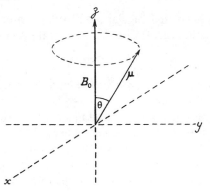

Fig. 2.2 The precession of the magnetic moment vector **μ** about the direction of the magnetic induction B_0.

the particle; no absorption of energy from B_1 can occur. In order for B_1 to change the magnetic energy, it has to rotate in the xy plane at the Larmor frequency, i.e. it must be in resonance with the precession about B_0. A rotating magnetic induction of this type arises from circularly polarized electromagnetic radiation of frequency ν_L; for most purposes, however, linear polarization is satisfactory, since linear oscillation may be regarded as the superposition of two counter-rotations. Only the component having the correct sense will be in resonance with the precession, the other having no effect. Energy may now be absorbed from B_1 since it is in resonance with the precession, and the effect is to tip **μ** towards the xy plane (i.e. to increase θ) by an amount depending on the strength of B_1.

It is now necessary to bring into the classical picture one of the consequences of quantization mentioned earlier, namely that the component of the nuclear magnetic moment vector along any given direction may only take up one of a discrete set of values. For a nucleus for which $I = \frac{1}{2}$, this means that the magnetic moment can precess about the $+z$ or $-z$ axis with a unique angle θ [corresponding to the $m_z = +\frac{1}{2}$ (α) and $m_z = -\frac{1}{2}$ (β) states respectively] which depends upon the value of B_0 and the nature of the nucleus. In an assembly of identical nuclei of spin $\frac{1}{2}$ all will precess with the same angle θ, and there will be slightly more precessing about $+z$ than about $-z$ since the Boltzmann distribution favours slightly the lower energy state. There is no way in which the nuclear moments distinguish the x and y directions, however, and so there is no phase coherence in the xy plane; the situation is depicted in Fig. 2.3.

The net result is an overall magnetization M, of the nuclear sample, in the direction of the z axis. As we saw before, at resonance the sample is able to absorb energy from a secondary applied field which rotates at the Larmor frequency in the xy plane; conveniently, the magnetic vector of electromagnetic radiation of frequency ν_L (the Larmor frequency) supplies the energy. The effect is to change the net magnetization M, of the sample, and an important set of equations due to Bloch [2.3, 2.4] shows how this varies as a function of time.

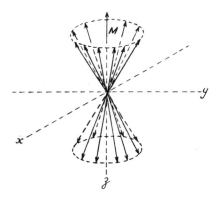

Fig. 2.3 The precession of the magnetic moments of an assembly of identical spin $\frac{1}{2}$ nuclei about the direction of an applied magnetic induction.

2.1.4 The Bloch equations

Combining equations (2.16) and (2.17), and multiplying both sides by the gyromagnetic ratio γ gives for a magnetic moment μ in any induction B comprising B_0 and B_1:

$$\gamma \frac{\mathrm{d}f}{\mathrm{d}t} = \gamma \mu \times B = \frac{\mathrm{d}\mu}{\mathrm{d}t} \qquad (2.22)$$

since $\mu = \gamma p$ by definition

M is the vector sum of the individual μ's, and summing Equation (2.22) over all μ gives:

$$\frac{\mathrm{d}M}{\mathrm{d}t} = \gamma M \times B \qquad (2.23)$$

Simple theory of vectors (see, for example, ref. 2.5) enables this equation to be expanded in terms of the components along the Cartesian axes x, y, and z, and unit vectors along these axes i, j, and k respectively:

$$\frac{\mathrm{d}M}{\mathrm{d}t} = \gamma \left[(M_y B_z - M_z B_y)i + (M_z B_x - M_x B_z)j + (M_x B_z - M_y B_x)k \right] \qquad (2.24)$$

The induction B in Equations (2.22) and (2.23) consists of B_0 together with the magnetic vector of the applied radiofrequency B_1 which is equivalent to an induction rotating in the xy plane at an angular frequency ω rad s^{-1}. The components of B can easily be deduced:

$$B_x = B_1 \cos \omega t, \; B_y = B_1 \sin \omega t, \; B_z = B_0 \qquad (2.25)$$

Equation (2.24) may now be written as three separate equations, each describing the time dependence of a component of M:

$$\frac{\mathrm{d}M_x}{\mathrm{d}t} = \gamma (M_y B_0 + M_z B_1 \sin \omega t) \qquad (2.26)$$

117

$$\frac{dM_y}{dt} = \gamma (M_z B_1 \cos \omega t - M_x B_0) \tag{2.27}$$

$$\frac{dM_z}{dt} = -\gamma (M_x B_1 \sin \omega t + M_y B_1 \cos \omega t) \tag{2.28}$$

These equations take no account of relaxation, the natural processes which tend to restore the Boltzmann equilibrium in opposition to the perturbing influence of B_1. These are first-order processes which tend to restore M_x and M_y to their equilibrium values of zero, and M_z to M_0. The full form of the Bloch equations then becomes:

$$\frac{dM_x}{dt} = \gamma (M_y B_0 + M_z B_1 \sin \omega t) - \frac{M_x}{T_2} \tag{2.29}$$

$$\frac{dM_y}{dt} = \gamma (M_z B_1 \cos \omega t - M_x B_0) - \frac{M_y}{T_2} \tag{2.30}$$

$$\frac{dM_z}{dt} = -\gamma (M_x B_1 \sin \omega t + M_y B_1 \cos \omega t) - \frac{M_z - M_0}{T_1} \tag{2.31}$$

Since T_1 is the time constant for the decay of the component of magnetization along the z axis (parallel to B_0) it is called the longitudinal relaxation time; T_2 describes the decay of magnetization in the xy plane and is therefore the transverse relaxation time.

Solution of the Bloch equations is straightforward but laborious. For the conventional n.m.r. experiment defined earlier, in which B_0 is varied slowly at constant applied radiofrequency, or the radiofrequency is varied slowly with constant B_0 in order to observe resonance, two resonance signals are predicted to be observable to a sensing device or detector operating in the xy plane; one is 90° out of phase with B_1 and has the shape of a Lorentzian curve (absorption or v-mode) while the other is in phase with B_1 and has a dispersive or derivative shape (u-mode).

2.1.5 Pulse methods: the rotating frame of reference

Many modern n.m.r. spectrometers, notably those designed for the study of low-sensitivity nuclei such as ^{13}C or ^{15}N, subject the spin system to a train of equally spaced pulses of r.f. energy. The experiment examines the decay of magnetization of the sample as a function of time, in between pulses, and in order to solve the Bloch equations under these conditions it is helpful to refer the motion of the magnetization to a coordinate system that rotates about B_0 in the same direction as that of precession of the nuclear moments, rather than one which is fixed in space and time. This is the *rotating frame of reference* first introduced by Ramsey et al. [2.6].

Initially, as before, we consider the effect of a magnetic induction of magnitude B_0 applied along the z axis. The reference frame of Cartesian coordinates rotates around the z axis with the Larmor frequency, $\boldsymbol{\omega} = -\gamma B_0$. The net magnetization vector $\boldsymbol{M}$ is invarient with time in this reference system. $\boldsymbol{B}_1$ is now applied in the xy plane and rotates in the laboratory (stationary) frame at ω rad s^{-1}; it can be shown that in the rotating frame, the net result is an effective induction vector:

$$\boldsymbol{B}_{\text{eff}} = \boldsymbol{B}_0 + \boldsymbol{\omega}/\gamma + \boldsymbol{B}_1 \qquad (2.32)$$

At resonance, the term $\boldsymbol{\omega}/\gamma$ (which has the dimensions of magnetic induction and arises from the effects of rotation) exactly cancels $\boldsymbol{B}_0$ along the z axis and leaves only $\boldsymbol{B}_1$ in the xy plane to interact with the net magnetization vector $\boldsymbol{M}$. $\boldsymbol{B}_1$ rotates at the same frequency as the frame, and we can therefore consider that its magnitude B_1 is directed along the rotating x axis (the x^1 axis); the result is to cause $\boldsymbol{M}$ to precess about x', in the $z'y'$ plane, with frequency γB_1. In time t_p seconds, the angle θ through which $\boldsymbol{M}$ precesses is given by the relationship:

$$\theta = \gamma B_1 t_p \text{ (rad)} \qquad (2.33)$$

This is schematically shown in Fig. 2.4.

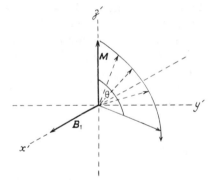

Fig. 2.4 The precession of $\boldsymbol{M}$ about $\boldsymbol{B}_1$ in the rotating frame.

In a pulsed n.m.r. experiment the r.f. induction B_1 is applied for t_p s; typically, B_1 may be about 0.01 tesla and t_p about 10–100 μs. The spectrometer is arranged to detect only the net component of $\boldsymbol{M}$ along the y' axis, and the 'intensity' of the absorption is therefore measured by the projection of $\boldsymbol{M}$ along y', and is proportional to $\sin \theta$. Maximum intensity is thus achieved when $\theta = \pi/2$ rad (the 90° pulse), and vanishes to zero when θ approaches π rad (180° pulse). The reduction of M_z to zero must not be confused with the phenomenon of saturation alluded to earlier, which occurs when too powerful a value of B_1 is used in the conventional continuous wave experiment and leaves no magnetization in any direction; the 90° pulse merely redirects it along y'.

In a typical pulse experiment, B_1 is returned to zero after time t_p seconds; ideally, the system is allowed to relax back to its equilibrium magnetization (i.e. M returns to the z' axis) and the pulse is applied again. This sequence may be repeated many times. It is instructive to consider what happens to the spin system during the 'switch-off' period, for upon this depends the behaviour of the system when the second and subsequent pulses are applied; it is also possible to illustrate the significance of T_1 and T_2.

Suppose that application of B_1 has caused M (the net magnetization due to the assembly of nuclear moments) to tip towards the y' axis by an angle θ. When B_1 is reduced to zero, there is a component of M along both the y' axis and the z' axis, but not along x'; however, because of spin–spin relaxation (i.e. processes which allow the nuclei mutually to exchange energy) the nuclear moments begin to 'fan out' in the $x'y'$ plane in both directions (i.e. towards $+x'$ and $-x'$). The component of M along y', $M_{y'}$, decays to zero with a time constant T_2, the spin–spin or transverse relaxation time. More accurately, T_2 applies for a sample of nuclei in a perfectly homogenous induction B_0; in practice slight inhomogeneities in B_0 cause some nuclei to precess at different frequencies, some faster than the rotating frame and some slower. This also causes $M_{y'}$ to diminish, and the true decay constant for transverse relaxation in an inhomogenous field is T_2^*.

The restoration of the z' component of M, $M_{z'}$, to its equilibrium value M_0 occurs with time constant T_1, the spin–lattice or longitudinal relaxation time; the energy is transferred to the lattice or surrounding medium by precesses which will be considered later in some detail. The net result of both relaxation processes is to cause the spins to perform a damped precession while the net magnetization components decay to their equilibrium values; since the nuclear moments are at this time precessing in the absence of the applied r.f. induction B_1, the residual magnetism in the $x'y'$ plane is said to be due to *free induction*.

2.1.6 Free induction decay

Let us imagine that a 90° pulse has been applied along the x' axis, and the magnetization vector M lies entirely along the y' axis. A spectrometer is arranged to detect magnetization in the xy plane, and so the magnitude of M_{xy}, the free induction signal, determines the strength of the signal. As transverse (spin–spin) relaxation occurs, the signal decays with time constant T_2^* in an exponential fashion. This is described pictorially in Fig. 2.5.

Figure 2.5(c) applies for the situation where the sample contains an assembly of nuclei, all of which have identical Larmor frequencies. We shall see later that in practice we more usually deal with a situation where, in a given sample, nuclei of the same type may have very slightly different Larmor frequencies due to phenomena known as *shielding* and *spin–spin coupling*. Quantum mechanically, this corresponds to a situation where there are many nuclear energy levels

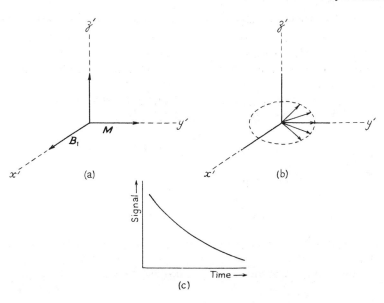

Fig. 2.5 (a) The 90° pulse transfers the net magnetization M from the z' to the y' axis in the frame rotating at the Larmor frequency. (b) During free induction decay, the net magnetization reduces to zero in the y' direction because the nuclear moments dephase, or fan out, in the $x'y'$ plane. (c) The signal detected in the xy plane decays to zero with exponential time constant T_2^*. This is free induction decay (f.i.d.).

between which transitions may be induced; classically, absorption of magnetic energy by a sample in an induction B_0 will occur at a variety of precession frequencies.

Application of a short intense pulse of electromagnetic energy at a frequency close to the range of Larmor frequencies of the sample can be shown to be capable of exciting all transitions simultaneously, i.e. it is equivalent to a band of radiation containing all Larmor frequencies characterizing the assembly of nuclei [2.7]. The f.i.d. is not now a simple exponential decay, but is an interferogram made up of all the resonance absorption frequencies.

2.1.7 Fourier transformation

Traditionally, spectroscopists are not familiar with the presentation of a spectrum in a way which studies the behaviour of a system as a function of time after the simultaneous excitation of all transitions; this *time domain* spectrum, however, does contain all the necessary information, i.e. the frequencies characterizing the resonances. Previously, we have examined spectra in the *frequency domain*, e.g. for n.m.r. observations we traditionally use *continuous wave* excitation in which B_0 is maintained and each individual resonance line of a sample is successively excited by sweeping through a range of radiofrequencies. It is possible, however, to transfer the information from the time domain into the frequency domain by a convenient mathematical analysis known as Fourier transformation, and this enables the spectroscopists to examine spectra in their more familiar form.

Mathematically, a function $f(y)$ in the time domain is related to a function $f(x)$ in the frequency domain by the general relationship:

$$f(x) = \int_{-\infty}^{+\infty} f(y) \exp(-ixy) \, dy \tag{2.34}$$

The inverse relationship also holds:

$$f(y) = \frac{1}{2\pi} \int_{-\infty}^{+\infty} f(x) \exp.(ixy) \, dx \tag{2.35}$$

$f(x)$ is said to be the *Fourier transform* of $f(y)$, and the n.m.r. problem of converting a time-domain spectrum (the f.i.d.) into its frequency-domain counterpart (the frequency absorption spectrum) may be stated in the form:

$$M(\omega) = \int_{-\infty}^{+\infty} M(t) \exp(-i\omega t) \, dt \tag{2.36}$$

where $M(\omega)$ is the magnetization as a function of frequency and $M(t)$ is the magnetization as a function of time after the application of a pulse (the f.i.d.). In a modern spectrometer, this transformation is carried out by a digital computer interfaced 'on line' to the instrument (see later); it is beyond the scope of this chapter to consider the process further, but reference may be made to the relevant sections of the book by Farrar and Becker [2.8] and the article by Gillies and Shaw [2.9].

2.1.8 Relaxation processes

The importance of the various processes whereby the magnetization of a sample of nuclei is restored to the equilibrium value, after a perturbation has been applied, has been emphasized in the preceding pages. It is convenient to discuss these processes now, although the methods for measuring the characteristic time constants will be dealt with later.

Both spin—lattice (longitudinal) relaxation and spin—spin (transverse) relaxation are caused by time-dependent magnetic or electric fields at the nucleus; they are not spontaneous processes. The origin of these fields lies in the random thermal motions present in any macroscopic assembly of molecules or atoms. The nucleus may experience such perturbations caused by the spins of other nuclei moving past it, from unpaired electrons in a paramagnetic species, or from magnetic fields generated by the rotation of the molecule in which the nucleus is present (spin—rotational interactions). If the shielding experienced by the nucleus due to its surrounding electronic environment (see later) changes as the molecule rotates, the magnetic field seen by the nucleus alternates and this too causes relaxation (by *chemical shift anisotropy*). If the nucleus has spin greater than $\frac{1}{2}$ it may be quadrupolar, and as the molecule vibrates and rotates the electric field gradients in the region of the nucleus also give rise to relaxation. Generally, for relaxation to occur, a *time dependent* perturbation must act directly on the nucleus.

A prerequisite for successful relaxation is that molecular motions have a suitable time scale, since fluctuations that occur much more rapidly than the n.m.r. frequencies (typically $10^7 - 10^8$ Hz) have little effect. Electronic motions and molecular vibrations are much faster than this, and do not cause relaxation; any process which induces rapid transitions between the α and β spin states, and fluctuates strongly at the resonance frequency, causes efficient relaxation. The time scale for n.m.r. is essentially slow, and of the same order as that for translation and rotation of molecules in a fluid medium, and so it is these processes which give rise to the necessary field [2.10]. It is convenient to discuss these motions in terms of an effective *correlation time*, τ_c which may be regarded as the time taken for the molecule to rotate through 2π radians, or translate through a distance comparable to its diameter.

Relaxation is best effected by frequencies comparable to the Larmor frequency of a given nucleus; for a ^{13}C nucleus in an applied induction of 2.35 tesla this is 2.5×10^7 Hz, or about 1.6×10^8 rad s^{-1}. The reciprocal of the resonance frequency may be taken as the correlation time (in this case τ_c is about 7×10^{-9}s) which leads to optimum relaxation. Figure 2.6 illustrates the consequences of different correlation times upon both T_1 and T_2 [2.11].

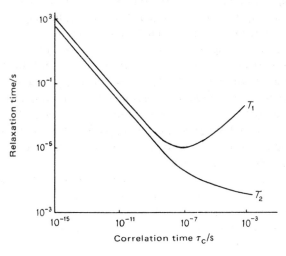

Fig. 2.6 The relationship between T_1, T_2 and τ_c.

The correlation times for most molecules fall below 10^{-9}s; for small molecules τ_c is typically $10^{-12} - 10^{-13}$s but may be as long as 10^{-10}s for some of the larger organic molecules (e.g. steroids) or 10^{-9} s for some polymers. From Fig. 2.5 it is clear that in general any change that shortens τ_c results in lengthening both T_1 and T_2. Thus lowering viscosity and raising temperature both decrease τ_c and consequently increase T_1 and T_2; this implies that the resonance is more easily saturated (the effect of longer T_1) but the line narrower (the effect of longer T_2). For molecules with $\tau_c = 10^{-10}-10^{-9}$ s, broad resonance lines are observed but these are difficult to saturate.

2.1.9 Mechanisms of spin–lattice relaxation

This process is the exchange of energy between nuclear spins and the surrounding medium (the lattice, which acts as a heat sink for the energy released when the nuclei decay to their ground state). In order to have efficient energy transfer, it is necessary to have physical interactions which 'couple' the spin system to the lattice; there are a number of these which are of importance: (a) magnetic dipole–dipole interactions (i.e. between the nucleus in question and other spin-$\frac{1}{2}$ nuclei, or with electrons of unpaired spin); (b) chemical shift anisotropy interactions; (c) scalar–coupling interactions; (d) spin–rotation interactions.

Frequently (especially, for example, in the case of ^{13}C nuclei bonded to hydrogen) the first of these mechanisms dominates. The *dipole–dipole* process originates in local magnetic fields associated with other nuclei in the same molecule. The nucleus 'sees' a net magnetic induction comprising B_0 and contributions from the local fields of the nuclei; at this stage, we ignore non-time-dependent fields due to the electronic structure of the molecule (i.e. shielding mechanisms) and those giving rise to spin–spin coupling since they do not contribute to relaxation. The strength and direction of the combined interactions depend on the magnetic moments of the nuclei, internuclear separation, and the relative orientation *within the molecule* with respect to B_0. The relative orientation with respect to B_0 is constantly changing in a fluid, as a result of rapid molecular motions (tumbling, internal rotations) and the net induction 'seen' by the observing nucleus rapidly varies. Such time-dependent fluctuations may have Fourier components of a frequency suitable to induce relaxation, and Fig. 2.6 gives the necessary relationship between τ_c for these processes and relaxation efficiency.

Generally, translational motion does not give rise to fluctuating fields of suitable frequency. If the sample contains paramagnetic species this is not the case, and intermolecular relaxation may become important. The magnetic fields arising from unpaired electron spins are so enormous that, in spite of an inverse-sixth-power dependence upon distance, they can nevertheless exert powerful forces at nuclei in other molecules. The random translational motions between the paramagnetic species and the molecule under examination gives rise to powerful fluctuating fields of frequency suitable for inducing relaxation. This is a dipole–dipole process (the electron is a 'spin-$\frac{1}{2}$' species) which invariably dominates all others, and T_1 values for materials containing paramagnetic ions, free radicals, or even dissolved oxygen, are very short.

As we shall see later, a nucleus is effectively shielded from the applied induction B_0 by the electrons in a molecule; since chemically different nuclei are shielded to different extents, they resonate at different Larmor frequencies, giving rise to the *chemical shift* phenomenon. The shielding is best discussed in terms of a tensor σ, the chemical shift tensor; the magnitude of this at a nucleus is dependent upon the orientation of the molecule with respect to B_0. Generally, in a fluid medium, rapid molecular motions average out the various components of σ and this value, σ, is that usually measured in the n.m.r. experiment (see later). If the components of σ are not all of equal magnitude, i.e. the chemical shift tensor is anisotropic, there will of course be rapid fluctuations of the effective field seen by the nucleus, and a mechanism for relaxation is evident, by the process of *chemical shift anisotropy*. It seems likely that this process is of little significance at values of B_0 less than about 2.5 tesla; at higher values (e.g. 6.0 tesla as used in some superconducting systems) it may well be as important as the dipole–dipole mechanism.

We shall discuss later the phenomenon of spin–spin coupling whereby there is energy of interaction between nuclei in the same molecule. The nucleus in which we are interested has spin $\frac{1}{2}$, but the nuclei to which it is spin–spin coupled may have spin greater than $\frac{1}{2}$; if the lifetime of the spin states of these nuclei is short compared with that of the nucleus whose resonance is being studied (i.e. they are relaxing more rapidly), the local field due to their magnetism fluctuates rapidly and the *scalar relaxation* mechanism operates. Quadrupolar nuclei have spin greater than $\frac{1}{2}$, and if they are in an asymmetric field environment (arising from the structure in which they are present) they have very short T_1. Nuclei of spin $\frac{1}{2}$ which are coupled to such species may not in fact show splittings in their spectra due to the phenomenon simply *because* the lifetime of spin states is so short; nevertheless the scalar coupling may cause them to have short T_1, although sometimes the mechanism is only effective in shortening T_2.

The phenomenon of *spin–rotation* relaxation arises from magnetic fields generated by the motion of *molecular magnetic moments* due to the electronic distribution in a molecule. The mechanism involves quantum rotational states of the molecule or of small symmetrical groups undergoing rapid internal rotation (e.g. CH_3 groups in a large molecule such as a steroid). It is possible to define an angular momentum correlation time; τ_J, for such species, which is related to τ_c (the molecular reorientation correlation time). Since τ_J becomes longer as temperature is increased, whereas τ_c decreases, the effect of higher temperatures is to make spin–rotation relaxation more efficient (shorter T_1) while dipole–dipole relaxation becomes less efficient (i.e. longer T_1). Generally, the spin–rotation mechanism is only of importance, compared with the other mechanisms, for small molecules particularly in the vapour or gas phase (since, again, τ_J is longer under these conditions than in the condensed phase where it may be very short).

125

2.1.10 Mechanisms of spin–spin
relaxation

Instead of transferring energy from the spin system to an external sink (the lattice), spin–spin relaxation involves the interchange of energy within the spin system. In the pulsed n.m.r. experiment this occurs when the pulse has been completed, and is shown by a 'fanning out' or dephasing of spins parallel to the xy plane after the pulse has transferred the net magnetization, out of alignment with the z' axis, towards the y' axis. Quantum mechanically, it is best described as being a process which limits the lifetimes of the nuclear spin states and consequently causes line broadening.

The scalar relaxation mechanism, described above for T_1, is important in determining T_2. At a nucleus, k, the local fields from neighbouring magnetic nuclei have oscillating components, and a nucleus j may therefore produce a magnetic induction oscillating at its Larmor frequency which induces a transition in nucleus k. The energy comes from j, and a simultaneous reorientation of both nuclei occurs with conservation of energy between the pair. Only nuclei that are identical may undergo this exchange of spin.

In a given sample, each nucleus 'sees' a steady magnetic induction which is modulated by a small fluctuating local induction produced by neighbouring nuclei; only nearest neighbours are actually significant. The modulation means that there is a broadening of the energy levels of the system and hence of the lines. If B_{loc} is the 'spread' of this local induction, the range of frequencies of Larmor precession is:

$$\Delta v = \frac{\mu B_{loc}}{Ih} \tag{2.37}$$

and if the nuclear dipoles are precessing in-phase at a particular instant of time they require $(\Delta v)^{-1}$ seconds to get out of phase; Δv is of the order of 10^{-4} s and contributes towards T_2. This process is of great importance when the nucleus is bonded to a quadrupolar species; thus, the absorption lines of protons bonded to ^{14}N $(I = 1)$ are often very broad.

Nuclei that are 'unlike' may contribute very significantly towards B_{loc}, and so the dipole–dipole relaxation mechanism may contribute to T_2. For this reason, the fluctuating field arising from unpaired electron spins in solutions containing paramagnetic species may also make B_{loc} very large and hence $(\Delta v)^{-1}$ very small, i.e. very short T_2 with consequently broad lines.

Generally, changes in T_1 are accompanied by parallel changes in T_2, and this is well illustrated in Fig. 2.6 which assumes domination by the dipole–dipole mechanism. For τ_c greater than 10^{-9} s, however, the behaviour is different, as shown clearly in Fig. 2.6.

2.2 THE EXPERIMENTAL TECHNIQUE

2.2.1 The spectrometer

The basic requirements of a spectrometer are simple:

(1) A *powerful magnet* into which the sample is placed. A net magnetization is induced in the direction of the applied magnetic induction B_0 (z' in the rotating frame).

(2) A *radiofrequency transmitter,* which supplies the (Larmor) frequency appropriate for the particular field strength and nucleus under examination. The linearly oscillating r.f. field is applied perpendicularly to the direction of B_0 (in the xy plane) and the magnetic vector of this field has a component rotating at the Larmor frequency, i.e. along the x' axis in the rotating frame.

(3) A *radiofrequency receiver,* which is arranged to detect resonance by registering only signals which appear in the y' direction of the rotating frame.

(4) A *recorder,* to provide a permanent record of the signals observed by the receiver.

Depending upon the particular spectrometer, a search is made for resonance by:

(a) sweeping the magnetic induction in the region of B_0 and keeping the radiofrequency constant; or

(b) sweeping the radiofrequency in the region of the Larmor frequency, while keeping B_0 constant; or

(c) applying a pulse of radiofrequency of high power for a very short interval of time, and observing free induction decay.

As the name implies, *continuous wave* (c.w.) *spectrometers* use a continuous train of radiofrequency; they will be discussed separately from the pulsed spectrometers, although the requirements for the magnet system are common to both types.

The Magnet

It is essential that the magnet should provide a very stable field which is highly homogeneous in the volume of space into which the sample is placed.

If very rigorously thermostated and shielded from stray magnetic influences, *permanent magnets* are capable of providing very stable fields. Some commercially successful spectrometers such as those produced by Varian Associates (T60, EM360) and Perkin-Elmer Ltd. (R10, R12, R24.) use permanent magnets. Most of these operate at *ca* 1.4 tesla (60 MHz resonance frequency for ^{1}H, 56.4 MHz for ^{19}F) and are designed for use in spectrometers of a 'routine' as opposed to a 'research' type. There are difficulties in using permanent magnets at very high fields, and the (now obsolete) Perkin-Elmer R14 spectrometer which operated at 2.35 T (100 MHz for ^{1}H, 94.1 MHz for ^{19}F) was unique; the present maximum field strength for such a system is realized in the Perkin-Elmer R32 spectrometer, operating at 2.1 T (90 MHz for ^{1}H, 84.7 MHz for ^{19}F), and this is probably the limit of practicability.

The advantage of *electromagnets* is flexibility, and field strengths in the range 1.0–2.5 T are easily obtainable. The technique of stabilizing the field is well established, and stability equivalent to that of the very best permanent magnets is routinely available. Commercial spectrometers use electromagnets and operate at 1.4 T (Varian A60, HA60, JEOL C6OHL), 2.1 T (Bruker HFX90, WH90), and 2.35 T (Varian HA100, XL100, JEOL MH100, PS100). These instruments find use both as routine and research spectrometers.

For operation at very high field strengths (up to about 7 T) *superconducting magnets* (cryogenic magnets) are used. Once the field has been established, they have the stability of a permanent magnet, and corrective devices such as those used with conventional electromagnets are unnecessary.

Magnetic field stabilization

The stability of an electromagnet depends largely upon the stability of the current that passes through the coils, and hence there are very exacting requirements for the power supply of the magnet. Highly regulated alternating currents are rectified, and the direct current which is supplied to the magnet is stabilized using a feed-back system. The current is passed through a resistance and the resultant voltage is compared with that from a reference source of high stability (e.g. a Weston cell); the difference voltage is amplified and fed as an error signal to the system through which the magnet current has passed and causes self-compensation. This first-order stabilization of the current produces field stability to about one part in 10^6.

In order to provide a field which is stable to better than one part in 10^8, a *flux stabilizer* system is used. Correction coils are wound around the pole pieces of the magnet, and these detect rapid changes in the total flux across the gap between the pole-pieces. The voltages induced in the coil are again used in a feed-back system with amplification, and correcting voltages are applied across suitably sited compensating coils (usually wound on the same former as the detection coils). This technique corrects short-term (rapid) fluctuations very well, but it cannot compensate for slower long-term drifts in the field strength.

There are two methods in use which provide the third-order corrections to field stability required to overcome long-term drift to better than one part in 10^6 or 10^7; both effectively 'lock' the value of B_0 to the radiofrequency by keeping in resonance a reference n.m.r. signal. In the earlier of these methods a control probe is used, in addition to the normal probe holding the sample under investigation. This holds the reference sample (usually H_2O) permanently in the magnetic field, albeit in a slightly different position from the sample; the resonance signal from this control probe generates an error signal if either the field or the radiofrequency drifts, and this is fed back either into the flux stabilizer or into the input of the amplifier of the magnet power supply.

A somewhat more sophisticated, and indeed superior, system using only the sample probe was developed by Primas [2.13] and is widely used. To the sample is added an internal reference compound having a sharp signal well away (*ca.* 0.5 p.p.m.) from resonances which occur in the sample (e.g. tetramethysilane for ^{1}H, hexafluorobenzene for ^{19}F). The dispersion mode (*u*-mode) signal from the reference is fed as an error signal to the flux stabilizer, which corrects any drift from resonance and maintains a constant field/frequency ratio very precisely over long periods of time. The correction signal may be derived from resonance of a nucleus of the same type as is being observed in the sample (homonuclear field/frequency lock, as used in Varian HA100 or JEOL MH100 spectrometers) or from a different type of nucleus (e.g. ^{2}D resonance, as in Varian XL100 and CFT-20, JEOL PS100, Bruker HFX-90 or WH90 spectrometers, which employ heteronuclear lock systems).

Field homogeneity

In order to achieve the resolution demanded in modern high-resolution n.m.r. spectroscopy, it is necessary to have magnetic fields which are homogeneous to within one part in 10^8 or 10^9 in the volume of space occupied by the sample. The basic requirement for such homogeneity lies in very precise pole-piece design; these must be metallurgically uniform, precisely parallel to each other, and free from machining marks. Traditionally, pole-pieces were almost optically flat but some modern systems use a logarithmically curved surface (the curve is very slight) with improved results.

Even the best pole-piece design and manufacture cannot produce the optimum homogeneity required for the n.m.r. experiment. The final results are obtained by utilizing a system of fine coils (shim coils, or Golay coils) in a certain pattern on each pole face; currents pass through these coils and may be varied by manual controls, and the complex field patterns generated used to correct the existing overall field inhomogeneity.

Further improvement in the homogeneity of the field is brought about by spinning the sample, at a rate of about 20 Hz, about the y axis of the magnet system (i.e. vertically, with B_0 applied horizontally). This averages out some of the effects of inhomogeneity in the xz plane.

The probe

The probe is the electrical device at the centre of the magnetic field into which the sample is placed in a suitable holder. The holder acts as a turbine, and air jets are arranged inside the probe to provide tangential forces which spin the sample about the vertical axis.

The probe may be of the *crossed-coil* variety, in which case it contains separate orthogonal coils, one of which acts as a transmitter and the other as a receiver of radiofrequency; it may alternatively be of the *single-coil* variety, in which case

129

it acts as one arm of a radiofrequency bridge.

In the crossed-coil probe, the radiofrequency is applied to the sample, via the transmitter coil, along the x' axis of the rotating frame of reference. At resonance the net magnetization is tipped away from the z' axis and has a component in the y' direction; this induces a voltage in the receiver coil, which detects magnetization in this direction, and this forms the basis for the n.m.r. signal which is finally displayed.

The single coil probe forms one arm of a radiofrequency version of a Wheatstone bridge. Radiofrequency is continuously supplied to the coil, and the bridge network balances out this signal. At resonance, absorption of r.f. energy causes the bridge to go out of balance and the resultant signal is displayed as the spectrum.

Although modern c.w. spectrometers may use either slow *field* or *frequency* sweep in order to produce the resonance spectrum, the probe is usually fitted with a device for rapidly and repetitively sweeping the magnetic field through resonance. This consists of a pair of Helmholtz coils mounted with their axes along the direction of the main magnetic field. To these *sweep coils* the output of a saw-tooth generator is fed, which allow the field to be swept through resonance in a uniform manner and in a recurrent fashion. The purpose is to allow oscilloscope presentation of the signal, for use when searching for the resonance of a particular nucleus or when making adjustments to the spectrometer such as probe-tuning or optimizing field homogeneity.

Figure 2.7 shows a schematic layout for a typical c.w. spectrometer. A feature to note is that the recorder is arranged to provide the voltage ramp necessary to sweep either the radiofrequency or the field as required; this ensures an exact correlation between the two and enables pre-calibrated recorder charts to be used.

Fourier transform (f.t.) *spectrometer*

The basic spectrometer system of magnet, probe, radiofrequency oscillator, receiver, and recorder is similar to that required for c.w. operation, and in fact most modern pulsed spectrometers may be operated in a c.w. mode. The differences lie in the requirement of units to pulse the radiofrequency, the need for high power amplifiers, and the necessity for a highly sophisticated data handling system; indeed the philosophy displayed by an instrument such as the Varian CFT-20 (a $^{13}C/^1H$ f.t. only system) is to design the whole system around the computer!

A schematic diagram of a typical f.t. spectrometer is shown in Fig. 2.8. The main features common to a c.w. system have been removed. The heart of the system is the computer and its interface to the spectrometer system, which controls the timing of the system and the acquisition and manipulation of data. The experimental sequence is as follows.

(1) At a command from the computer, the pulsing unit allows high radiofrequency

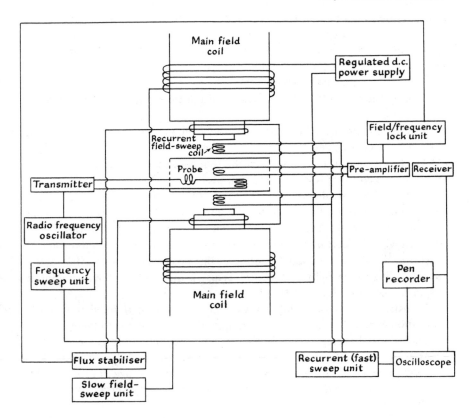

Fig. 2.7 Schematic diagram of a typical continuous-wave spectrometer.

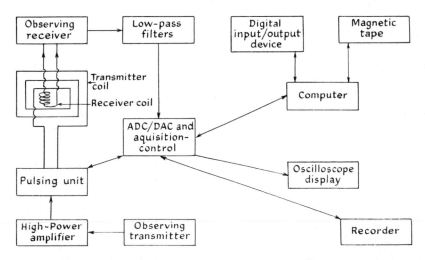

Fig. 2.8 Schematic diagram of a typical Fourier–transform spectrometer.

power (close to the Larmor frequency for the nucleus under investigation) to pass to the transmitter coil. After a short interval of time (typically $10-100\,\mu s$) the computer issues a command to shut down the radiofrequency source (i.e. an r.f. 'gate' or switch is opened and closed).

(2) The f.i.d. is observed by the observing receiver for an interval of time as instructed by the computer. This is the acquisition time, and is typically about $1-5$ s.

(3) After going through a low-pass filter, the analogue signal is converted to digital form (by an a.d.c.) and stored in the computer.

(4) The pulsing unit is reactivated and the sequence repeated. The new f.i.d. is added to that already present in the computer.

(5) When a previously determined number of f.i.d.'s have been added coherently the computer applies a Fourier transform analysis to the stored (time domain) spectrum and produces the frequency domain spectrum in digital form.

(6) The digitzed spectrum is converted to analogue form (d.a.c.) and output on a chart recorder is the usual way. A listing of line positions and intensities corresponding to the digitized spectrum is also printed out via the digital input/output device (typically a teletype).

All instructions for operating the system (e.g. pulse length in microseconds, acquisition time, spectral width, display parameters) are input to the computer via the teletype, and during the whole process the computer controls everything via a master program which is permanently in residence in core. A magnetic tape or disc unit provides a convenient back up facility for the storage of programmes, spectra, or unprocessed data and considerably enhances the versatility of the system.

Further discussion of Fourier transform systems is beyond the scope of this chapter. For further information, the reader is referred to the excellent accounts by Farrar and Becker [2.14] and by Gillies and Shaw [2.15].

2.2.2 Useful accessories

Spin decoupling

We shall see later that many energy levels in a spin system (and hence many of the resonance lines observed) arise because of mangetic interactions between spin-$\frac{1}{2}$ nuclei in the same molecule. These occur via the bonding electrons of the spin system, and the phenomenon results in the appearance of fine structure on resonance lines and is known as spin−spin coupling. Often, it is desirable to observe the spectrum without this multiplicity of lines, and application of the correct radiofrequency in the form of a secondary applied field at the origin of a multiplet will effectively 'decouple' the spin of the particular nucleus concerned from the rest of the spectrum; the spectrum may then be observed in the absence of coupling with this nucleus.

Most spectrometers are equipped with apparatus which can supply the secondary irradiating frequency at the desired position, a *spin decoupler*. Sometimes it is necessary to irradiate individual lines in the spectrum, as opposed to a group of lines, with very great precision; this is, for example, for the observation for *spin–spin tickling* (see later). In order for this to be successful, the secondary oscillator has to be of the same specification as the basic r.f. oscillator for the spectrometer; ideally both the observing field (B_1) and the secondary irradiating field (B_2) should be derived from the same source, as in a frequency synthesizer.

Spin decoupling, is *homonuclear* when, as the name implies, it decouples nuclei of the same type as are being observed in a given experiment; *heteronuclear* decoupling has recently achieved considerable importance, since it is most convenient to observe ^{13}C n.m.r. spectra which are completely decoupled from ^1H spins. In order to irradiate all ^1H resonances simultaneously, it is necessary to use very high powers in the B_2 field (about 10 W) and to *noise modulate* the frequency; this has the effect of spreading the useful power over a wide range (which may be selected manually), hence irradiating all lines in the spectrum.

INDOR Spectroscopy

Spin decoupling and spin–spin tickling are special cases of the double irradiation experiment. A useful variation is known as *INDOR spectroscopy* (Inter-*N*uclear *DO*uble *R*esonance) and this forms a useful accessory to a basic spectrometer. The requirement is to be able to monitor continuously the intensity of a chosen spectral line, using B_1, while sweeping through all other resonances with the B_2 field. The INDOR spectrum is displayed as the intensity of this line as a function of the B_2 frequency.

Variable-temperature accessory

In order to study chemical and physical rate processes (and sometimes merely to take up solids into solution) an essential feature of a spectrometer is the presence of a device to control accurately the temperature of the sample over as wide a range as possible. It is necessary to maintain fixed temperatures for considerable periods of time, and for temperatures to be reproducible with some precision. This subject is adequately covered in the literature [2.16].

2.2.3   Experimental procedures

Sample preparation

Samples are almost invariably examined as liquids or dilute solutions in inert solvents. Gases under pressure may be observed, but solids cannot usually be examined because of extreme line broadening. The solution is contained in a cylindrical glass tube which is inserted into a turbine which fits into the probe. In order to achieve the best results from spinning the tube, it is necessary to have high-quality tubing of precision bore; typical diameters in use are 5, 8, 10, and 12 mm o.d. Samples should be free from suspended particles and paramagnetic impurities which cause line broadening (but may help to prevent saturation); removal of dissolved oxygen often aids the production of high-quality spectra.

Referencing

It is usual to quote frequencies of lines in n.m.r. spectra by referring to an arbitrarily chosen standard. This is selected to have a single, sharp resonance which occurs at a frequency outside the range of frequencies typifying the sample; for ^1H, tetramethylsilane in convenient, and the ^{13}C absorption of this same molecule provides a useful reference for ^{13}C studies, although early studies with ^{13}C used CS_2. The references may be dissolved in the sample under investigation (internal standard) or contained separately in a glass capillary which is mounted centrally in the sample tube (external standard); in the latter situation, it is necessary to correct the frequency differences for the different magnetic susceptibilities of the sample and reference, The *frequency measurements* are usually made directly upon the pre-calibrated charts of modern spectrometers. For accurate work, however, it is often desirable to measure frequency differences directly using a frequency counter of high precision. In a Fourier transform spectrometer, the frequencies are given directly as part of the computer's output.

Integration

In a conventional (c.w.) spectrum, the relative areas under the absorption lines in a given spectrum are proportional to the number of equivalent nuclei contributing to them. The area under a peak is given by the integral of the trace, and all spectrometers are arranged to output the spectrum as an integral when required. In a Fourier transform experiment there are certain difficulties in obtaining accurate integrals owing to incomplete (and unequal) relaxation between pulses, and digitization limitations caused by small core size in the computer [2.14, 2.15]; assuming steps have been taken to overcome these problems, a simple routine enables direct output of integrals from the computer.

Spectrometer operation

In principle this is straightforward. In a typical c.w. experiment, operation of the field/frequency lock circuit places the resonance of the reference line at the right hand end of the recorder chart. The operator then decides over what range the frequency should be varied, and how much power should be used in the B_1 field; amplifier gain settings enable spectra to be increased in size to 'fill' charts vertically, filters remove unwanted noise, and a suitable 'phase' setting is chosen to select exclusively the v-mode signal from the u-mode. Operation of the recorder, from its left-hand extremity, causes the field to be swept synchronously (low field to high field) or the frequency (high frequency to low) to be swept, according to which operation is chosen.

In a typical f.t. experiment, the whole operation is under computer control; instructions are input via the teletype in the form of simple mnemonics. The manual settings of the spectometer consist of adjusting the level of power in the pulse, and selecting its frequency. A typical set-up of instructions then defines pulse length (microseconds), acquisition time, interval between pulses, spectral width, instruction for digitization and manipulation of the f.i.d. before transformation, and display and output parameters.

Measurement of T_1 and T_2

The pulse spectrometer lends itself very well to the measurement of relaxation times. The experimental technique is to decide upon a suitable multi-pulse sequence, the details of which are given to the computer which sets up the necessary experiment.

For T_1 measurement, a $180°-\tau-90°$ pulse sequence is widely used. First, a $180°$ pulse inverts the magnetization along the $-z'$ axis. Only longitudinal relaxation occurs, and the system would eventually return to its equilibrium state with M_0 aligned along $+z'$; however, after a time τ, a $90°$ pulse is applied along the z' axis and the magnetization is rotated along the y' axis. The free induction signal which results is allowed to decay in the usual way, and upon Fourier transformation yields lines whose intensities reflect the extent to which spin–lattice relaxation has occurred in the time τ; if the sequence is to be repeated many times for signal enhancement purposes (e.g. for ^{13}C n.m.r.), it is necessary to wait for a considerable time (of the order of five times the longest T_1 in the sample spectrum) between sequences.

The whole operation is repeated with a sequence of different values of τ, and a plot of the intensity of each line against τ shows an experimental increase (from a maximum negative value with very short τ, through zero, to a maximum positive value) with time constant T_1, which may then be evaluated.

For T_2 measurements, rather more complex pulse sequences are required. These include the spin-echo sequence of Hahn [2.17] ($90°-\tau-180°-2\tau-$ observe f.i.d. echo), the Carr–Purcell modification of this [2.18] ($90°-\tau-$ $180°-2\tau-180°-2\tau-180°-2\tau$, etc.), and other developments such as that of Meiboom and Gill [2.19]. These methods are fully described elsewhere [2.20] and are beyond the scope of the present account.

2.3 THE ANALYSIS OF HIGH RESOLUTION N.M.R. SPECTRA

In the preceding sections, we discussed the theoretical and experimental basis of the technique. Unlike other spectroscopic methods, however, it is not sufficient

135

in n.m.r. merely to measure line positions and intensities; it is necessary to 'translate' this information into fundamental parameters known as *chemical shifts* and *coupling constants,* and this is the process of spectral analysis.

The precise value of the Larmor frequency for a given nucleus depends upon its electronic environment, i.e. upon the chemical properties of the atom in which it is present. Thus, all ^1H nuclei do not have precisely the same Larmor frequency, and different compounds have absorption signals at different frequencies for a given applied field. For example, water, benzene, and cyclohexane each show a single resonance at a different frequency, This phenomenon is called the chemical shift and is particularly important because chemically different nuclei *in the same molecule* also have different chemical shifts; ethanol, for example, shows three distinct ^1H absorption signals, of relative intensities $1 ; 2 ; 3$ (OH, CH_2, CH_3). Chemical shifts are usually measured from an arbitrary reference, as discussed in the previous section. It is found that they are dependent upon the field strength at which the measurement is made, and in fact are proportional to B_0. They originate in the fact that nuclei are screened from B_0 by the electrons within a molecule, and the extent of this is proportional to B_0; the net field 'seen' by a nucleus is then:

$$B_{eff} = B_0(1 - \sigma) \qquad (2.38)$$

where σ is the screening constant. The difference in the screening constants of two nuclei is the chemical shift between them.

It is conventional to measure chemical shifts in frequency units (Hz) irrespective of whether the experiment was carried out by field or frequency sweep (the interconversion is obvious). They are, however, reported in terms of a dimensionless parameter, *parts per million* or p.p.m., which is field-independent, by using the following relationship:

$$\text{Chemical shift (p.p.m.)} = \frac{\text{Chemical shift (Hz)} \times 10^6}{\text{Observation frequency (Hz)}} \qquad (2.39)$$

Thus, a chemical shift of 90 Hz from a certain reference, measured at 60 MHz observing frequency, is equivalent ot a shift of 150 MHz measured at 100 MHz:

$$\frac{90 \times 10^6}{60 \times 10^6} = \frac{150 \times 10^6}{100 \times 10^6} = 1.5 \text{ p.p.m.}$$

Frequently it is observed that chemically shifted lines show multiplet structure, arising from the phenomenon of *spin–spin coupling.* This originates in mutual interactions between nuclei within a molecule, which are transmitted via the bonding electronic structure. Two nuclei which are coupled in this way show a fine splitting, in their spectra, which is common to them both and depends upon their spin quantum number I.

If two spin-$\frac{1}{2}$ nuclei are coupled to, and chemically shifted from, each other they each appear as a doublet, the separation of which is called the spin–spin coupling constant and is independent of field strength. Spin–spin coupling constants, J, are measured in units of linear frequency, Hz.

If several nuclei are chemically equivalent to each other, for instance as in a methyl group (see later), they show no mutual splittings due to coupling. However, 'external' interactions with this group cause resonances to which the nuclei are coupled to be split into a multiplet structure of constant separation J. We shall see later how this arises, but it should be remembered at this stage that such a group will cause splitting into $(n + 1)$ lines where n is the number of spin-$\frac{1}{2}$ nuclei in the group.

It might be imagined that these simple rules would enable chemical shifts and coupling constants to be measured directly from a spectrum. This is indeed true if the chemical shifts between nuclei are significantly larger than the coupling constants between them (by a factor for 10 or more), but when this is not the case, i.e. when chemical shifts (Hz) and coupling constants (Hz) are comparable, then line separations do not correspond to the required parameters, and other phenomena (e.g. distortion of intensities, appearance of unexpected lines, disappearance of expected lines) occur. In these circumstances, it is necessary to resort to spectral analysis in order to extract the required information from measured line positions and intensities.

2.3.1 Nomenclature

In order to discuss spectral analysis, it is first necessary to understand the meaning of certain well-used terms and the conventional method of descriptively labelling a particular spin system.

Chemical equivalence

This occurs when two or more nuclei have identical chemical shifts; broadly speaking it may arise from three separate causes.

(1) The symmetry properties of the molecule may dictate this; e.g. in benzene (I) all six protons are chemically equivalent, as are the three protons in 1,3,5-trichlorobenzene (II).

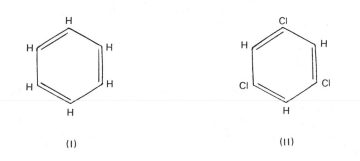

(I)   (II)

137

(2) An *effective* symmetry element may be introduced into the molecule by an averaging process which is 'fast' on the n.m.r. time scale (see later). Cyclohexane (III), for example, exists as a rapidly equilibrating mixture (III) of conformational isomers of equal energy.

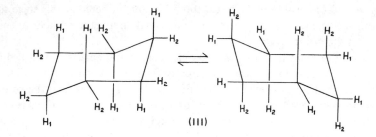

(III)

In either isomer, the protons fall into two different groups of six chemically equivalent nuclei (by symmetry), the axial and equatorial protons; the interconversion causes the identity of the two groups to interchange, and the result is chemical equivalence of all twelve protons *by averaging*. If the process is made slow by lowering the temperature, the separation into two groups of equivalent nuclei becomes apparent.

An interesting example is provided by the CH_3 protons of an ethyl group in a molecule such as ethyl chloride (IV). Rapid equilibration between conformations of equal energy causes chemical equivalence of the three nuclei; the CH_2 protons are of course equivalent by symmetry in each conformation irrespective of the conformational equilibrium.

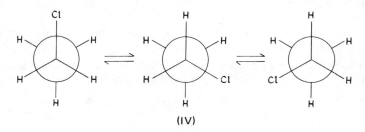

(IV)

If two of the hydrogens in the CH_3 group are replaced by other substituents, the $CH_2 Cl$ protons find themselves in an asymmetric environment, i.e. they are diastereotopic. In each of the three staggered formations they are non-equivalent, as is shown in the Newman projections (V).

The conformations are now of different energy, and hence unequally populated, and the spectrum of the $CH_2 Cl$ group protons is the weighted average of the individual spectra of all three; the two protons H_a and H_b have different average environments and are chemically non-equivalent. It is not immediately obvious, however, that even if all three conformations were of equal energy, and hence equally populated, the two protons would still be in different environments.

If the environment of each proton H_a and H_b is described in terms of its *gauche* (g) or trans (t) relationship to the substituents H, X, and Y, the average environment of each (when equally weighted between the three conformations) appears to be identical:

$$\text{For } H_a = \frac{(H_g + X_g + Y_t)_i + (H_t + X_g + Y_g)_{ii} + (H_g + X_t + Y_g)_{iii}}{3}$$

$$\text{For } H_b = \frac{(H_g + Y_g + X_t)_i + (H_g + X_g + Y_t)_{ii} + (H_t + X_g + Y_g)_{iii}}{3} \quad (2.40)$$

$$(2.41)$$

However:

$$(H_g + X_g + Y_t)_i \neq (H_g + X_g + Y_t)_{ii} \quad (2.42)$$

$$(H_t + X_g + Y_g)_{ii} \neq (H_t + X_g + Y_g)_{iii} \quad (2.43)$$

$$(H_g + X_t + Y_g)_{iii} \neq (H_g + X_t + Y_g)_i \quad (2.44)$$

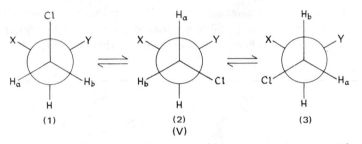

To understand why these inequivalences hold, it is necessary to consider secondary interactions involving the groups H, X, and Y, Thus $(H_g + X_g + Y_t)_i$, for example, involves H_g which is *gauche* to H_a and H_b and *trans* to Cl; in the environment $(H_g + X_g + Y_t)_{ii}$, however, H_g is *gauche* to H_b and *trans* to H_a. These differences hold of course for $(X_g + Y_t)_i$ and $(X_g + Y_t)_{ii}$ and similar considerations apply to the other situations. Diastereotopic protons are therefore *intrisically non-equivalent* because of the asymmetric environment in which they exist. The inequalities of conformational populations which almost inevitabily accompany such a situation may either increase or decrease the degree of non-equivalence which is observed in practice.

(3) Accidental equivalence may sometimes occur in situations where there is no obvious reason why this should be so; the electronic environments of the nuclei just happen to be identical. In this case it is usually possible to resolve the equivalence by changing the solvent in which the observation is made, or by adding small quantities of a so-called lanthanide 'shift reagent' which complexes with certain functionalities (if present in a molecule) and induces differential shielding at geometrically non-equivalent sites. For example, in 4-fluoro-4-aminobibenzyl (VI) the two groups of CH_2 protons are equivalent in all solvents;

139

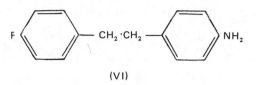

(VI)

in the presence of a europium complex (which coordinates with the NH_2 function) the spectrum shows the expected signals arising from two non-equivalent pairs of CH_2 protons.

Magnetic equivalence

This is a concept somewhat more difficult to grasp than chemical equivalence. Magnetically equivalent nuclei are always chemically equivalent, but not all chemically equivalent nuclei are magnetically equivalent; they must have identical spin–spin coupling constants to each different nucleus in the system. In a molecule such as 1-chloro-3,5-dibromobenzene (VII) the symmetry demands that there are two different chemical environments for the protons (a, b) and c, and

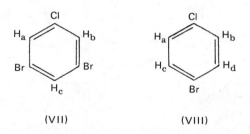

(VII)   (VIII)

H_a and H_b are chemically equivalent. H_a and H_b each have the same coupling to H_c (i.e. *meta* coupling via an intervening C–Br group) and they are therefore also magnetically equivalent. In *para*-chlorobromobenzene (VIII), however, the situation is different. By symmetry, H_a and H_b are chemically identical, and H_c and H_d are also equivalent to each other. However, H_a has an *ortho* coupling to H_c and a *para* coupling to H_d, while H_b has an *ortho* coupling to H_d and a *para* coupling to H_c; this seemingly trivial differentiation means that the protons H_a and H_b do not show magnetic equivalence and this has important consequences in the form taken by the spectrum.

In the case of a molecule such as CH_3CH_2Cl, the two chemically different sets of protons show *effective* magnetic equivalence within each set because of averaging (i.e. the three CH_3 protons are all magnetically equivalent to each other, as are the two CH_2 protons).

It is very useful to have a short-hand notation to describe a system of spin-$\frac{1}{2}$ nuclei present in a molecule (a spin system) in such a way as to identify chemical and magnetic equivalence or non-equivalence, and also to indicate which nuclei have a small or large chemical shift (δ) between them compared with their mutual coupling constant J. Conventionally, this is done by allocating to each nucleus a letter of the alphabet. Chemically equivalent nuclei are given the same letter, nuclei with a large chemical shift between them are given letters at opposite ends of the alphabet (or as far apart as possible), and nuclei between which there is a small chemical shift are given letters close together in the alphabet. If nuclei are chemically but not magnetically equivalent they are differentiated by using a prime notation (').

A two-spin system may now be described as AB (if $J_{AB} \simeq \delta_{AB}$), or AX (if $J_{AB} \ll \delta_{AB}$), or even A_2 (if $\delta_{AB} = 0$).

Three-spin systems may take any of several possible forms: ABC (if $J_{AB} \simeq \delta_{AB}$, $J_{AC} \simeq \delta_{AC}$, $J_{BC} \simeq \delta_{BC}$), ABX (or AXY) (if $J_{AB} \simeq \delta_{AB}$, $J_{AX} \ll \delta_{AX}$, $J_{BX} \ll \delta_{BX}$), AMX (if $J_{AM} \ll \delta_{AM}$, $J_{AX} \ll \delta_{AX}$, $J_{MX} \ll \delta_{MX}$), or AX_2 (or A_2X) (if $J_{AX} \ll \delta_{AX}$ and there are two equivalent spins X).

Four-spin systems are even more complex, and the possibility of chemical equivalence with magnetic non-equivalence arises: ABCD, ABCX, ABMX, ABXY, A_2B_2, A_2X_2, AA'BB', AA'XX', etc., where AA' signifies a pair of chemically equivalent nuclei which are magnetically non-equivalent.

The extension of this nomenclature to even larger spin systems is obvious, and need not concern us further. In situations where the chemical shifts, δ, are considerably larger than the coupling constants, J, the spin system is said to be first-order, but when the J's and δ's are comparable in magnitude a second-order spin system arises. Thus, AMX is first order, whereas ABC is second order; mixed systems such as ABX must be considered to be second order. The differentiation is important, for in a first order spectrum it is possible to measure chemical shifts and coupling constants directly in terms of the observed line separations, but not in a second order case where a full quantum mechanical analysis may be necessary.

Before we can proceed further, it is perhaps necessary to consider how the splittings from spin—spin coupling arise. Consider a pair of non-equivalent nuclei, H_A and H_X. In the absence of spin—spin coupling all A nuclei are divided between two nuclear energy levels, as are all X nuclei.

In the presence of spin—spin coupling, the two different spin states in which A nuclei may find themselves exert different magnetic fields at nucleus X. These are independent of B_0 and additional to it; they may reinforce or oppose B_0 and therefore cause a splitting of the levels shown in Fig. 2.9. 'X' exerts an exactly equal influence on 'A', and the situation is exemplified in Fig. 2.10.

The splitting of the levels, when converted to units of linear frequency (Hz), is equal to $J/2$, where J is the spin—spin coupling constant. For each nucleus, only two transitions are observed and these are symmetrically disposed about the single transition which would occur in the absence of coupling; they are

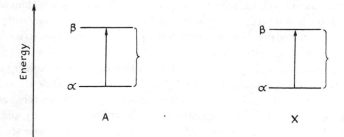

Fig. 2.9 Energy levels of two nuclei in the absence of spin—spin coupling

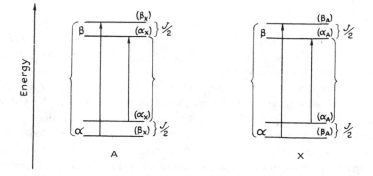

Fig. 2.10 Energy levels of two nuclei in the presence of spin—spin coupling

separated by J (Hz). In the example shown, the β spin state of X has lowered the energy of the α state of A and raised the energy of the β state of A; conversely, the α spin state of X raises the energy of $\alpha(A)$ and lowers the energy of $\beta(A)$. Because of the mutual nature of the interactions, the same effects, due to the spin states of A, occur on the X spin; the spin—spin coupling is said to be of positive sign. If, on the other hand, the α spin state of X *lowers* the energy of the α spin state of A and increases the energy of the β spin state, the coupling is said to be of negative sign. It is important to realize that the two transitions shown for A and the two for X are the only ones observable; this is because: (a) $J/2$ is very small compared with the major splittings (Hz compared with 10^8 Hz) and no suitable frequency is available for excitation between the closely spaced levels; (b) the spin state of X cannot change simultaneously with that of A, and hence the transition $\alpha_A\beta_X$ to $\beta_A\beta_X$ is allowed, but $\alpha_A\beta_X$ to $\beta_A\alpha_X$ is not.

2.3.2 First order spectra

In a first order situation, any spin-$\frac{1}{2}$ nucleus which is coupled to another chemically non-equivalent spin will be split into a doublet; the other nucleus will also

142

have a doublet splitting of the same magnitude. Independently, these nuclei may be spin coupled to others in the same way and extra splittings occur. In an AMX (first order 3-spin system) there will thus be three groups of four lines. The separation between the centres of the multiplets corresponds to chemical shifts, δ, and the separations within the multiplets to coupling constants, J. Such a spectrum is shown schematically in Fig. 2.11

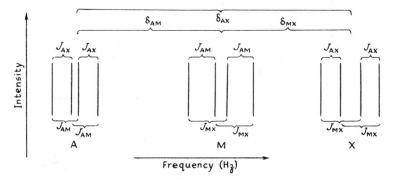

Fig. 2.11 Schematic AMX spectrum; $J_{MX} > J_{AM} > J_{AX}$.

Any first order system which does not involve equivalent spins may now be easily analysed. If the system contains a group (or groups) of chemically and magnetically equivalent nuclei, and is first order (e.g. $A_3 X_2$ or $A_2 X_2$) a simple rule enables the spin–spin splitting pattern to be rationalized. A group of such spins which is coupled to another (either single nucleus or another group) will split the resonance of the latter into $(n + 1)$ lines, where n is the number of nuclei in the equivalent groups. Thus, an $A_3 X_2$ spectrum has the form shown in Fig. 2.12; note, however, that the intensities of the lines are not now all equal but are in the ratio of the binomial coefficients.

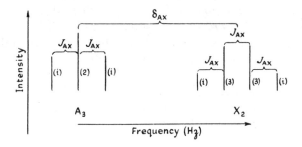

Fig. 2.12 The spectrum of an $A_3 X_2$ spin system. The total intensity of A_3 is 3/2 times that of X_2, and the relative intensities within each group are given by the numbers in parentheses.

143

In order to understand how this arises, it is necessary to consider the number of ways in which the spins of one group, say A_3, can combine together to perturb the energy levels of the other, say X_2. In all there are eight different ways of combining together the three A spins, since each may be only either α or β, and there are 2^n ways of combining n spins. The energy of interaction with the rest of the spin system (X_2 in this case) depends upon the sum of the m_z values for each combination, F_z; in fact, only four different values of F_z are possible and this is shown schematically in Table 2.2.

Table 2.2. The different ways of combining together the three equivalent spins of an A_3 group of nuclei; they are grouped together according to F_z ($=$ sum of m_z)

Spins	F_z
$\beta\beta\beta$	$-3/2$
$\beta\beta\alpha, \beta\alpha\beta, \alpha\beta\beta$	$-1/2$
$\beta\alpha\alpha, \alpha\beta\alpha, \alpha\alpha\beta$	$+1/2$
$\alpha\alpha\alpha$	$+3/2$

Since the two nuclei of the X_2 spectrum are chemically and magnetically equivalent, we may consider the effect of coupling upon just one of them; the lines due to the other superimpose *exactly* on these, and merely double the intensity. The α and β spin states of X are perturbed by the eight different spin states of the A_3 group, and split as shown in Fig. 2.13, where the allowed transitions are indicated. Since the transitions fall into two sets of three degenerate lines and two separate lines, the resultant spectrum shows four lines of intensities $1 : 3 : 3 : 1$.

A similar rationale may be used for any group of equivalent spins. In the first order case of three non-equivalent spins, A M X, the system may be described in terms of eight energy levels. These correspond to each of the 2^3 ($=8$) ways of combining together the α and β spin states of the component spins. In general for a *first order* system of n spins there will be 2^n energy levels.

The transition frequencies are given by the differences between these energy levels, but not all transitions are allowed; *selection rules* enable the correct predictions to be made
(1) Transitions are only allowed between states which differ in F_z by ± 1 ($+1$ corresponds to absorption, -1 to the corresponding stimulated emission).
(2) The transition must involve the change of orientation of the spin of only one nucleus.

Figure 2.14 illustrates the situation for the A M X case, and shows the allowed transitions. Thus, for example, a transition is allowed between energy levels corresponding to states $\alpha\alpha\alpha$ and $\alpha\alpha\beta$ (F_z 3/2 and 1/2 respectively) but not between $\alpha\alpha\alpha$ and $\alpha\beta\beta$ ($F_z = 3/2$ and $-1/2$). The transition $\alpha\beta\alpha$ to $\beta\alpha\beta$ (F_z $+1/2$ to $-1/2$) is not allowed because all three nuclei have change spin simultaneously.

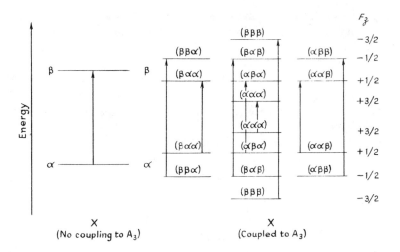

Fig. 2.13 The spin energy levels of an X nucleus coupled to an A_3 group of nuclei. The spin combinations of A_3 which are responsible for each energy level are shown in parentheses, and the F_z values are shown at the right.

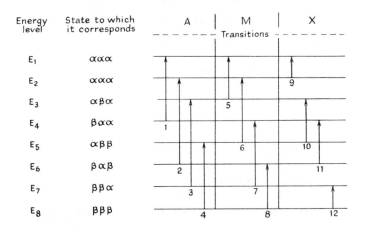

Fig. 2.14 Allowed transitions between energy levels for the case of an AMX molecule.

2.3.3 Second order spectra

In order to analyse second order spectra, it is necessary to formulate the problem in quantum-mechanical terms, the details of which are beyond the scope of this chapter. The consequences (which arise when the coupling constants in a system become comparable to the chemical shifts) are to distort the line intensities and to introduce extra (weak) lines into the spectrum. Line separations change and

145

are not simply related to chemical shifts and coupling constants. Generally, any spin system in which the coupling constant between a pair of nuclei is greater than about one-tenth of their mutual chemical shift gives a spectrum which requires second order analysis.

In order to show briefly some of the consequences, let us consider the AB spin system. This is characterized by a coupling constant J_{ab} Hz, and the chemical shift between the two nuclei δ_{AB} Hz. The form of the spectrum is solely a function of the ratio J_{AB}/δ_{AB}.

The analysis is straightforward. Referring to the schematic diagram, Fig. 2.15, the chemical shift δ_{AB} is given in terms of the separation in Hz of the couter lines (1 and 4) and the inner lines (2 and 3):

$$\delta_{AB} = [(1 + 4)(2 + 3)]^{\frac{1}{2}} \qquad (2.45)$$

J_{AB} is equal to the separation of the outer lines:

$$J_{AB} = (1 - 2) = (3 - 4) \qquad (2.46)$$

Also, the intensity of the inner lines is related to that of the outer lines:

$$\frac{\text{Intensity of 2 (or 3)}}{\text{Intensity of 1 (or 4)}} = \frac{(1 - 4)}{(2 - 3)} \qquad (2.47)$$

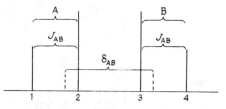

Fig. 2.15 Schematic AB spectrum

Spectra arising from larger spin systems are complex, but in some cases explicit algebraic formulae are available in order to calculate the parameters in terms of measured line positions and intensities. For further reading concerning the practice theory of spectral analysis and the form taken by various types of spectra, the reader should consult the excellent monograph by Abraham [2.22], and references [2.21] and [2.23–2.29].

2.4 DOUBLE RESONANCE METHODS

It is convenient to discuss at this stage the various double resonance methods, since they are often of great value in spectral analysis and may in some cases provide the only means of obtaining information upon relative signs of coupling constants.

Spin decoupling

The most widely used technique consists of applying a *powerful* secondary r.f. field at the chemical shift position of one nucleus in a spin system. This causes frequent transitions to occur between the energy levels of this spin, and decouples it from the remaining spins. Thus, if the secondary field is applied at ν_X in an ABX system the AB part reduces from a complex eight-line system to an instantly analysable AB quartet structure. This technique was developed by Bloch [2.30], Royden [2.31], and Bloom and Shoolery [2.32].

Limitations of the experiment arise because the secondary irradiating power required for decoupling increases rapidly with the second order nature of a spectrum. Incomplete decoupling results from too low power and the spectra are very complex. Further difficulty may occur if its desired to remove several couplings by simultaneous irradiation at several chemical shifts, because the decoupling field has a finite bandwidth. This may be overcome by employing discrete frequencies from separate sources (multiple resonance) or, in the case of heteronuclear double resonance, by *noise modulation* of the secondary irradiation [2.33]. Noise modulation has the effect of spreading the useful power across a wider, definable, bandwidth. Thus it is possible to examine ^{13}C spectra while simultaneously decoupling all protons over a range of, say, 10 p.p.m. in the ^{1}H spectrum.

Irradiation with weak r.f. fields

An interesting experiment involves irradiation at selected points in the spectrum with a secondary field of medium strength. This was the first technique to give information about the relative signs of J's in first order spin systems. It will be illustrated by reference to an AMX system, in which nucleus A absorbs to low field, M is midfield, and X is to high field; for convenience $J_{AM} > J_{MX} > J_{AX}$, and the spectrum is illustrated schematically in Fig. 2.11. The consequences of this partially selective double irradiation depend upon the relative signs of the coupling constants, and may be summarized as follows: "If, in a first order spin system, the *low* field half of a multiplet (say A) is irradiated the effects are observed in a second multiplet (say X), then if the two couplings to the third nucleus (J_{AM}, J_{MX}) are of the same sign, the *low* field half of X will collapse; if J_{AM} and J_{MX} are of opposite sign, the *high* field half will collapse." The converse statement is also true, and it is never possible to obtain information about the sign of the J between the irradiated and observed nuclei.

It is simply necessary, then, in a three-spin system to carry out two experiments to determine the relative signs of the three J's. Irradiation of half of A and observing X gives the relative signs of J_{AM} and J_{MX}, and irradiation of M and observing X gives the relatives sign of J_{AM} and J_{AX}. Analogous experiments may be designed for more complex spin systems, but in no case is it possible to obtain the *absolute* signs of the J's.

The place of partial irradiation studies has been largely taken by *spin-tickling experiments* for the determination of relative signs of J's. The experiment consists of applying a *weak* secondary field precisely *upon* the resonance frequency of individual lines and observing the effects of this mild perturbation elsewhere in the spectrum. In order to understand what happens, it is helpful to consider the example of an AMX system once more, together with the energy level diagram in Fig. 2.14. Each transition has four others which are said to be 'connected' to it, because they have one or other of its energy levels in common. Thus, line 1 is connected to lines 5 and 9 (upper level E_1 in common) and to lines 7 and 11 (lower level, E_4, in common). For any connected pair of transitions, if the common energy level is intermediate between the lowest and highest of the three they are said to be *progressively* connected. If the common level is the highest *or* lowest of the three they are said to be *regressively* connected. Thus, transition 1 is regressively connected to 5 and 9 and progressively connected to 7 and 11.

The consequences of a tickling experiment may now be summarized, and the rules apply to a first order *or* a second order spectrum.

(a) A transition which has an energy level in common with (i.e. is connected to) the irradiated transition will be split into a doublet.

(b) If the connection is regressive (R) the doublet will be well-resolved; if progressive (P) the doublet will be poorly resolved.

(c) The size of the doublet splitting is proportional to the strength of the secondary irradiating field and to the square root of the intensity of the irradiated transition.

The value of the experiment lies in the ability to identify transitions. Thus, irradiation of line 1 causes 5 and 9 to give a deep splitting and 7 and 11 a shallow splitting. If J_{AX}, and J_{MX} are all positive, this corresponds to the first and third lines of M (R and P respectively) and the first two lines of X (R and P respectively). If J_{AX} is opposite in sign to the others, however, this corresponds to the same situation in M but the first two lines of X are now P and R respectively. Obviously many such combinations exist, but it is possible with a little thought to design a few simple experiments which will give the relative signs of all the J's. Extensions to other, more complicated, spin systems are possible. For a more detailed discussion, see Abraham [2.22], pp. 302 *et seq.*

INDOR spectroscopy is a useful double irradiation experiment, designed by Baker and Burd [2.34] and exploited well by Freeman and Anderson [2.35] and many others (e.g. [2.36, 2.37]). It is analogous to spin-tickling in that it enables connectivity relationships between energy levels to be deduced. Experimentally, the *intensity* of an individual transition is monitored while the weak secondary field is swept through resonance of the other transitions. This is plotted as a function of the irradiation frequency, and can be arranged to coincide with the normal absorption spectrum. When a line is brought into resonance, which is regressively connected to the observing line, a negative signal is observed (corresponding to a decrease in intensity) and when a progressive transition is in resonance an increase in intensity of the observing line causes a positive signal. The application to the determination of relative signs then follows.

The variations in intensity of the observed line arise from changes in energy level population brought about by the (sweeping) secondary r.f. field. Referring again to Fig. 2.14, consider transition 1 as an observed line. Passage of the secondary irradiation through line 5 (R) increases the population of the *upper* level E_1 and hence reduces the intensity of line 1; passage through 6 has no effect, but passage through 7 (P) causes an increase in the population of E_4 (the lower level of 1) and therefore an increase in the intensity of line 1. Irradiation of 7, 10, and 12 obviously has no effect, but 9 (R) is accompanied by a decrease and 11 (P) by an increase in the intensity of line 1.

2.5 THE CHEMICAL SHIFT

The chemical shift of a nucleus in a molecule is a sensitive measure of its electronic environment, and therefore has great potential as a structural tool. We shall discuss, in simple terms, the various factors which influence this parameter.

The simplest situation occurs for an isolated atom in a ^1S state, e.g. ^3_2He which has spin-$\frac{1}{2}$ and natural abundance of 10^{-4} per cent. In the presence of the external magnetic induction, B_0, the entire spherical electron distribution precesses about the direction of application z. This produces a secondary magnetic induction which is proportional to B_0 but *opposed* to it along the z direction. The induction which is 'seen' by the nucleus, B_{eff}, is given by:

$$B_{eff} = B_0(1 - \sigma) \tag{2.48}$$

where σ is the screening constant due to this diamagnetic circulation of electrons. (This process is exactly analogous to the classical situation of a current in a closed loop of wire in an applied magnetic field, where Lanz's law applies.) The constant σ is given simply by the Lamb formula:

$$\sigma = \frac{e^2}{3m_ec^2} \int \frac{\rho(r)}{r} \, dv \tag{2.49}$$

where e is the electronic charge, m_e is the mass of the electron, c is the velocity of light, $\rho(r)$ is the probability density of the electrons distributed over the atom, r is a distance vector from the nucleus, and the integration is carried out over the volume of space occupied by the atom.

If the nucleus is part of a molecule, the electrons are no longer completely free to rotate about the direction of B_0, and this has two important consequences. First of all, the secondary field is not necessarily antiparallel to B_0 and the constant σ must be replaced by an *anisotropic shielding tensor* $\bar{\sigma}$; secondly the expression for shielding involves an additional term which is usually of opposite sign to the Lamb formula. The full treatment of this problem is due to Ramsey [2.38, 2.39] and applies to polyatomic molecules which, in the absence of B_0,

149

Spectroscopy   *Volume One*

have no resultant electron spin or electron orbital angular momentum; for an
elementary discussion, see [2.10, p. 57].

The three principal components of the tensor $\bar{\sigma}$ may be deduced, and an
average isotropic screening constant σ_0 calculated. This is what is measured
experimentally, and chemical shifts are differences in values of σ_0. The two
parts of σ_0 are called the diamagnetic (d) and paramagnetic (p) terms, and we
may write:

$$\sigma_0 = \sigma_d + \sigma_p \tag{2.50}$$

σ_d depends upon the electronic distribution in the electronic ground state only,
and is fairly easy to evaluate. The term σ_p, however, depends also upon excited
states, and although it is zero for electrons in s-orbitals (which have zero angular
momentum) it may be very large when there is an asymmetric distribution of
p or d electrons close to the nucleus and these electrons have low-lying excited
states. Thus, it is generally found that for ^{1}H nuclei in molecules σ_d is dominant,
but for others (e.g. ^{13}C, ^{19}F, ^{15}N, ^{31}P) σ_p is of overwhelming importance and can
give rise to very large variations in σ, i.e. large chemical shifts.

For the present discussion, it is convenient to separate the expression for σ
into terms which have a 'physical significance' to the chemist, and which may be
considered separately. This useful approach is based upon that of Pople [2.40,
2.41] which was extended by Zürcher [2.42] and considers that the shielding
constant for a nucleus in an isolated molecule, σ_0, is given by:

$$\sigma_0 = (\sigma_d + \sigma_p)_{loc} + (\sigma_d + \sigma_p)_{neighbour} + \sigma_E + \sigma_v \tag{2.51}$$

where $(\sigma_d + \sigma_p)_{loc}$ represents the screening due to (local) induced circulation of
electrons on the atom of which the nucleus is a part, and $(\sigma_d + \sigma_p)_{neighbour}$ arises
from induced circulation of electrons upon neighbouring atoms or groups
present in the molecule. σ_E arises from electric fields due to the presence of
permanent electric dipoles in the molecule, and σ_v represents the influence of
van der Waals fields arising from fluctuating electron distributions (time-dependent dipoles).

For ^{1}H, $(\sigma_p)_{loc}$ is negligible, $(\sigma_d)_{loc}$ may be calculated from an expression
closely analogous to the Lamb formula [Equation (2.49)]. An increase in
electron density in the hydrogen 1s orbital will cause an increased shielding, and
so the presence of electronegative substituents will tend to deshield a nucleus.
This is shown, for example, by the protons in a CH_3 group bonded to an electronegative atom (chlorine) and an electropositive group (trimethylsilyl); compared
with CH_4, the chemical shifts are given by Table 2.3. Here, as elsewhere, we
adopt the convention that shifts to low field (deshielding) are positive.

There have been attempts to relate chemical shifts, in such situations, directly
to the electronegativity of substituents (e.g. [2.43]). These are fraught with
difficulty, however, because of the importance of the other contributions to
shielding which may oppose the effect. Nevertheless, it is a good rule of thumb
that electron-withdrawing groups tend to *deshield* while electron-releasing groups

150

tend to *shield*. This idea holds true also in the case of protons in aromatic compounds; thus, the protons in nitrobenzene absorb some 1.5 p.p.m. to low field of those in aniline.

Table 2.3. Chemical shifts of protons in a CH_3 group

$CH_3 \cdot Cl$	+2.1 p.p.m.
$CH_3 \cdot H$	0 p.p.m.
$CH_3 \cdot Si(CH_3)_3$	-1.0 p.p.m.

It is interesting to compare, at this stage, the observed shielding in the hydrocarbons ethane, ethylene, benzene, and acetylene. The hydrogen 1s orbital density in these molecules may be deduced simply by considering the hybridization of the carbon in the C–H bonds. Higher s-character in the hybrid means stronger overlap with H(1s), and the shorter bond length which results is accompanied by removal of electrons from the region of the ^{1}H nucleus. Thus, the more acidic protons should absorb to lower field, and the expected order of shielding would be:

$$HC \equiv CH < H_2C = CH_2 \approx C_6H_6 < H_3C - CH_3$$

In practice, the chemical shifts are in the order:

$$C_6H_6 < H_2C = CH_2 < HC \equiv CH < H_3C - CH_3$$

The reason for this apparent anomaly lies in the importance of the second term in Equation (2.51), $(\sigma_d + \sigma_p)_{neighbour}$.

The term $(\sigma_d + \sigma_p)_{neighbour}$ shows its biggest contribution to σ_0 if the electrons in a near-neighbouring group or atom have a large anisotropic magnetic susceptibility. It is particularly important for linear closed-shell molecules such as acetylene or for large flat molecules with mobile π-electrons such as benzene. The term combines both the diamagnetic and paramagnetic currents on the other atoms, and since the paramagnetic term is much the more anisotropic of the two, the neighbour contribution is largest from atoms with low-lying excited states, such as the halogens. It is possible to calculate this term fairly simply, if the nucleus is at a distance R from an axially symmetric magnetically anisotropic group whose magnetic susceptibility along the z axis is $\chi_\parallel$ and perpendicular to it is $\chi_\perp$. A field B_0 along the z axis induces a magnetic dipole $B\chi_\parallel$ in the group, and if the nucleus lies at an angle θ off the axis in the plane perpendicular to it (xy) the dipole's magnetic field has a component $B\chi(3\cos^2\theta - 1)/R^3$ in the direction of the axis. This is a contribution to the shielding tensor $\bar{\sigma}$ along the direction, σ_{zz}, and consideration of the field along the three axes in turn gives:

$$\sigma_{xx} = \chi_\parallel(1 - 3\sin^2\theta)/R^3 \tag{2.52}$$

$$\sigma_{yy} = \chi_\perp/R^3 \tag{2.53}$$

$$\sigma_{zz} = \chi_\parallel(1 - 3\cos^2\theta)/R^3 \tag{2.54}$$

151

The averaged contribution to σ_0 is then given by the McConnell equation:

$$(\sigma_d + \sigma_p)_{neighbour} = \chi_\parallel - \chi_\perp (1 - 3\cos^2\theta)/3R^3 \qquad (2.55)$$

We may now examine the consequences of the magnetic anisotropy of the C–C, C=C, and C≡C bonds by evaluating this function in the surrounding volume of space. The results are illustrated in Fig. 2.16 in terms of cones in which the effect is shielding (−) or deshielding (+).

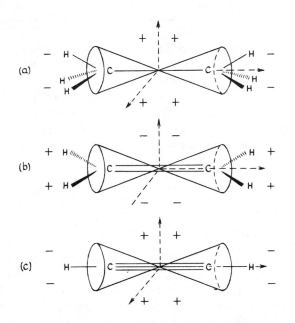

Fig. 2.16 Regions of shielding (−) or deshielding (+) due to the diamagnetic anisotropy of the C–C bond in (a) ethane, (b) ethylene, and (c) acetylene.

Other double bonds, such as C=O and C≡N, behave similarly to C=C, while a triple bond such as C≡N behaves as does the C≡C. The reason for the unexpected relative positions of the proton absorptions in ethylene and acetylene is now obvious; the deshielding of ethylene, relative to ethane, is *reinforced* by the protons' lying in the *deshielding* cone (Fig. 2.16) while that of acetylene is *opposed* by the protons' lying in the *shielding* cone. Their positions are consequently reversed.

Special consideration should be given to the very large anisotropic diamagnetic susceptibility of benzene rings, which causes large shielding or deshielding effects upon neighbouring protons in a molecule. This is the so-called *ring current effect,* which has received much attention. In benzene, the simplest system to consider, there are six mobile π-electrons which behave rather like charged particles moving on a circular wire. If an induction B_0 is applied perpendicularly to the ring, the electrons circulate with an angular frequency eB_0/m_e

leading to a current i given by:

$$i = \frac{3e^2 B_0}{2\pi m_e c} \qquad (2.56)$$

This gives rise to a diamagnetic moment opposed to the primary field, with lines of force as shown in Fig. 2.17.

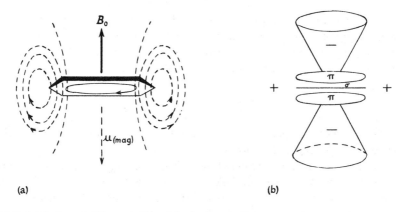

(a)   (b)

Fig. 2.17 (a) induced diamagnetism in a benzene ring.
     (b) shielding and deshielding cones for benzene.

  Figure 2.17 also shows the volume of space in which the effect is shielding and deshielding upon a suitably sited proton, i.e. regions where the lines of force oppose and reinforce the applied induction B_0. Obviously, in benzene itself, the protons around the periphery will be in a deshielding environment, and this is why benzene absorbs to low field of ethylene. In 1,4-polymethylenebenzenes (IX), some of the CH_2 protons will be in the shielding region and absorb at a relatively high-field position; this is observed in practice [2.44]. A further interesting example is provided by 18-annulene (IX), a Hückel $(4n + 2)$ $(n = 4)$ aromatic system which can also sustain a ring current. In this planar molecule, six protons project inwards and are in the shielding region of the induced magnetic field while twelve more project outwards, as in benzene, and are deshielded. The result is therefore a spectrum of two lines (intensity 2 : 1), at 9.28 and −2.99 p.p.m. from tetramethylsilane (TMS).

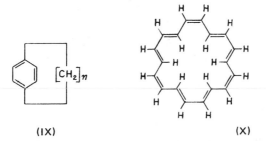

(IX)   (X)

In benzene, a simple estimate of the ring-current effect may be obtained
[2.45]. In Equation (2.56), which applies for B_0 perpendicular to the ring, i
may be replaced by a magnetic dipole at the centre of the ring of radius a,
and perpendicular to it. If R is the distance of a proton from the ring centre,
the secondary magnetic field at this point may be approximated by:

$$B' = \frac{i\pi a^2}{cR^3} = \frac{3eB_0 a^2}{2m_e c^2 R^3} \tag{2.57}$$

The screening constant is obtained by dividing by $-B_0$. If the applied field is
in the *plane* of the molecule, there is no induced ring current, and when averaged
by tumbling of the molecule, the ring current contribution to the average screening constant σ is obtained by dividing Equation (2.57) by 3:

$$\sigma_{\text{(ring current)}} = \frac{-e^2 a^2}{2m_e c^2 R^3} \tag{2.58}$$

Substitution of the approximate values in this equation indicates a contribution
of about +1.75 p.p.m. in benzene.

So far we have only considered the anisotropic magnetic properties of *groups*
in molecules; *atoms* with p or d electrons may show considerable anisotropy,
however, especially if they have low-lying excited states. Thus, there have been
suggestions that iodine may exert effects which are up to half as much as that
due to C=0, with the other halogens being somewhat less effective [2.46].
Oxygen and nitrogen may also exert significant effects [2.47, 2.48].

It is difficult, for these functionalities, to evaluate quite how important this
effect is, because their electronegativity also must affect $(\sigma_d + \sigma_p)_{\text{local}}$. Addition-
ally, however, the bond of which they are part has a permanent electric dipole
moment, μ. The electric field due to this can distort the local electronic environment of a proton and change its shielding, hence the term σ_E in Equation (2.51).
σ_E was first considered by Buckingham [2.49] who used it to explain the differences between the shielding of *o*-, *m*-, and *p*-protons in substituted benzenes.
Briefly, the value of σ_E is related to the component E_z of the electric field, due
to the permanent dipole moment, along the direction of the C−H bond containing the ^{1}H under examination, and to the value of E^2 at that nucleus:

$$\sigma_E = -AE_z - BE^2 \tag{2.59}$$

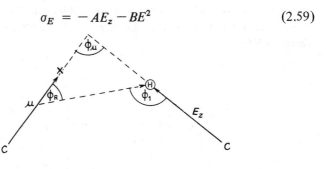

Fig. 2.18 Definition of terms for Equation (2.60)

The constant A is a measure of the ease of distortion of the C–H bond electrons along the bond direction, and B is a measure of the polarizability perpendicularly to this. The negative signs ensure that a field in the direction C–H
causes increased shielding. Zürcher has developed the earlier work [2.42] and
gives a simple method for calculating E_z:

$$E_z = \mu(3 \cos \phi_1 \cos \phi_R - \cos \phi_\mu)/R^3 \qquad (2.60)$$

where the dipole moment arises from a point dipole in the centre of the C–X
bond and the angles ϕ_1, ϕ_R, and ϕ_μ, and the distance vector R are defined in
Fig. 2.18.

In general, Zürcher suggests that the second term in Equation (2.59), $-BE^2$,
is negligible, and simply by considering the $-AE_z$ term he very successfuly
accounts for the chemical shifts of the C–18 and C–19 angular methyl groups in
steroids with a variety of polar substituents [2.42]. The value of A is about
$10^{-12} \, m^2 c^{-1}$.

The term involving R is of importance in determining the term σ_v in Equation (2.51). Time-dependent dipoles arising from fluctuating electronic distributions in atoms have a non-vanishing time average of E^2, $\langle E^2 \rangle$, even though the
value of E or its component E_z may average to zero. This is the van der Waals
field, and clearly:

$$\sigma_v = -B \langle E^2 \rangle \qquad (2.61)$$

$\langle E^2 \rangle$ can be calculated approximately from the expression:

$$\langle E^2 \rangle = 3\alpha I/r^6 \qquad (2.62)$$

where α is the polarizability of the perturbing atom from which the field arises
and I is its first ionization potential; r is the distance from the atom of the nucleus under consideration. These interactions are only likely to be of importance
at short distances (because of the r^{-6} dependence), and although the value of
B obtained by independent methods (for ^1H, $10^{-19} \, m^4 c^{-2}$) suggests that the
influence may be significant, there is little evidence to support this for ^1H
shielding [2.42].

Nuclei other than ^1H have been widely studied, particularly ^{19}F, ^{13}C, ^{31}P,
^{15}N, and ^{14}N. The influence of the terms other than $\sigma_{p,local}$ is rather similar to
that discussed for ^1H, although there is some evidence to suggest, for ^{19}F at
least, that van der Waals fields can exert significant effects (i.e. σ_v is important).
The $\sigma_{p,local}$ term is almost wholly dominant, and responsible for the very large
ranges of chemical shifts observed with these nuclei. The evaluation of this
term is very difficult, since it requires knowledge of excited-state wavefunctions.
Most progress has been made for ^{19}F and ^{15}N, and the reader should consult
the relevant review articles which give a comprehensive account of the theoretical
and empirical approaches [2.50, 2.51].

Solvent effects

We have so far discussed the factors which, in an isolated molecule, affect the average shielding constant of a nucleus, σ_0. In reality, the molecule is examined as an assembly in the liquid phase or in solution. Solvents can exert powerful influence upon shielding and hence upon chemical shifts. Again, it is convenient to break down the observed shielding constant into a number of contributions arising from different sources:

$$\sigma_{obs} = \sigma_0 + \sigma_B + \sigma_A + \sigma_{v'} + \sigma_{E'} + \sigma_c \qquad (2.63)$$

σ_0 is the shielding constant for the nucleus in the isolated molecule as discussed above; σ_B is a term which arises from the bulk diamagnetic susceptibility of the medium [2.52]; σ_A is important only for solvent molecules which are markedly anisotropic in structure (such as CS_2 or benzene) and tend to associate with solute molecules in a non-random fashion [2.53]; $\sigma_{v'}$ is shielding due to van der Waals dispersion forces between solvent and solute [2.54]; $\sigma_{g'}$ arises from a reaction electric field which is set up in the solvent parallel to the electric dipole of the solute and is dependent upon solvent polarity [2.55]; σ_c is a very general term which encompasses the many chemical-type interactions that may occur (e.g. hydrogen-bonding, charge-transfer, etc.). The significance of these quantities is discussed in full in the literature [2.56, 2.57]. A detailed examination shows that if shielding in a molecule of interest is measured with reference to a chemically and structurally similar molecule which is used internally, then the only solvent contributions that make σ_{obs} differ significantly from σ_0 are likely to be $\sigma_{E'}$ and σ_c. These contributions are minimized by utilizing a non-polar solvent which is chemically inert, such as n-hexane or CCl_4.

A comprehensive review of solvent effects upon shielding has been made by Laszlo [2.58]. This work also comments authoritatively upon the special effects of aromatic solvents which are of particular importance; further studies in this area are summarized by Ronayne and Williams [2.59] and Engler and Laszlo [2.60].

The relationship between structure and
chemical shift

The great value of n.m.r. spectroscopy, so far as the practical chemist is concerned, lies in the empirical way in which parameters such as chemical shifts may be related to structure. Many of the contributions to shielding discussed above behave in an additive way, and the chemical shifts induced in a molecule by introduction of a substituent (s.c.s. — substituent chemical shifts) may be used as an additive property. The 'translation' of such data into structural terms is largely a matter of experience, but in order that the reader may consider them in the light of the preceding discussion, we list some sources of compilations of chemical shifts: ^{1}H [2.61, 2.16, 2.62, 2.63]; ^{19}F [2.50, 2.64]; ^{31}P [2.65];

^{13}C [2.66, 2.67]. One of the sources [2.16] also contains data for the many other magnetic nuclei of importance, e.g. ^{10}B, ^{11}B, ^{15}N, ^{17}O, ^{27}Si.

2.6   SPIN—SPIN COUPLING

We have seen that this arises because the magnetic energy of a nucleus in a system can be affected by magnetic interactions with the other nuclei. These interactions are mutual, and are transmitted by the bonding electrons of the molecule of which the nuclei are part. The basic theory is well established, and was originally developed by Ramsey, [2.68, 2.69] McConnell [2.70] and Karplus [2.71, 2.72]. Useful reviews have been given by Murrell [2.73], Barfield and Grant [2.74], and Barfield and Chakrabati [2.75].

The simplest approach is to consider that the coupling constant between two spin-$\frac{1}{2}$ nuclei A and B, J_{AB}, is given as the sum of four terms:

$$J_{AB} = J_{1\,AB} + J_{2\,AB} + J_{3\,AB} + J_{4\,AB} \qquad (2.64)$$

$J_{1\,AB}$ is only of importance for a stationary molecule, i.e. one with a fixed orientation with respect to the applied induction B_0. It is a direct dipole—dipole interaction by means of which the magnetic field produced by one nucleus affects the other directly, through space. It is very large (e.g. for two protons 2 pm apart it may be as large as 30 kHz!) and depends upon the relative orientation of the two nuclei with respect to B_0. In solids, the width of lines will be of this order and obscure chemical shifts between protons which are usually (at the most) 1 kHz. In gases and liquids, the rapid tumbling and translation of the molecules averages $J_{1\,AB}$ to zero, however, and we need not consider it further.

The three remaining factors in Equation (2.64) are field-independent and arise from mechanisms involving the molecular electrons — they are electron-coupled interactions.

$J_{2\,AB}$ occurs because the nuclear moment of one nucleus, say A, interacts with the orbital electron currents which in turn interact with the moment of B and vice versa.

$J_{3\,AB}$ arises from a dipole interaction between the nuclear magnetic moment of A and the electron spin magnetic moments; the electron dipoles interact with the magnetic dipole of B and thus pass on the spin-coupling information.

$J_{4\,AB}$ is the best characterized of all. It is the Fermi contact term (so-called by analogy with that introduced by Fermi [2.76] to explain hyperfine structure in atomic spectra). It is an interaction between the nuclear magnetic moments, and electron spins in s-orbitals which have a non-zero value at the nucleus. An electron associated primarily with one nucleus has a finite probability of being at that nucleus and transmitting spin information via the other bonding electrons to the other nucleus.

For proton—proton coupling (J_{HH}) the Fermi contact term is all-important and accounts for up to 98 per cent of the magnitude of any observed coupling constant. $J_{2\,AB}$ and $J_{3\,AB}$ will therefore be neglected, and the statements made below concerning J_{HH} have been deduced for the Fermi contact mechanism only. We shall see later that for other nuclei the situation is not so simple.

Proton–proton couplings

The hydrogen molecule presents the simplest possible situation. Because both protons are chemically equivalent, it is not possible to observe $^1J_{HH}$ directly (the superscript refers to the number of bonds formally separating the two nuclei.) However, the spectrum of HD (in which a hydrogen atom has been substituted by its deuterium isotope) contains the necessary information. $D(= {}^2_1H)$ is a spin-1 nucleus, and splits the hydrogen resonance into a $1:1:1$ triplet of separation J_{HD}. J_{HH} is related to J_{HD} by the expression:

$$J_{HH} = J_{HD}(\gamma_H/\gamma_D) = J_{HD}/0.154 \tag{2.65}$$

where γ_H and γ_D are the gyromagnetic ratios of the proton and deuteron respectively. It is found that $J_{HD} = 43.5$ Hz; therefore $J_{HH} = 282.5$ Hz. (In general, such coupling constants may be compared by using the idea of a *reduced coupling constant* which is isotope-independent. The reduced coupling constant between two nuclei A and B, K_{AB}, is given by:

$$K_{AB} = J_{AB}(2\pi/\gamma_A\gamma_B\hbar) \tag{2.66}$$

and K_{AB} has the dimension of cm^{-3}.)

The Fermi-contact contribution to this can be calcualted, and it is found that about 260 Hz (of the 282.5 Hz) arises from this mechanism. Pictorially, it may be represented thus:

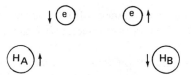

H_A interacts with the 1s electron by the contact mechanism, which favours the antiparallel orientation of the nuclear and electron spins. The Pauli exclusion principle causes the pair of electrons in the bond to line up antiparallel as shown, and the other electron interacts with nucleus H_B to give the preferred orientation shown and a *positive* coupling constant. In the same way, $^1J_{CH}$ is predicted to be positive and is used as a standard to which the signs of all other J's are referred.

The vast majority of available data concern spin-coupling between non-directly bonded nuclei, and it is convenient to consider three different categories: $^2J_{HH}$(geminal), $^3J_{HH}$(vicinal), and $^nJ_{HH}$(long-range).

$^2J_{HH}$ (*Geminal coupling*) *across carbon*

The original picture of $^2J_{HH}$ was given in terms of a valence-bond formalism, by Karplus and Anderson [2.71] and Hiroike [2.77]. It is likely that the reader will be more familiar with molecular orbital ideas, however, and the necessary picture was developed by Pople [2.78, 2.79].

A $-CH_2-$ fragment may be described in terms of four molecular orbitals, by forming linear combinations of the hydrogen 1s functions and carbon sp^3 hybrids. The resulting MO's, ψ_1 to ψ_4, are depicted in Fig. 2.19; two are bonding and two antibonding.

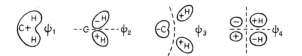

Fig. 2.19 Four MO's of a CH_2 group. The signs are algebraic, and ψ_1 is symmetric bonding, ψ_2 antisymmetric bonding, ψ_3 symmetric antibonding, and ψ_4 antisymmetric antibonding.

In the ground state, ψ_1 and ψ_2 are doubly filled. Coupling between the two hydrogens involves contributions from the pairs of electrons in each orbital separately. The magnitude and sign of the two contributions depends upon the singlet→triplet excitation energy (to states of the same symmetry), and the following rules may be deduced.
(1) Withdrawal of electrons from the symmetric bonding orbital will cause an *algebraic increase* in $^2J_{HH}$, i.e. will make it more *positive*.
(2) Withdrawal of electrons from the antisymmetric bonding *MO* will cause an algebraic *decrease* in $^2J_{HH}$ (i.e. will make it more *negative*.)
(3) As the H–C–H bond angle increases, $^2J_{HH}$ will also increase algebraically (become more positive).

Because of their different symmetry the MO's ψ_1 and ψ_2 have different overlap properties with substituents on the carbon.

ψ_1 can overlap strongly with orbitals of σ-symmetry, e.g. a p$_\sigma$ orbital of substituent X overlaps with ψ_1 as shown in (XI).

Thus, electronegative substituents withdraw electrons inductively from ψ_1. ψ_2 cannot overlap with such an orbital, however, and such a substituent has no effect upon this orbital, as shown in (XII). The ++ overlap is balanced by the +− repulsion; therefore there is no net overlap.

159

Rule 1 therefore suggests that in a fragment $X-CH_2-Y$ $^2J_{HH}$ will become more positive with increased electronegativity of X and Y. This is confirmed by the data in Table 2.4. ψ_2 can overlap with orbitals of π-symmetry, for example when in allylic or benzylic situations (XIII). This is hyperconjugation, and ψ_1 does not have the correct symmetry for this to occur, as shown in (XIV). The ++ overlap is balanced by the +− repulsion; hence there is no net overlap.

Table 2.4 $^2J_{HH}$ in a variety of methyl compounds [2.80]

Molecule	$^2J_{HH}$(Hz)
CH_4	−12.4
CH_3Cl	−10.8
CH_3OH	−10.8
CH_3F	− 9.6

(XIII)

(XIV)

Rule 2 enables some predictions to be made, since the hyperconjugation may either withdraw electrons from ψ_2 or donate electrons to it. Thus, in toluene there is hyperconjugative withdrawal from the CH_3 group with an algebraic decrease in $^2J_{HH}$ to −14.4 (cf. −12.4 Hz for CH_4). An even more pronounced example is afforded by malonitrile, $CH_2(CN)_2$ in which $^2J_{HH}$ is −20.3 Hz [2.81].

Rule 3 is illustrated by the data in Table 2.5, which shows clearly that as the H−C−H angle increases $^2J_{HH}$ becomes more positive.

A very interesting example is afforded by formaldehyde, in which the effects described by rules 1, 2, and 3 combine together to result in $^2J_{HH} = +41$ Hz [2.85] which is some 53 Hz more positive than that for CH_4. The change from sp^3 to sp^2 hybridization is responsible for about 15 Hz of this (rule 3). In addition, the electronegative oxygen atom removes electron density from ψ_1 to further increase $^2J_{HH}$ (rule 1) while hyperconjugative overlap of ψ_2 with a lone-pair of electrons in a suitably oriented p-orbital on oxygen results in an increase in electron density in ψ_2 with a further positive contribution to $^2J_{HH}$ (rule 2) (see Fig. 20).

Table 2.5 Data to illustrate the dependence of $^2J_{HH}$ upon the H–C–H bond angle

Molecule	$^2J_{HH}$ (Hz)	Ref.
$R_1 R_2 C=CH_2$	+10 to −3 (depending upon R_1 and R_2)	[2.82]
(structure with Cl Cl, H, H, R_1, R_2)	−7.1 to −8.38 (depending upon R_1 and R_2)	[2.83]
(structure with AcO, CH₃, OAc, H, H)	−12.36	[2.84]
CH_4	−12.4	

Fig. 2.20 The formation of the π-bonding orbital of formaldehyde, together with hyperconjugative overlap of ψ_2 (CH_2) and p (oxygen)

Because of the wide variety of structural situations in which a CH_2 group may be found, $^2J_{HH}$ can vary from about −30 to +40 Hz.

$^3J_{HH}$ (*Vicinal coupling*) *across a* C–C *bond*

These are the most commonly observed couplings in organic chemistry, and since they vary with the nature of substituents on the carbon atoms and with geometry and conformation, they are important in structure elucidation.

The most successful rationale has been developed in valence bond theory by Karplus [2.72] and leads to the following suggestions.

(1) $^3J_{HH}$ is always positive. Experimentally, values range from zero to +25 Hz.

(2) In olefinic systems, $^3J_{HH\,trans}$ is larger than $^3J_{HH\,cis}$. The π-system is *not* involved to a significant extent, however (about 10 per cent arises from σ–π interactions).

(3) In a saturated fragment, the coupling is approximately proportional to $\cos^2\phi$ where ϕ is the dihedral angle between the two C–H bonds.

(4) The presence of electron-withdrawing substituents upon the carbon atoms tends to decrease $^3J_{HH}$. Quite simply, fewer electrons to transmit the interactions results in smaller coupling.

161

The simplest situation to examine is that of monosubstituted ethylenes, $CH_2 = CHX$. $^3J_{trans}$ and $^3J_{cis}$ may be related to the electronegativity of the substituent, E_X, by the expressions due to Banwell and Sheppard [2.86, 2.87]:

$$^3J_{trans} = 19.0 - 3.2(E_X - E_H) \qquad (2.67)$$

$$^3J_{HHcis} = 11.7 - 4.0(E_X - E_H) \qquad (2.68)$$

Since the effects of substituents are additive, these may easily be extended to 1,2-disubstituted olefins.

An interesting observation, not predicted by theory, concerns $^3J_{HHcis}$ in cyclic olefins. It is found that the coupling *decreases* with ring size from a norm of 10−11 Hz in six- and seven-membered rings to *ca*.1 Hz in cyclopropene. It is possible to regard *ortho* couplings in aromatic compounds as a special case of $^3J_{HHcis}$, and these show a similar dependence upon ring size. Thus, in the seven-membered ring of azulenes it is about 10−12 Hz, in benzenes about 5−9 Hz, and in the five-membered aromatic heterocycles (furan, thiophen, etc.) it is about 2−5 Hz.

The conformational mobility of C−C single bonds makes it rather more difficult to describe $^3J_{HH}$ in saturated fragments >CH−CH<. Both the dihedral angle between the C−H bonds and the nature of any substituents are important in determining the magnitude, and a convenient expression to describe the trends is:

$$^3J_{HH} = A + B \cos \phi + C \cos 2\phi \qquad (2.69)$$

where ϕ is the dihedral angle, defined in the Newman projection of the general molecule abCH−CHde (XV). A, B, and C are empirical constants which depend upon the electronegativity and orientation of substituents and also upon the hybridization of the carbons (i.e. whether they are part of an open chain or ring, and upon the size of rings). Equation 2.69 is an adaptation of the well-known Karplus relationship which is perhaps easier to visualize in its original form:

$$^3J_{HH} = J^0 \cos^2 \phi - 0.3 \text{ (for } 0° \leqslant \phi < 90°) \qquad (2.70)$$

$$^3J_{HH} = J^{180} \cos^2 \phi - 0.3 \text{ (for } 90° \leqslant \phi < 180°) \qquad (2.71)$$

Depending upon the values of J^0 and J^{180} (which are functions of the same properties as A, B, and C above, and are empirically determined) this leads to a family of curves with minima of −0.3 Hz at $\phi = 90°$ and $270°$ and maxima of 5−16 Hz at $\phi = 0°$ and $180°$.

The difficulties of evaluating J^0 and J^{180} in a given situation are illustrated by the work of Abraham and Gatti [2.88] and Phillips and Wray [2.89]. The case of 1,2-distributed ethanes, $XCH_2 CH_2 Y$, was considered. Two different conformational isomers are possible, with $^3J_{HH}$ couplings as defined (XVI).

The various J's are related to the electronegativities (on the Huggins scale) of X and Y by the expressions:

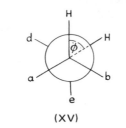

(XV)

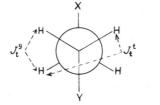 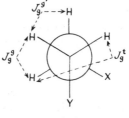

trans isomer   (XVI)   *gauche* isomer

$$J_t^g = 3.51 + (E_y - 0.91)(0.367E_X - 0.402) \qquad (2.72)$$

$$J_t^t = 18.07 - (E_X + E_y) \qquad (2.73)$$

$$J_g^g = 3.40 + (E_y - 1.93)(2.950 - 1.003E_X) \qquad (2.74)$$

$$J_g^{g'} + J_g^t = 26.92 - 2.03(E_X + E_y) \qquad (2.75)$$

The behaviour of $J_g^{g'}$ and J_g^g with respect to changes in the electronegativity of X and Y will serve to highlight some of the problems revealed by these expressions. $J_g^{g'}$ is a coupling between two protons in a *gauche* relationship to each other. The group X is *trans* coplanar to one proton and Y is *trans* coplanar to the other; the magnitude of $J_g^{g'}$ decreases linearly with the sum of E_X and E_Y. On the other hand, J_g^g is a coupling between two *gauche* protons, only one of which has a *trans* substituent X, while the other is in a *gauche* relationship to Y; the magnitude of J_g^g is much more sensitive to changes in X than to changes in Y. This is consistent with suggestions made elsewhere [2.90, 2.91] that substituents exert a maximum effect when they are in a *trans*-coplanar relationship with respect to each other. A direct consequence of this is that, for a given pair of substituents X and Y, J_g^g is invariably smaller than J_t^g; thus, in 1,2-difluoroethane, J_t^g is 6.3 Hz while J_g^g is 1.43 Hz, even though the dihedral angle is the same in each case!

The situation of $^3J_{HH}$ in molecules which have no conformational preference, such as CH_3CHXY, is much easier to define. A value of $^3J_{HH}$ averaged by rotation is obtained, and related linearly to the sum of the electronegativities of the first atoms of the substituent groups [2.92].

$$^3J_{av} = 18.0 - 0.80 \, \Sigma E \qquad (2.76)$$

163

In rigid systems, such as conformationally fixed six-membered rings, the statements made above concerning 1,2-disubstituted ethane apply. Thus, axial—axial couplings are similar to J_{trans} in substituted ethanes (J^{180}) and are much greater than the equatorial—axial or equatorial—equatorial couplings which are comparable to J_{gauche} (J^{60}). Electronegative substituents have the expected effects.

$^nJ_{HH}$ (*Long-range coupling*)

We have stated that $^3J_{HH}$ in olefins and aromatic compounds do not involve the π-electrons but are transmitted almost wholly by the σ-bonded framework. Long-range couplings *may* involve interactions with π-electrons, however, and so it is convenient to discuss them in two separate categories.

(*a*) *Coupling via σ-bonds only.* These are invariably positive, and perhaps the best known examples are *meta* and *para* couplings in aromatic compounds. Because the C—H bonds are orthogonal to the π system there is negligible overlap and the couplings occur solely via the σ-bonds. The coupling mechanisms arise because of *delocalization of the σ-electrons* as shown in (XVIII).

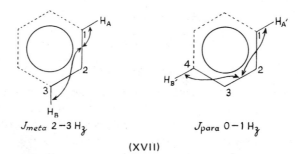

J_{meta} 2—3 H$_3$ J_{para} 0—1 H$_3$

(XVII)

For $J_{HH\,meta}$, *geminal* delocalization occurs between the C_1—H_A and C_1—C_2 bonds; *vicinal* delocalization occurs between the C_1—C_2 and C_3—H_B bond, and hence the Fermi contact mechanism can operate to give $J_{H_A H_B}$. For $J_{HH\,para}$, the mechanism involves vicinal delocalization between the $C_1 H_A{}'$ and $C_4 H_B{}'$ bonds and the C_2—C_3 bond. This picture, essentially due to Pople, and Sentry [2.93, 2.94] accounts for the fact that in conformationally mobile systems analogous couplings are very stereoscopic. This is because vicinal delocalization between bonds follows an approximately $\cos^2 \phi$ relationship (cf. vicinal couplings) with maxima at $\phi = 0°$ (*cis*) and $180°$ (*trans*) and minima at $\phi = 90°$ and $270°$ (ϕ is the dihedral angle between the bonds).

It is found that $^4J_{HH}$ is usually finite (about 1—3 Hz) when there is a planar-W arrangement of bonds, and negligible otherwise. Thus, in derivates of bicyclo[(2.2.1)] heptane (XVIII) coupling is observed between H_A and H_X but not between H_A and $H_X{}'$; similarly, the H_B—H_Y coupling is observed but not H_B—$H_Y{}'$ [2.95].

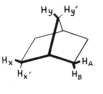

(XVIII)

There are very many examples which illustrate this well; e.g. in certain carbo-hydrate derivatives (XIX) which exist in solution in the conformation shown, a $^4J_{HH}$ is invariably seen between H_1 and H_3 in the α-anomer but not in the β-form.

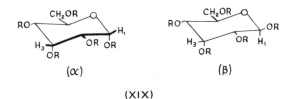

(XIX)

It does not seem to make any significant difference if the coupling pathway involves only carbon atoms or if a heteroatom intervenes (e.g. O, N, or S). The occurrence of strain, as in five-membered rings, does increase $^4J_{HH}$ significantly, however, as shown for the compound (XX) [2.96] where $^4J_{H_A H_B}$ is 6–8 Hz while $^4J_{H'_A H'_B}$ and $^4J_{H_A H_B'}$ are unresolved.

(XX)

Longer-range couplings occur invariably across all *trans-coplanar* zig-zag pathways. Examples are afforded by the compounds (XXI) – (XXIII) in which $^5J_{HH}$ is approximately 1 Hz.

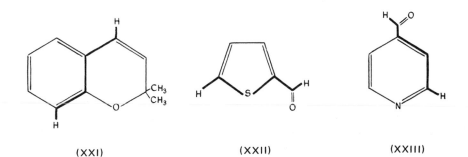

(XXI) (XXII) (XXIII)

Further examples of long-range couplings are discussed in a review by Sternhall [2.97] and in the relevant chapter of the book by Jackman and Sternhall [2.61].

(*b*) *Couplings involving σ–π interactions.* The two most common types are the allylic (HC–C=C–H) and homoallylic (HC–C=C–C–H) couplings.

Allylic coupling relies upon the fact that there is a small but finite degree of overlap between the electrons of the vinylic C–H bond and the π-bonding MO. The allylic C–H bond may overlap more extensively (hyperconjugation) and hence a coupling mechanism is available (XXIV)

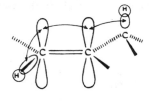

(XXIV)

The overlap between the π-bond and the allylic C–H bond is dependent upon their relative orientation, and is shown at its maximum in (XXIV). If the C–H bond is in the *plane* of the olefin and not perpendicular to it, the overlap decreases to zero; in fact, the size of *J* (which is almost always negative in sign) varies as $\cos^2 \phi$ where ϕ is the dihedral angle between the π-bond and the C–H bond. As a general rule, *transoid* coupling (as illustrated) tends to lie in the range −1 to −4 Hz while *cisoid* coupling is about −2 to −4 Hz. In a given compound, J_{cisoid} tends to be slightly larger than $J_{transoid}$.

The homoallylic $^5J_{HH}$ arises from hyperconjugative overlap between both allylic C–H bonds and the π-bonding MO (XXV). It is always positive in sign, but since two dihedral angles ϕ and ϕ' are involved it varies as $\cos^2 \phi \cos^2 \phi'$. Consequently, if *either* C–H bond is orthogonal to the π-system, this coupling will be zero. It normally lies in the range 0 to +4 Hz, and is comparable for both *cisoid* and *transoid* situations.

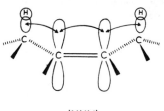

(XXV)

Closely related to allylic coupling is the benzylic coupling observed between α-protons on the side-chains of alkylbenzenes, and ring protons. In this case the coupling to *ortho* and *para* protons may be as high as 1 Hz, with coupling to the *meta* proton considerably smaller.

*Spin–spin coupling involving nuclei other
than hydrogen*

When *both* nuclei which are coupled together are in atoms with p-electrons (e.g. $^{19}F–^{19}F$, $^{13}C–^{19}F$, $^{19}F–^{31}P$, etc.) the mechanisms other than Fermi contact are often very important. The terms in Equation (2.64) are often large and may be opposite or similar in sign. The consequence is that such couplings may be very large or very small and there may not be a regular attenuation with increasing number of bonds separating the nuclei; long range J's may be bigger than short-range J's, and signs may vary unpredictably. In addition, it is difficult to rationalize the effects of substituents and changes in geometry in a way analogous to that given earlier for J_{HH}.

Fluorine–fluorine couplings are the best documented, and a brief considera-tion of these will highlight some important general features. $^2J_{FF geminal}$ is very susceptible to the effects of substituents, and these are difficult to cate-gorize. One trend which is apparent, in spite of this, is that it is always positive and tends to increase with decrease in the F–C–F bond angle [2.98]. Thus, in olefins $^2J_{FF}$ can vary from $+16$ to $+87$ Hz, in three-membered rings it is about $+157$ Hz, and in six- and seven-membered rings it is about $+237$ to $+284$ Hz. In fluoroalkenes, depending upon substituents, it may be in the range $+150$ to $+350$ Hz.

In fluoroalkanes, $^3J_{FF vicinal}$ is usually small, about $1–16$ Hz. This is very much smaller than some longer-range couplings, but was explained by Evans [2.99] who observed that such a coupling was the weighted average of several conformations and that J_{gauche} was opposite in sign to J_{trans}, resulting in a small net coupling. In rigid systems, $^3J_{FF}$ may be numerically higher ($15–40$ Hz). In fluoroalkanes, $^3J_{FF}$ is a sensitive indicator of geometry since the *trans* coupling is always very much larger than the *cis* (typically, $^3J_{FF trans}$ is about -130 Hz, $^3J_{FF cis}$ is about $+10$ to $+30$ Hz). The reason for these larger values is that the coupling mechanism involves the π-system.

Longer-range couplings, $^nJ_{FF}$, are well known and may exist over many bonds. The pathway may be entirely of σ-bonds, or involve π-participation, and the value depends upon geometrical factors. Thus, Peake and Thomas [2.100] have shown that diequatorial $^4J_{FF}$ involving the CF_3 group in perfluoro-methylcyclohexanes is about 9 Hz, axial–equatorial values are about 1 Hz, and axial–axial couplings are about $25–30$ Hz. In addition, couplings analogous to allylic and homoallylic J_{HH}'s are known and show stereospecificity [(XXVI)–(XXVIII)].

Fluorine–fluorine couplings in aromatic compounds involve the π-structure, and consequently are very susceptible to substituent effects and variable in size. J_{ortho} is of opposite sign to the others, and typical values are: $J_{FF ortho}$ about 20 Hz, $J_{FF}(meta)$ about $0–7$ Hz, and $J_{FF}(para)$ about $0–15$ Hz.

If one of the coupled nuclei is hydrogen, however, it seems that the Fermi contact mechanism is predominant and the situation is somewhat easier to

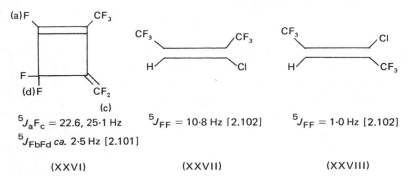

$^5J_{a}F_c = 22.6, 25.1$ Hz $^5J_{FF} = 10.8$ Hz [2.102] $^5J_{FF} = 1.0$ Hz [2.102]

$^5J_{FbFd}$ *ca.* 2.5 Hz [2.101]

(XXVI)   (XXVII)   (XXVIII)

discuss. Most work has involved $^1J(^{13}C-^1H)$ directly bonded) and various $J(^{19}F-^1H)$.

In hydrocarbons, $^1J_{CH}$ has been related to the amount of s-character contributed by carbon to the C–H bond. Thus, sp^3-hybridized carbons have the smallest values, about 125 Hz, with sp^2 about 160 Hz and sp about 250 Hz. Substitution on the carbon by electronegative atoms increases $^1J_{CH}$ significantly. $^1J_{CH}$ is always positive, and is taken as the standard to which all other signs may be related.

$^2J_{HF}$, rather like $^2J_{HH}$, tends to become more positive with increase of the H–C–F bond angle, in marked contrast with $^2J_{FF}$. Thus, $^2J_{HF}$ varies from +72 to +85 Hz in fluoro-olefins (H–C–F angle about 120°) to +47–+57 Hz in cyclohexyl and alkyl fluorides (H–C–F angle about 109°). Exactly as for $^2J_{HH}$, electron-withdrawing substituents increase $^2J_{HF}$ (Table 2.3); the same situation holds in silicon derivatives, and the subject has been examined in detail by Frankies [2.103] and by Pendlebury, Phillips, and Wray [2.104]. There is some evidence to suggest that β-substituent effects upon $^2J_{HF}$ may be stereospecific, and this has been discussed by Phillips and Wray [2.105].

$^3J_{HF}$ is, again, rather similar to $^3J_{HH}$ in its dependence upon structure. Thus, Williamson *et al.* have shown that $^3J_{HF}$ is a well-behaved function of the dihedral angle ϕ between the C–H and C–F bonds [2.106]. At $\phi = 0$, it has a value of about +31 Hz; at $\phi = 90°$ it decreases to zero, and at $\phi = 180°$ it increases again to about +44 Hz. The issue is complicated by the fact that $^3J_{HF}$ probably depends upon bond lengths and bond angles also [2.106]. These observations verify the generality of the findings by Hall *et al.* for fluorinated carbohydrates [2.107–2.109] and of Mousseron-Canet for fluorinated glycyrrhetic acids [2.110]. Electronegative substituents reduce the magnitude of this parameter, and the effects are stereospecific; the situation is more complicated than for $^3J_{HH}$, however, since the influence depends to some extent upon which carbon (i.e. α to H or to F) the substituent is place [2.111].

Further consideration of coupling involving nuclei other than hydrogen is beyond the scope of this chapter; for further information, reference [2.16] should be consulted. Typical values of *J* for the situations considered in this section are tabulated [2.16, 2.61–2.63, 2.66, 2.67].

2.7 THE EFFECTS OF MOLECULAR CONFORMATIONAL MOTION

N.m.r. spectroscopy has made great contributions to the study of certain rate processes in chemistry. Because of the long time-scale inherent in n.m.r., many phenomena which occur in shorter times may affect the spectrum. In principle, any equilibrium process involving exchange of nuclei between chemically different or equivalent sites may be studied, provided that it has first-order rate constants in the range $1-10^3\,\mathrm{s^{-1}}$. Processes that have been studied include electron and group transfer reactions, bond switching, molecular and ionic association, ion solvation, and molecular conformational motion (particularly in organic compounds). These aspects are covered in detail in chapter 9 of reference [2.16] and elsewhere [2.112–2.119]. We shall concern ourselves here with the last aspect only, specifically the way in which n.m.r. may be used to study internal rotations about bonds, and ring inversions.

The simplest possible situation arises when a nuclear spin has an equal probability of being in two chemically different environments, and jumps at random from one to the other with rate constant k. If it gives rise to a single line in each environment, and the frequency of the jumping process is comparable to the frequency separation between these lines, then profound effects may occur.

There are two important limiting situations: the fast exchange limit, and the slow exchange limit. In the first case, the exchange frequency ν_e is much greater than the frequency separation between the lines, $\Delta\nu$. The result is a single line at the average position, of width:

$$\delta\nu \simeq (\Delta\nu)^2/\nu_e \qquad (2.77)$$

At the other extreme, $\nu_e \ll \Delta\nu$ and the two lines remain separate. They are each broadened, however, and of width $\delta\nu \simeq \nu_e$. It is in the region between these limits, that of 'intermediate exchange', that the spectrum holds useful kinetic information since the shape of the resonance line then depends most critically upon the rate of the process. Frequently, it is possible to pass from one extreme to the other by varying the temperature of the exchanging system; a modern spectrometer may have a variable-temperature probe operating in the range 103–437 K, and processes with energy of activation in the range 20–100 kJ mol^{-1} may be studied.

By considering what happens to the Bloch equations when the exchange occurs, an expression for the line-shape may be derived in terms of the lifetime in each site, τ, and the frequencies of the two resonance lines at the slow exchange limit, ν_A and ν_B.

The shape function, $g(\nu)$ is:

$$g(\nu) = K\,\frac{\tau(\nu_A - \nu_B)^2}{\left[\tfrac{1}{2}(\nu_A + \nu_B) + \nu\right]^2 + 4\pi^2\tau^2(\nu_A - \nu)^2(\nu_B - \nu)^2} \qquad (2.78)$$

K is a normalizing constant, and τ may simply be considered as the reciprocal of the rate constant k. In a typical experiment, the line shape is studied as a function of temperature over the whole region from slow to fast exchange. ν_A and ν_B are measured directly at the slow exchange limit, and k is estimated, at each temperature, from the line shape. The process may then be discussed in terms of the Arrhenius activation parameters, using the relationship:

$$k = A \exp(-E_a/RT) \qquad (2.79)$$

Alternatively, the process may be discussed in terms of the transition-state theory of reaction rates using the Eyring equation:

$$k = (kT/h)[\exp(-\Delta H^*/RT)\cdot\exp(\Delta S^*/R)] \qquad (2.80)$$

where the terms have their usual meaning. The ΔS^* and ΔH^* parameters should be obtained from the temperature dependence of ΔG^* using the relationship:

$$k = (kT/h)[\exp(AG^*/RT)] \qquad (2.81)$$

$$\Delta G^* = \Delta H^* - T\Delta S^* \qquad (2.82)$$

Sometimes the approximation:

$$\Delta H^* = E_a - RT \qquad (2.83)$$

is used and ΔG^* and ΔS^* are obtained from k at the chosen standard temperature.

The various methods of analysing the line shape, and the inherent errors in the estimation of thermodynamic parameters, are fully discussed in reference [2.116] which provides a very useful bibliography of recent work.

REFERENCES

2.1   Bloch, F., Hansen, W.W. and Packard, M., *Phys. Rev.*, **69**, 37 (1946).

2.2   Ramsey, N.F., *Nuclear Moments*, Chapman and Hall. London (1953)

2.3   Bloch, F., *Phys. Rev.*, **70**, 460 (1946).

2.4   Bloch, F., Hansen, W.W. and Packard, M., *Phys. Rev.*, **70**, 474 (1946)

2.5   Farrar, T.C., and Becker, E.D., *Pulse and Fourier Transform n.m.r.*, Academic Press, New York, pp. 4, 5(1971)

2.6   Tabi, I.I., Ramsey, N.F. and Schwinger, J., *Revs. Mod. Phys.*, **26**, 167 (1954).

2.7   Ref. 5, p.66 *et seq.*

2.8   Ref. 5, pp. 14–16, 29, 66–78.

2.9   Gillies, D.G. and Shaw, D., *Annual Report on n.m.r.* (ed. F.F. Mooney), Academic Press, London, 5A, p.557 (1972)

2.10   Carrington, A. and McLachlan, A.D., *Introduction to Magnetic Resonance*, Harper and Row, New York, pp. 185, 257 (1967)

2.11   Bloembergen, N., Purcell, E.M., and Pound, R.V., *Phys. Rev.*, **73**, 679 (1948)

2.12   Ref. 5, Ch.4.

2.13   Primas, H., 5th. European Congress on Molecular Spectroscopy, Amsterdam (1961).

2.14 Ref. 5, Ch.3.

2.15 Ref. 9, pp.580—606.

2.16 Emsley, J.W., Feeney, J. and Sutcliffe, L.H., *High Resolution N.M.R. Spectroscopy,* Pergamon Press, London, 1st. ed., pp. 212—217 (1967)

2.17 Hahn, E.L., *Phys. Rev.* **80**, 580 (1950).

2.18 Carr, H.Y. and Purcell, E.M., *Phys. Rev.,* **94**, 630 (1954).

2.19 Meiboom, S. and Gill, D., *Rev. Sci. Instrum.,* **29**, 688 (1958).

2.20 Ref. 5, Ch. 5, 6

2.21 Ref. 16, Ch. 8.

2.22 Abraham, R.J. *Analysis of High Resolution N.M.R. Spectra,* Elsevier, London (1971).

2.23 Mathieson, D.W. (Ed.), *Nuclear Magnetic Resonance For Organic Chemists,* Academic Press, London, Ch. 5—7 (1967).

2.24 Diehl, P., Harris, R.K. and Jones, R.G., Progress in N.M.R. Spectroscopy, (Ed. Emsley, J.W., Feeney, J. and Sutcliffe, L.H.), Pergamon, London, **3**, p. 1 (1967).

2.25 Corio, P.L. *Chem. Rev.* **60**, 363, (1960)

2.26 Castellano, S. and Bothner-By, *J. Chem. Phys.,* **41**, 3863 (1964).

2.27 Swalen, J.D. and Reilly, C.A., *J. Chem. Phys.,* **37**, 21 (1962).

2.28 Haigh, C.W., *Annual Reports on N.M.R. Spectroscopy,* (Ed. E.F. Mooney), Academic Press, London, 4, p.311 (1971).

2.29 Ferguson, R.C. and Marquardt, D.W., *J. Chem. Phys.,* **41**, 2087 (1964)

2.30 Bloch, F., *Phys. Rev.* **93**, 944 (1954).

2.31 Royden, V., *Phys. Rev.,* **97**, 1261 (1954).

2.32 Bloom, A.L. and Shoolery, *J. Phys. Rev.,* **97**, 1261, (1955)

2.33 Ernst, R.R., *J. Chem. Phys.,* **45**, 3854 (1966)

2.34 Baker, E.B. and Burd, L.W., *Phys. Rev. Sci. Instrum.,* **34**, 238 (1963).

2.35 Freeman, R. and Anderson, W.A., *J. Chem. Phys.,* **39**, 806 (1963).

2.36 Kowalewski, V.J., *Progress in N.M.R. Spectroscopy* (Ed. Emsley, J.W., Feeney, J. and Sutcliffe, L.H.), Pergamon, London (1969).

2.37 Freeman, R. and Gestblom, B., *J. Chem. Phys.,* **47**, 2774 (1967).

2.38 Ramsey, N.F., *Phys. Rev.,* **78**, 699 (1950).

2.39 Ramsey, N.F., *Phys. Rev.,* **86**, 243 (1952).

2.40 Pople, J.A., Schneider, W.G. and Bernstein, H.J. *High-resolution Nuclear Magnetic Resonance.* McGraw-Hill, New York, Ch.12 (1959).

2.41 Pople, J.A., *Proc. Roy. Soc.,* **A239**, 541 (1957).

2.42 Zürcher, R.F., *Progress in N.M.R. Spectroscopy* (Ed. Emsley, J.W., Feeney, J. and Sutcliffe, L.N.,), Pergamon, London, **2**, p. 205 (1967).

2.43 Cavanaugh, J.R. and Dailey, B.P., *J. Chem. Phys.,* **34**, 1094 (1961).

2.44 Waugh, J.S. and Fessenden, R.W., *J. Amer. Chem. Soc.,* **79**, 846 (1957).

2.45 Pople, J.A., *J. Chem. Phys.,* **24**, 1111 (1956).

2.46 McConnell, H.M., *J. Chem. Phys.,* **27**, 226 (1957).

2.47 Pople, J.A., *J. Chem. Phys.,* **37**, 60 (1962).

2.48 Gil, V.M.S. and Murrell, J.N., *Trans. Faraday Soc.,* **60**, 248 (1964).

2.49 Buckingham, A.D., *Canad. J. Chem.,* **38**, 300 (1960).

2.50 Emsley, J.W. and Phillips, L., *Progress in N.M.R. Spectroscopy* (Ed. Emsley, J.W., Feeney, J. and Sutcliffe, L.H.), Pergamon, London, 7, p. 1 (1971).

2.51   Witanowski, M. and Webb, G.A., *Annual Reports on N.M.R. Spectroscopy* (Ed. E.F. Mooney), Academic Press, London, 5A, p. 395 (1972).

2.52   Bothner-By, A.A. and Glick, R.E., *J. Chem. Phys.*, 25, 362 (1956); 26, 1647 (1957).

2.53   Buckingham, A.D., Schaeffer, T. and Schneider, W.G., *J. Chem. Phys.* 32, 1227 (1960).

2.54   Howard, B.B., Linder, B. and Emerson, M.T., *J. Chem. Phys.*, 36, 485 (1962)

2.55   Buckingham, A.D., *Canad. J. Chem.*, 38, 300 (1960).

2.56   Ref. 16, Ch. 11.

2.57   Emsley, J.W. and Phillips, L., *Mol. Phys.*, 11, 437 (1966).

2.58   Laszlo, P., *Progress in N.M.R. Spectroscopy* (Ed. Emsley, J.W., Feeney, J. and Sutcliffe, L.H.), Pergamon, London, 3, p. 231 (1967).

2.59   Ronayne, J. and Williams, D.H., *Annual Review of N.M.R. Spectroscopy* (Ed. E.F. Mooney), Academic Press, London, 2, p. 83 (1969).

2.60   Engler, E.M., and Laszlo, P., *J. Amer. Chem. Soc.*, 93, 1317 (1971).

2.61   Jackman, L.M. and Sternhell, S. *Applications of N.M.R. Spectroscopy in Organic Chemistry*, 2nd. Ed., Pergamon, London (1969).

2.62   Mathieson, D.W., Ed., *Nuclear Magnetic Resonance For Organic Chemists*, Academic Press, London (1967).

2.63   Chapman, D. and Magnus, P.D., *Introduction to Practical High Resolution N.M.R. Spectroscopy*, Academic Press, London (1966).

2.64   Dungan, C.H. and Van Wazer, J.R., *Compilation of Reported F-19 Chemical Shifts*, Wiley-Interscience, New York (1970).

2.65   Critenfield, M.M., Dungan, C.H., Letcher, J.H., Mark, V. and Van Wazer, J.R., *P-31 N.M.R. Spectroscopy*, Wiley-Interscience, New York, (1969).

2.66   Stothers, J.B., *Carbon-13 N.M.R. Spectroscopy*, Academic Press, New York (1972).

2.67   Levy, G.C. and Nelson, G.L., *Carbon-13 Nuclear Magnetic Resonance for Organic Chemists*, Wiley, New York (1972).

2.68   Ramsey, N.F. and Purcell, E.M., *Phys. Rev.*, 85, 143 (1952).

2.69   Ramsey, N.F., *Phys. Rev.*, 91, 303 (1953).

2.70   McConnell, H.M., *J. Chem. Phys.* 24, 460 (1956).

2.71   Karplus, M. and Anderson, D.H., *J. Chem. Phys.*, 30, 6 (1959).

2.72   Karplus, M., *J. Chem. Phys.*, 30, 11, (1959).

2.73   Murrell, J.N., *Progress in N.M.R. Spectroscopy* (Ed. Emsley, J.W., Feeney, J. and Sutcliffe, L.H.), Pergamon, London, 6, p. 1 (1971).

2.74   Barfield, M. and Grant, D.M., *Advances in Magnetic Resonance*, (Ed. Waugh, J.S.), Academic Press, New York, 1, p. 149 (1965).

2.75   Barfield, M. and Chakrabati, B., *Chem. Revs.*, 69, 757 (1969).

2.76   Fermi, E., *Z. Physik*, 60, 320 (1930).

2.77   Hiroike, E., *J. Phys. Soc. Japan*, 15, 270 (1960).

2.78   Pople, J.A., *Abstr. 3rd National Meeting Society of Applied Spectroscopy*, Cleveland, Ohio, paper 83 (1964).

2.79   Pople, J.A. and Bothner-By, A.A., *J. Chem. Phys.*, 42, 1339, (1965).

2.80   Banwell, C.A. and Sheppard, N., *Discuss. Faraday Soc.*, 34, 115 (1962).

2.81   Gutowsky, H.S., Karplus, M. and Grant, D.M., *J. Chem. Phys.*, **31**, 269 (1958).
2.82   Schaeffer, T., *Canad., J. Chem.*, **40**, 1 (1962).
2.83   Graham, J.D. and Rogers, M.T., *J. Amer. Chem. Soc.*, **84**, 2249 (1962).
2.84   Musher, J.I., *J. Chem. Phys.*, **34**, 594 (1961).
2.85   Shapiro, B.L., Kopchik, R.M. and Ebersole, S.J., *J. Chem. Phys.*, **39**, 3154 (1963).
2.86   Banwell, C.N. and Sheppard, N., *Discuss. Faraday Soc.*, **34**, 115 (1962).
2.87   Banwell, C.N., Sheppard, N. and Turner, J.J., *Spectrochim. Acts.*, **11**, 794 (1960).
2.88   Abraham, R.J. and Gatti, G., *J. Chem. Soc. (B)*, 961 (1969).
2.89   Phillips, L. and Wray, V., *J. Chem. Soc. Perkin II*, 536 (1972).
2.90   Booth, H., *Tetrahedron Letters*, 41 (1965).
2.91   Bhacca, N.S. and Williams, D.H., *Applications of N.M.R. Spectroscopy in Organic Chemistry*, Holden Day, New York (1964).
2.92   Abraham, R.J. and Pachler, K.G.R., *Mol. Phys.*, **6**, 165 (1963)
2.93   Pople, J.A. and Santry, D.P., *Mol. Phys.*, **7**, 269 (1963).
2.94   Pople, J.A. and Santry, D.P., *Mol. Phys.*, **8**, 1 (1964).
2.95   Musher, J.I., *Mol. Phys.*, **6**, 93 (1963).
2.96   Wiberg, K.B., Lampman, G.M., Guila, R.P., Connor, D.S., Schertler, P. and Lavanish, J., *Tetrahedron*, **21**, 2749 (1965).
2.97   Sternhall, S., *Revs. Pure Appl. Chem.*,   **4**, 15 (1964).
2.98   Phillips, L., *Elucidation of Organic Structures by Physical Methods, pt.1.* (Ed. Bentley, K.W. and Kirby, G.W.), Wiley-Interscience, New York, p. 323 (1972).
2.99   Evans, D.F., *Discuss. Faraday Soc.*, **34**, 139 (1962).
2.100 Peake, A. and Thomas, L.F., *Trans. Faraday Soc.*, **62**, 2980 (1966).
2.101 Banks, R.E., Barlow, M.G., Deem, W.R., Haszledine, R.N. and Taylor, D.R., *J. Chem. Soc. (C)*, 981 (1966).
2.102 Burton, D.J., Johnson, R.L. and Bogan, R.T., *Canad. J. Chem.*, **44**, 635 (1966).
2.103 Frankiss, S.G., *J. Phys. Chem.*, **71**, 3418 (1967).
2.104 Pendlebury, M.H., Phillips, L. and Wray, V., *J. Chem. Soc. Perkin II*, 787 (1974).
2.105 Phillips, L. and Wray, V., *J. Chem. Soc. Perkin II*, 928 (1974).
2.106 Williamson, K.L., Li Hou, Y.F., Hall, F.H., Swager, S. and Coulder, M.S., *J. Amer. Chem. Soc.*, **90**, 6717 (1968).
2.107 Hall, L.D., and Manville, J.F., *Canad.J. Chem.*, **45**, 1299 (1967); **47**, 17 (1969).
2.108 Hall, L.D., Manville, J.F., Foster, A.B. and Hems, R., *Chem. Comm.* 158 (1968).
2.109 Hall, L.D. and Manville, J.F., *Chem. Comm.* 37 (1968).
2.110 Mousseron-Canet, M. and Crouzet, F., *Bull. Soc. Chim. France*, 3023 (1968).
2.111 Phillips, L. and Wray, V., *J. Chem. Soc. (B)*, 1618 (1971).
2.112 Johnson, C.S., *Advances in Magnetic Resonance* (Ed. Waugh, J.H.), Academic Press, New York, **1**, p. 33 (1965).

2.113   Booth, H., *Progress in N.M.R. Spectroscopy* (Ed. Emsley, J.W., Feeney, J. and Sutcliffe, L.H.), Pergamon, London, 5. p. 149 (1969).

2.114   Thomas, W.A. *Annual Reports on N.M.R. Spectroscopy* (Ed. E.F. Mooney), Academic Press, London, 3, p. 92 (1970).

2.115   Harris, R.K., *Specialist Periodical Reports, (The Chemical Society) N.M.R. no. 2.,* p. 216 (1972).

2.116   Sutherland, I.O., *Annual Reports on N.M.R. Spectroscpy,* (Ed. E.F. Mooney), Academic Press, London, 4, p. 71 (1971).

2.117   Gutowsky, H.S. and Holm, C.H., *J. Chem. Phys.,* 25, 1228 (1956).

3 Nuclear quadruple resonance (n.q.r.) spectroscopy

3.1 INTRODUCTION

Nuclear quadrupole resonance (n.q.r.) is a branch of radio-frequency spectroscopy which is closely related to nuclear magnetic resonance (n.m.r.). Both techniques involve the coupling of radio-frequency radiation with a nuclear magnetic moment to bring about transitions between nuclear orientations of different energies. The difference between the two lies in the origin of the external nuclear energy levels. In the case of nuclear magnetic resonance, the energy levels are governed by the interaction of the nuclear magnetic dipole moment with an externally applied magnetic induction*, whereas in nuclear quadrupole resonance the levels are governed by an interaction of the nuclear electric quadrupole moment with the electric field gradient produced at the nucleus by the charge distribution to its environment.

The way in which the quadrupole interaction is observed depends upon the relative magnitudes of the nuclear magnetic and nuclear quadrupole interactions. When the magnetic interaction is large in comparison with the quadrupole, the latter causes a splitting or broadening of the n.m.r. lines. When the quadrupole interaction is dominant, the transition frequencies are largely determined by the electric field gradients at the nucleus, and the magnetic interaction (in non-zero field) is seen as a splitting or broadening of the n.q.r. lines. In the absence of a magnetic induction, there is no magnetic interaction and the unperturbed resonance lines are observed. Here we are primarily concerned with the latter case which is alternatively known as 'zero field' or 'pure' quadrupole resonance, or just 'n.q.r.'.

With nuclear magnetic resonance the transition frequencies are proportional to the applied magnetic induction, so transitions between the magnetic levels are possible using a fixed-frequency oscillator while sweeping the magnetic field. With n.q.r. the electric induction gradient is a fixed property of the molecule or crystal and is considerably larger than any practical externally applied field

* Please see Page vi for an explanation of the use in this book of the terms *magnetic field* and *magnetic induction*

gradient. This means that a variable-frequency detection system must be used. The range of nuclear quadrupole interactions is such that transition frequencies can occur anywhere between 100 kHz and 1 GHz (1000 MHz), making detection by a single spectrometer very difficult.

3.1.1 The fundamental requirements of nuclear quadrupole resonance spectroscopy

(1) The sample must contain a nucleus with a quadrupole moment, in an unsymmetrical environment. This necessarily implies that the nuclear spin quantum number, I, should be greater than $\frac{1}{2}$, and that the site symmetry of the atom containing the quadrupolar nucleus should be less than cubic or tetrahedral.

(2) The effect is observed only in the solid state, where the field gradient axes are fixed in space. In a liquid the molecular vibrations and tumbling motions, which occur at a rate much greater than the radio-frequency time scale, cause the average field gradient experienced by the nucleus to be reduced to zero. The effect could in principle be seen in gases (if the collision rate is low) or in liquid crystals which have been oriented.

(3) To make detection possible, the chosen nuclear isotope should be present in reasonably high natural abundance, and in addition about 2 g or more of polycrystalline material are usually required. The resonance lines in general are much broader by a factor of 10^3 or 10^4 than those observed in high resolution n.m.r. and are therefore more difficult to detect.

(4) A sensitive radio-frequency detection system is required, the operating frequency of which is variable over the region of the spectrum of interest. For example ^{35}Cl n.q.r. frequencies fall in the range 0–60 MHz and most of this region can be covered by one spectrometer. Details of these will be given in Section 3.3.

For the above reasons the most popular n.q.r. nuclei are: ^{14}N, ^{35}Cl, ^{37}Cl, ^{79}Br, ^{81}Br, and ^{127}I. Other suitable nuclear isotopes are ^{10,11}B, ^{27}Al, ^{33}S, ^{55}Mn, ^{59}Co, ^{63,65}Cu, ^{69,71}Ga, ^{75}As, ^{93}Nb, ^{115}In, ^{121,123}Sb, ^{135}Ba, ^{175}Lu, ^{185}Re, ^{157}Au, ^{201}Hg, and ^{209}Bi.

3.2 GENERAL PRINCIPLES

First of all we shall consider the nature of the nuclear quadrupole moment and then see how this interacts with its electronic environment. The quadrupole moment of the nucleus is a consequence of its spin and spin degeneracy; it is unlike the quadrupole moment of molecules. A nucleus for which $I \geqslant 1$ behaves as if it does not have spherical symmetry but is distorted along the axis of the spin. In a simple model in which the nuclear charge distribution is related to that of a molecule we may imagine the charge to be compressed or extended along this axis, according to the sign of its quadrupole moment. A positive moment has a prolate spheroidal shape (extension along the spin axis) and a negative moment has an oblate spheroidal shape (contraction along the spin axis) see Fig. 3.1. The magnitude of the quadrupole moment, *eQ*, corresponds to the deviation of the nucleus from spherical symmetry and is defined as:

$$eQ = \int \rho r^2 (3\cos^2 \theta - 1) \, d\tau \qquad (3.1)$$

where *e* is the absolute value of the electronic charge, ρ is the charge density in a volume element $d\tau$ inside the nucleus at a distance *r* from the centre, and θ is the angle which the radius vector *r* makes with the nuclear spin axis (nucleus in the spin state $M_I = I$).

In many molecules the electronic distribution around a nucleus is not spherically symmetrical, unless the atom containing the nucleus happens to be at a site of tetrahedral or cubic symmetry. If the atom in question has a nucleus with zero *eQ*, the potential energy associated with the nuclear–electronic interaction will be independent of the nuclear orientation. If, however, the nuclear electric quadrupole moment is non-zero, the energy of the system will depend upon the nuclear orientation. Obviously there will be a particular nuclear orientation in

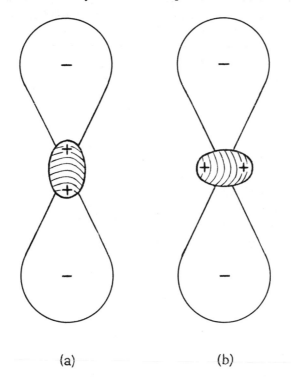

(a)　　　　　　　　(b)

Fig. 3.1 Interaction of a quadrupolar nucleus with a p-orbital charge distribution. Interaction energy of (a) is less than that of (b).

which the potential energy will have a minimum value. On a classical (non-quantized) level, consider the orientations shown in Fig. 3.1. This shows two possible orientations of a quadrupolar nucleus with respect to an electronic charge distribution similar to that generated by a filled p-orbital. In Fig. 3.1(a) the energy of interaction will be lower because there is less separation between opposite charges. This could be equated with the ground state configuration, and Fig. 3.1(b) with an excited state, since energy has to be expended to turn the nucleus, which moves the opposite charges further apart. On the molecular level, of course, the nuclear orientations will be quantized and the nucleus will precess, like a gyroscope, about the direction of maximum field gradient (if this has axial symmetry), rather than as shown in Fig. 3.1. There is here a further analogy with n.m.r., where the nuclear magnetic dipole moment precesses about the direction of the applied induction (see Chapter 2).

For a quadrupolar nucleus, therefore, different nuclear orientations give rise to a set of energy levels. Transitions between these levels may be brought about by the fact that the quadrupolar nucleus still possesses a magnetic dipole moment, and if an oscillating magnetic induction is applied from a coil round the sample, coupling occurs between the magnetic moment of the nucleus and the magnetic components of the radiation. In this way the nuclear orientation is altered with respect to the direction of maximum electric field gradient. Note that the energy levels are set by an electric quadrupolar interaction but that the transitions are magnetic dipole in type.

When the electric field gradient is symmetrical about an axis, the expression for the energy levels is:

$$E = \frac{e^2 Qq}{4I(2I-1)} \ [3M_I^2 - I(I+1)] \tag{3.2}$$

where eQ is the nuclear quadrupole moment defined in Equation (3.1), eq is the magnitude of the electric field gradient in the direction of the axis of symmetry, I is the nuclear spin quantum number, and M_I is the nuclear magnetic quantum number, which takes the values $+1, I-1, \ldots, -I$. Note that there is degeneracy in $+M_I$ and $-M_I$ states because of the M_I^2 term.

The electric field gradient is a tensor quantity and this can be defined in a diagonal form so that all off-diagonal terms, e.g. $\partial^2 V/\partial x \partial y$, are zero, when the three axes x, y, and z are called the principal axes of the tensor. According to the Laplace equation, the sum of the field gradients in these directions, which are the second derivatives of the electrostatic potential, V, is equal to zero:

$$\frac{\partial^2 V}{\partial x^2} + \frac{\partial^2 V}{\partial y^2} + \frac{\partial^2 V}{\partial z^2} = 0 \tag{3.3}$$

The convention is to choose the principal axes in such a way that:

$$\left| \frac{\partial^2 V}{\partial z^2} \right| \geqslant \left| \frac{\partial^2 V}{\partial y^2} \right| \geqslant \left| \frac{\partial^2 V}{\partial x^2} \right|$$

and to define:

$$\frac{\partial^2 V}{\partial z^2} = eq \tag{3.4}$$

An asymmetry parameter, η, is also defined:

$$\eta = \left(\frac{\partial^2 V}{\partial x^2} - \frac{\partial^2 V}{\partial y^2}\right) \bigg/ \frac{\partial^2 V}{\partial z^2} \tag{3.5}$$

η measures the departure of the field gradient from axial symmetry, and can take values from 0 to 1. When $\partial^2 V/\partial x^2$ equals $\partial^2 V/\partial y^2$ the field gradient is symmetric about the z axis and η is equal to zero.

3.2.1 Half integral spins

Firstly we shall consider the case of common quadrupolar nuclei, ^{35}Cl, ^{79}Br, and ^{81}Br, which have a spin $I = 3/2$. From Equation (3.2) there will be only two levels of energies:

$$E_{\pm 1/2} = -\tfrac{1}{4} e^2 Qq \tag{3.6}$$

$$E_{\pm 1/2} = \tfrac{1}{4} e^2 Qq$$

The selection rule for magnetic dipole transitions is:

$$\Delta M_I = \pm 1 \tag{3.7}$$

Only one frequency is therefore observed:

$$\nu = \tfrac{1}{2} \frac{e^2 Qq}{h} \tag{3.8}$$

The expression $e^2 Qq/h$ is termed the nuclear quadrupole coupling constant and has the units of frequency (MHz). It can be of either sign, according to the sign of eQ or eq, but this is not found from a simple measurement of the transition frequency since a reversal of sign has the effect of inverting the order of the levels, leaving the frequency unchanged.

For nuclei of spin $I = 5/2$, for example ^{127}I or ^{121}Sb, transitions between three levels are possible [see Fig. 3.2(a)]. The magnetic dipole selection rule (3.7) ensures that only two transition frequencies are observed:

$$\nu_{(1/2 \rightarrow 3/4)} = \frac{3}{20} \frac{e^2 Qq}{h}$$

$$\tag{3.9}$$

$$\nu_{(3/2 \rightarrow 5/2)} = \frac{3}{10} \frac{e^2 Qq}{h}$$

In this case of an axially symmetric field gradient the frequencies are in the ratio

179

of 1 : 2; it will be seen later that the effect of a finite η term is to cause departure from this situation.

For the spin 7/2 case, e.g. ^{59}Co or ^{123}Sb, three transitions between four energy levels are possible [see Fig. 3.2(b)].

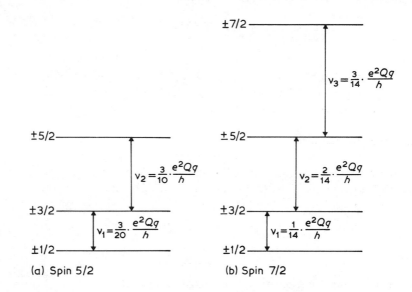

(a) Spin 5/2   (b) Spin 7/2

Fig. 3.2 Zero-field quadrupole energy levels and transitions for (a) $I = \frac{5}{2}$ and (b) $I = \frac{7}{2}$ ($\eta = 0$).

When the asymmetry parameter is non-zero, the degeneracy in $\pm M_I$ remains but the transition frequencies are modified. For spin 3/2 the transition frequency is given by:

$$\nu = \tfrac{1}{2} \frac{e^2 Qq}{h} \left(1 + \frac{\eta^2}{3} \right)^{\frac{1}{2}} \tag{3.10}$$

In this case the quadrupole coupling constant cannot be derived from the n.q.r. frequency ν unless η is known. However, in practice η is often small for singly coordinated atoms (e.g. chlorine), and the error in neglecting η values below 0.1 is less than 0.16%, so in this instance the coupling constant may be taken as twice the resonance frequency.

For spin 5/2 and 7/2 the transition freqencies for η values of < 0.25 are given in Table 3.1.

The main effect of a non-zero asymmetry parameter is to remove the harmonic relationship between successive transitions. From a ratio of two resonance frequencies, both $e^2 Qq/h$ and the asymmetry parameter η can be determined.

Table 3.1

Spin	Transition frequency ($\eta \neq 0$)
1	$\nu_0 = \dfrac{1}{2} \left(\dfrac{e^2 Qq}{h}\right) \eta$
	$\nu_- = \dfrac{3}{4} \left(\dfrac{e^2 Qq}{h}\right) \eta \left(1 - \dfrac{\eta}{3}\right)$
	$\nu_+ = \dfrac{3}{4} \left(\dfrac{e^2 Qq}{h}\right) \eta \left(1 + \dfrac{\eta}{3}\right)$
$\dfrac{3}{2}$	$\nu_1 = \dfrac{1}{2} \left(\dfrac{e^2 Qq}{h}\right) \eta \left(1 + \dfrac{\eta^2}{3}\right)^{\frac{1}{2}}$
$\dfrac{5}{2}$	$\nu_1 = \dfrac{3}{20} \left(\dfrac{e^2 Qq}{h}\right) \eta \, (1 + 0.09259\eta^2 - 0.63403\eta^4)$
	$\nu_2 = \dfrac{3}{10} \left(\dfrac{e^2 Qq}{h}\right) \eta \, (1 + 0.20370\eta^2 + 0.16215\eta^4)$
$\dfrac{7}{2}$	$\nu_1 = \dfrac{1}{14} \left(\dfrac{e^2 Qq}{h}\right) \eta \, (1 + 3.63333\eta^2 - 7.26070\eta^4)$
	$\nu_2 = \dfrac{2}{14} \left(\dfrac{e^2 Qq}{h}\right) \eta \, (1 - 0.56667\eta^2 + 1.85952\eta^4)$
	$\nu_3 = \dfrac{3}{14} \left(\dfrac{e^2 Qq}{h}\right) \eta \, (1 - 0.1001\eta^2 - 0.01804\eta^4)$

3.2.2 Integral spins

For nuclei of spin $I = 1$, the most important example of which is ^{14}N, the resonance frequencies are given below. The effect of a finite asymmetry parameter is to lift the degeneracy due to $\pm M_I$, so three levels exist (see Fig. 3.3):

$$\nu_+ = \tfrac{3}{4} \frac{e^2 Qq}{h} \left(1 + \frac{\eta}{3}\right)$$

$$\nu_- = \tfrac{3}{4} \frac{e^2 Qq}{h} \left(1 - \frac{\eta}{3}\right) \tag{3.11}$$

$$\nu_0 = \tfrac{1}{2} \frac{e^2 Qq}{h} \eta$$

ν_0 is rarely observed since a large value of η is needed to bring it into a frequency range where detection is possible. When η is zero, a single transition is observed:

$$\nu = \tfrac{3}{4} \, \frac{e^2 Qq}{h} \tag{3.12}$$

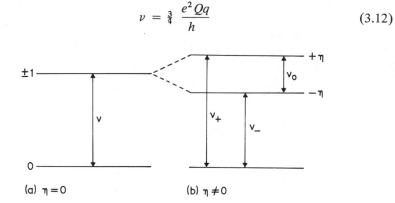

Fig. 3.3 Zero-field quadrupole energy levels and transitions for $I = 1$: (a) $\eta = 0$; (b) $\eta \neq 0$.

The transition frequencies which are measured from a nuclear quadrupole resonance spectrum of a polycrystalline sample can provide (with the exception of the spin 3/2 case) values for the quadrupole coupling constant $e^2 Qq/h$ and the asymmetry parameter η using the above equations. To determine η for spin $I = 3/2$, it is necessary to resort to Zeeman splitting studies, in which a weak magnetic field is applied to the sample (see section 3.5).

3.3   EXPERIMENTAL DETECTION OF N.Q.R. FREQUENCIES

The detection of n.q.r. signals is dependent upon the coupling of the nuclear magnetic moment with a suitably applied rotating magnetic field B_1. In this respect it resembles detection in nuclear magnetic resonance but the following differences exist:

(1) As indicated earlier (p. 176) the n.q.r. frequency is determined by the electric field gradient within the crystal, so a variable-frequency detector is necessary. In the case of n.m.r. the oscillator frequency used is determined largely by the magnitude of the applied field.

(2) Relaxation times of quadrupolar nuclei tend to be shorter than those of nuclei which possess only a magnetic moment, because of the greater efficiency of quadrupolar relaxation processes in comparison with dipolar processes. This means that a high radio-frequency (r.f.) power is required in order to observe resonance.

(3) The degeneracy of the quadrupole spin states $(\pm M_I)$ causes a net cancellation of the nuclear induction in a direction perpendicular to the axis of the transmitting coil. Therefore crossed-coil detection is not possible in zero magnetic field and single-coil systems are used. (A crossed coil system could

be used in the presence of a weak magnetic field since this removes the degeneracy.)

The variable-frequency requirement rules out the bridge form of detection often used in n.m.r. since this system requires complicated adjustments.

There are three usual methods for n.q.r. detection: (1) super-regenerative oscillators; (2) continuous wave oscillators; (3) pulsed r.f. or spin-echo method. The first two methods both use an r.f. oscillator to act as a detector as well as an exciter of the nuclei. In the third method these functions are carried out by a separate receiver and transmitter. Each of these methods will be discussed and their relative merits compared.

Alternative methods for determining nuclear quadrupole coupling constants other than by pure quadrupole resonance are given in Chapters 1, 2, 5 of this Volume.

3.3.1 The super-regenerative oscillator

The super-regenerative oscillator (s.r.o.) was the first circuit to be used in the detection of n.q.r. in solids by Dehmelt and Kruger in 1950 [3.1]. Essentially it consists of an r.f. oscillator which is turned off periodically at a rate which is a small fraction ($\sim 1/100$) of the r.f. frequency. When turned on again, the oscillations build up either from noise (this is termed the incoherent mode of operation) or from the tail of the preceding pulse (coherent mode) until they reach their limiting value (see Fig. 3.4). The sample is situated in the inductance coil of the tank circuit of the oscillator (this is a parallel combination of an inductance and a variable capacitance which sets the running frequency of the oscillator), and the circuit parameters are adjusted for coherent mode operation. If an input signal due to the exchange of r.f. energy with the sample appears

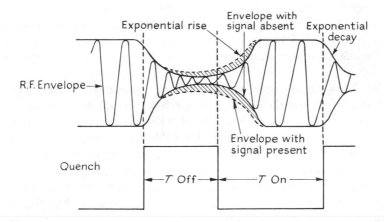

Fig. 3.4 Oscillation envelope of a super-regenerative oscillator (Courtesy of Professor J.A.S. Smith).

in the tank circuit of the oscillator at the tuned circuit frequency, the oscillations build up sooner and the overall area of the r.f. envelope is increased (see Fig. 3.4). This increases the level of the oscillator valve anode current, which is detected, either directly on an oscilloscope or by other means. The rate at which the oscillator is 'quenched' determines the phase relationship between successive bursts of r.f. and also, in the extreme case, determines whether the oscillator is operating in the linear or logarithmic mode. The former mode requires a rapid quench frequency so that the oscillations never reach their limiting value. In this (the linear) mode the ouput is a linear function of input. When the oscillations are allowed to limit, with a slower quench frequency the relationship between input and output is logarithmic, and it is this mode of operation which is useful for n.q.r. detection. When in this mode, the s.r.o. can be made to have variable coherence by changing the quench frequency. A change in coherence has a considerable effect on the power spectrum (plot of amplitude versus frequency) of the oscillator. As the coherence decreases, the linewidth of the radiation increases sharply. Since the quenching of the oscillator is an extreme form of amplitude modulation, the power spectrum of the radiation will contain side-bands separated from the carrier frequency by integral multiples of the quench frequency. The important feature of the s.r.o. as an n.q.r. detector is that the width of the lines in the power spectrum can be adjusted to be comparable with the natural linewidth of the nuclear resonance. By n.m.r. standards n.q.r. lines are very broad, perhaps of the order of 1–30 kHz. The advantage of the s.r.o. is that all the nuclei can be excited simultaneously, rather than just a proportion of these, as with continuous wave oscillators (see Section 3.3.2). A disadvantage is that each n.q.r. absorption can also be excited by the s.r.o. side-bands and hence a frequency-swept spectrum will contain multiple lines from a single n.q.r. signal. However, these side-band responses can be suppressed by varying the quench frequency at a rate which is much slower than the quench frequency itself but faster than the recording time constant of the instrument. This has the effect of moving the side-bands in and out from the fundamental signal, and when a recording time constant of about 10 seconds is used the side-bands become smeared out and only the fundamental signal is recorded.

A block diagram of a super-regenerative n.q.r. spectrometer is shown in Fig. 3.5. This is based upon the Decca NQR Spectrometer, which is used for the detection of resonances in the region 5–60 MHz. The sample, usually 2 or 3 g of polycrystalline material, is contained in a thin-walled glass tube which fills the r.f. coil. The coil is usually situated in a copper or brass can and the whole probe assembly is immersed in a Dewar vessel of coolant, usually liquid nitrogen. N.q.r. measurements are often made at low temperature because line intensities are usually (but not always) greater than at room temperature. This follows from a more favourable Boltzmann distribution of energy level populations, and in addition there is less broadening from molecular torsional vibrations or hindered rotations at low temperature. However, large changes in the relaxation

times with temperature can in some cases result in an increase of signal strength
with temperature.

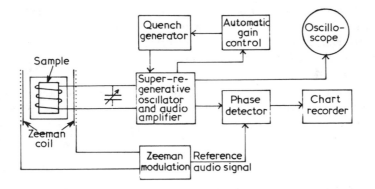

Fig. 3.5 Block diagram of a super-regenerative n.q.r. spectrometer using Zeeman
modulation.

A frequency sweep is carried out by a motor drive to the variable capacitor
in the tank circuit. An optimum quench frequency is selected and during the
run an automatic gain control unit maintains a constant coherence level (i.e. a
constant sensitivity) while the r.f. frequency is swept. This is accomplished by
changing the mark/space ratio of the quench pulses while keeping the quench
frequency constant. Signals are detected by changes in the anode voltage level
and can be observed on an oscilloscope or more usually using a phase-sensitive
detector. This is a device for improving the signal-to-noise ratio of a signal
which is in phase with a particular reference frequency. Usually this is an
audio-frequency with which the n.q.r. signal is modulated. The easiest way of
accomplishing this is to include a 'varicap' diode (a voltage dependent capacitor) in the tank circuit. The phase detector then measures the component of
the signal which is present at this audio-frequency and the output is fed to a
chart recorder.

This frequency modulation method has the disadvantage that signals other
than those due to n.q.r. absorption may be detected (e.g. from radio trans-
mitters, piezoelectric responses). For n.q.r. observation a form of magnetic
modulation is useful because only true n.q.r. signals can respond to it. With
magnetic or Zeeman modulation a magnetic field of up to 0.02 T is periodically applied to the sample at a low audio-frequency (<200 Hz). The
magnitude of line splitting caused by a magnetic field is a sensitive function
of the orientation of the crystal with respect to the field. For a polycrystalline
sample, therefore, application of the field causes the signal to be broadened
and almost erased for the period during which the magnetic field is applied. A
large coil surrounding the samples supplied with a bisymmetric wave-form
(⎍⎍⎍⎍) to provide the modulating field. This wave-form is used to

185

ensure that no residual magnetic field is present in the coil during the 'off' period. The reference frequency supplied to the phase detector is in fact at twice the frequency of this bisymmetric waveform since there will be two 'field off' and 'field on' periods per cycle. When an n.q.r. signal is present the phase detector then detects the component of it which is in phase with the 'field off' period, and this is recorded on a chart recorder in the normal way. Some calibration of the chart is necessary and this can be provided by a frequency marker unit.

A typical n.q.r. spectrum from this type of spectrometer is shown in Fig. 3.13.

Super-regenerative oscillators can operate successfully at several hundred MHz, but in the low frequency range, below 5–10 MHz, their sensitivity is often reduced. It is in this region that continuous wave oscillators have been most commonly used.

3.3.2 Continuous wave oscillators

A continuous wave oscillator is somewhat simpler than an s.r.o. since there is no quenching of the oscillations. The system can be self-detecting in that the oscillator both excites and detects the r.f. absorption at the resonance condition.

An oscillator may be regarded as an amplifier with sufficient positive feedback to supply its own input, and thus maintain continuous oscillation. When there is only just enough feedback to sustain oscillations the circuit is very sensitive to changes in the shunt impedance of the tank circuit. When the oscillator frequency equals that of the nuclear quadrupole resonance, the level of amplitude of the oscillator falls and this is detected, usually by grid rectification and then by an audio-frequency modulation technique, as described in Section 3.3.1. This type of oscillator (known as a marginal oscillator) is capable of r.f. levels somewhat lower than those of the s.r.o. At low level its sensitivity is greatest, and its noise factor is also more favourable. A typical circuit of this type of oscillator is given by Pound and Knight [3.2].

Another type of oscillator, due to Robinson [3.3], has a different feedback arrangement. This uses a second valve (or transistor) which provides a limited output, giving a constant amount of feedback. This system (the limited oscillator) has the advantage that sensitivity is maintained over a wider frequency range without the need for constant adjustment of the feedback.

Both of these oscillator-detectors give better line-shapes than the s.r.o. type of detector, and have in fact been well used in n.m.r. applications where low levels of r.f. are often needed. For n.q.r. they are usually used for nuclei whose resonances are more easily saturated, for example ^{14}N. The frequency range for this nucleus is 1–5 MHz, a difficult region for s.r.o. detection.

A typical ^{14}N n.q.r. spectrometer of this type has a simpler form to that shown in Fig. 3.5. No quench generator is of course required and indeed an a.g.c. system is not absolutely necessary. The frequency sweep is again made by a motor drive to the tank circuit capacitor. The output from the oscillator is usually passed to an audio amplifier before phase detection. Both frequency modulation or Zeeman modulation can be used for detection when the asymmetry parameter is small, but the latter form does not work well with ^{14}N signals when η is large, since a high field strength is required to cause an appreciable splitting of such signals.

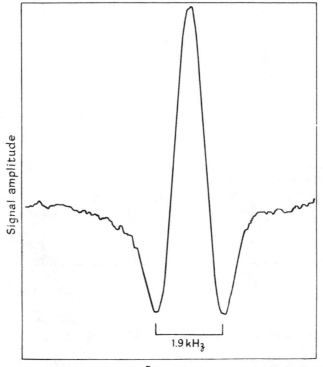

Fig. 3.6 ^{14}N quadrupole resonance signal from hexamethylenetetramine. $\nu = 3.308$ MHz at 296 K.

Fig. 3.6 shows the ^{14}N n.q.r. signal from 2 g of hexamethylenetetramine at room temperature, obtained with a Robinson type oscillator-detector.

Chlorine resonance is also possible with this type of oscillator, provided that the linewidth is small (< 5 kHz); Fig. 3.7 shows a resonance signal from ^{35}Cl in gallium trichloride at room temperature. This compound exists as a dimer in which there are two bridging chlorines and four terminal ones. The molecule has two-fold symmetry, so only three lines are observed, at 20.22 and 19.08 MHz due to the terminal chlorine atoms, and at 14.67 MHz due to the bridging chlorine atoms (frequencies at 306 K). Notice the large frequency difference between resonances of bridging and terminal chlorine atoms (see Section 3.4.2).

3.3.3 Pulsed r.f. detection

The use of pulsed methods for the detection of n.q.r. has not been as widely

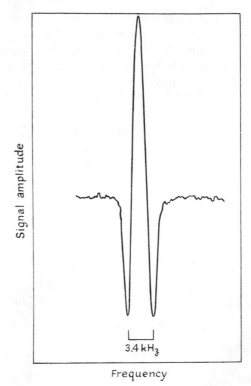

Fig. 3.7 One of the ^{35}Cl quadrupole resonance signals from gallium trichloride using a c.w. spectrometer; $\nu = 20.274\,\mathrm{MHz}$ at 294 K.

adopted as those described in the two previous sections, owing largely to the more complex circuitry and high r.f. powers required. However, the amount of information obtainable is greater; for example, nuclear relaxation times can be studied by this method.

The action of the r.f. pulses on the nuclear spins is similar in both n.q.r. and n.m.r. When in a magnetic field the nuclei will precess about the field direction at a rate of γB, where γ is the gyromagnetic ratio of the nucleus and B is the magnetic induction. In zero field, the quadrupolar nuclei may be regarded as precessing about the direction of the principal axis of the electric field gradient in the crystal (the z axis, at least if $\eta = 0$). The rate of precession is dependent upon their orientation, θ; in classical terms this is equal to $3e^2 Qq \cos \theta / 4\,\Omega$, where Ω is the nuclear angular momentum. In a polycrystalline sample the field gradient z axes will lie in all possible directions with respect to the r.f. coil, so there will always be a proportion of crystallites so arranged in the sample that their z axes coincide with a particular direction in the laboratory frame of reference, to be compared with the field direction in n.m.r. When an intense radio-frequency field B_1 is applied to the sample via an r.f. coil, the nuclei (in both n.m.r. and n.q.r. cases) begin in addition to precess about B_1 at a rate of γB_1. If this field is applied at right angles to B, say along the x axis in a rectangular coordinate system, xyz (with B along the z axis), the new rotation will be in the yz plane. For n.q.r. of polycrystalline samples there is no directional requirement of the r.f. field since the choice of a laboratory frame z direction is arbitrary. The length of time for which B_1 is switched on controls the amount of rotation of the macroscopic nuclear magnetization vector. If the pulse is switched off when this rotation is 90°, the magnetization is then in a position to induce an r.f. voltage in the coil. A single coil or two coaxial coils may be used for n.q.r. detection but otherwise the apparatus is similar to that used for n.m.r. The relaxation times T_1 and T_2 are also measured by the same techniques (Chapter 2). The absence of a magnetic induction requirement allows larger samples to be used than is usual in n.m.r., which aids signal detection. Where a narrow-band transmitter or receiver is used, searching over a wide frequency range is tedious because constant retuning is necessary to maintain sensitivity. Searching for an n.q.r. signal is more difficult than for magnetic resonance, where the frequency range of any particular nucleus is relatively narrow and can be predicted from the value of γB. Pulsed methods are useful for detecting very broad resonances which may not be observable by any other technique, or signals with long T_1 (which saturate easily in a c.w. spectrometer).

Figure 3.8 shows the free induction decay signal (following a 90° pulse) from ^{35}Cl resonance in CsDCl$_2$ at 290 K, using a pulsed r.f. spectrometer. The marker spacing in $10\,\mu s$. And the time constant of the decay, T_2^*, is $50\,\mu s$.

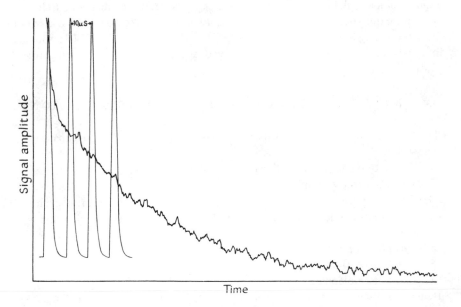

Fig. 3.8 Free induction decay signal for ^{35}Cl resonance in CsDCl$_2$; $\nu = 21.28$ MHz at 290 K (Courtesy of Prof. J.A.S. Smith).

Double resonance

A highly sensitive method of n.q.r. detection using pulsed r.f. is a double resonance technique first used by Hahn [3.4]. The method of spin mixing in the
laboratory frame involves the use of an indicator nucleus A which is either bonded
to, or is in proximity with, the quadrupolar nucleus B. Nucleus A should be
capable of generating an easily observed n.m.r. or n.q.r. signal, usually seen as a
free induction decay (f.i.d.) signal. Suppose that A is a 'good' n.m.r. nucleus.
The sample is placed in a magnetic field and the f.i.d. signal observed. The sample
is then rapidly removed from the magnetic field into a coil of a continuous-wave
r.f. transmitter, the operating frequency of which can be swept. When coincident
with an n.q.r. frequency of nucleus B this changes the level populations of B.
The sample is now returned to the magnetic field; as it does so, at some intermediate value of the field, the 'A' and 'B' transitions come into coincidence
('level crossing'). Energy therefore passes from the A system to the B; the
populations of the A levels are now altered, and thus a change in this signal is
noticed when the sample reaches the original magnetic field and the pulse
sequence repeated. This method has been used very successfully to determine ^{14}N
n.q.r. frequencies in a number amino-acids, simple peptides, and other molecules
of biological interest [3.5]. In these experiments, ^{1}H was used as the indicator
nucleus (A). As an indication of the inherent sensitivity of this method, it has
been possible to measure n.q.r. frequencies due to ^{17}O in quinolines. The natural
abundance of this isotope is only 0.037% [3.6]. Alternatively, nucleus A can be
a quadrupolar nucleus and the experiment carried out in zero magnetic field.
Deuterium quadrupole resonance in CD_2Cl_2 has been detected at 127.2 kHz (77 K)
using the strong n.q.r. signal from ^{35}Cl [3.7]. ^{39}K and ^{41}K quadrupole resonance
in potassium chlorate has also been observed, again using ^{35}Cl as the indicator
[3.8].

3.4   INTERPRETATION AND CHEMICAL APPLICATIONS

The main uses of nuclear quadrupole resonance spectroscopy are summarized
below.
(1) Information about chemical bonding in the solid state.
(2) Molecular structure information.
(3) Characterization of molecular or ionic species (i.e. as a 'fingerprint' technique)
(4) Test for electronic wave-functions used in calculating theoretical coupling
    constants.
(5) Crystallographic and molecular symmetry information.
(6) Detection of phase transitions.
(7) Solid state molecular motion studies.
Item (1) is discussed in detail in Section 3.4.1; (2) and (3) in Section 3.4.2; (5),
(6), and (7) are covered in Section 3.5. Some examples of theoretical coupling
constant calculations have been given by Lucken [3.9].

3.4.1 Chemical bonding

In order to be able to interpret nuclear quadrupole coupling constants in terms of chemical bonding parameters, we shall examine the origin of the electric field gradient first of all in an atom, A, and then in a hypothetical molecule, AB. Atom A contains the quadrupolar nucleus under investigation.

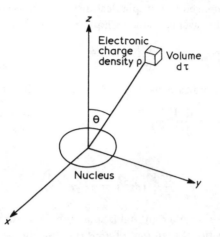

Fig. 3.9 Axes for calculation of charge distribution around nucleus A.

Consider an element of volume $d\tau$ and charge density ρ, distant r from the nucleus of A (Fig. 3.9). If θ is the angle between the spin axis of the nucleus (the z axis) and the position of the charge element, the potential due to the charge element at the nucleus of A is:

$$V = \frac{\rho d\tau}{r}$$

Therefore;

$$\frac{dV}{dz} = \frac{-\rho d\tau}{r^2} \cos\theta \text{ (since } dr/dz = \cos\theta\text{)}$$

and

$$\frac{d^2 V}{dz^2} = \rho\left(\frac{3\cos^2\theta - 1}{r^3}\right) d\tau \tag{3.13}$$

For the field gradient due to the total electron cloud around the nucleus, Equation (3.13) must be integrated over all space:

$$\frac{d^2 V}{dz^2} = \int \rho\left(\frac{3\cos^2\theta - 1}{r^3}\right) d\tau \tag{3.14}$$

In quantum-mechanical terms the charge density ρ can be replaced by $-e\psi^*\psi$:

$$\frac{d^2 V}{dz^2} = -e \int \psi^* \left(\frac{3\cos^2\theta - 1}{r^3} \right) \psi \, d\tau \qquad (3.15)$$

This is the total field gradient exerted by the electronic charge distribution at the nucleus of A. To proceed further we need to know the contributions from electrons in different orbitals. Taking the case of the s-orbital first, the value of $(3\cos^2\theta - 1)/r^3$, averaged over a spherical distribution, is equal to zero. For a p_z-orbital, however, the average value of Equation (3.15) is:

$$\frac{d^2 V}{dz^2} = -\tfrac{4}{5} e \left\langle \frac{1}{r^3} \right\rangle \qquad (3.16)$$

where the brackets indicate 'average value'. Similarly for a p_x or p_y orbital:

$$\frac{d^2 V}{dz^2} (p_x) = \frac{d^2 V}{dz^2} (p_y) = \tfrac{2}{5} e \left\langle \frac{1}{r^3} \right\rangle \qquad (3.17)$$

For a d_{z^2} orbital:

$$\frac{d^2 V}{dz^2} (dz^2) = -\tfrac{4}{7} e \left\langle \frac{1}{r^3} \right\rangle \qquad (3.18)$$

The value of $\langle 1/r^3 \rangle$ for a d-orbital is smaller by a factor of eight (using hydrogen-like wave-functions) than that of a p-orbital of the same principal quantum number. Therefore a 3d electron, for example, will exert considerably less field gradient at the nucleus than a 3p electron. Also, the inverse cubed dependence on r ensures that the field gradient due to an electron in a $4p_z$ orbital will be less than that from an electron in a $3p_z$ orbital. It follows from Equations (3.16) and (3.17) that:

$$\frac{d^2 V}{dz^2} (p_z) = -\frac{2d^2 V}{dz^2} (p_x) = -\frac{2d^2 V}{dz^2} (p_y) \qquad (3.19)$$

This equation is used extensively in the Townes–Dailey theory for an *approximate* treatment of nuclear quadrupole coupling constants in molecules [3.10], which we now examine.

Townes–Dailey method

Now consider that nucleus A has become part of a diatomic molecule A–B. Suppose that atom A is a halogen, e.g. chlorine. Contributions to the field gradient at the nucleus of A will arise from:

(1) The two electrons in the covalent bond, A–B.
(2) The $3p_x$ and $3p_y$ electrons on A.
(3) The electrons which previously occupied the 3s orbital in the free atom A, and now occupy an sp-hybridized orbital. (see Equations 3.20 and 3.21).
(4) The inner-shell electrons on atom A, $1s^2 2s^2 2p^6$.
(5) The electrons of atom B, with the exception of the one involved in the

covalent A–B bond.

(6) The nuclear charge of B (field gradient of opposite sign).

Townes and Dailey [3.10] showed, from calculations on the molecule ICl, that the positive nuclear contribution from B approximately equals the negative contribution from the electrons of atom B, including the one involved in the A–B bond. This is reasonable provided that the bond between A and B is fairly covalent. This implies that the contribution from electron density in the orbital overlap region of the σ bond is neglected. In ICl it was shown that the overlap term was responsible for less than 5 per cent of that due to a $3p_z$ electron. The main source of the electric field gradient at nucleus A is generated by the electron from A involved in the A–B bond, plus the $3p_x$ and $3p_y$ electrons on A, with due allowance for any sp-hybridization of the bonding atomic orbital, including contribution (3) above.

The contributions to the field gradient, eq (or $d^2 V/dz^2$), can now be calculated. For a singly coordinated halogen the z axis is collinear with the bond direction. If the bonding orbital of A, ψ_b, is sp-hybridized:

$$\psi_b = \sqrt{s}\, \psi_{3s} + (1 - s)^{1/2} \psi_{3p_z} \qquad (3.20)$$

where s is the degree of sp-hybridization. The other hybridized orbital will be:

$$\psi_c = (1 - s)^{1/2} \psi_{3s} - \sqrt{s}\psi_{3p_z} \qquad (3.21)$$

Let the populations of $\psi_b = b$, $\psi_c = c$, $3p_x$ and $3p_y = 2$. Then the total field gradient in the z direction is, using Equation (3.19):

$$eq = [b(1 - s) + cs - 2] eq_p \qquad (3.22)$$

where $eq_p = (d^2 V/dz^2)(3p_z)$, the field gradient due to a $3p_z$ electron. For the chlorine atom case, $b = 1$, $s = 0$, and $c = 2$, so $eq_{atom} = eq_p$. Therefore, for the general case the molecular field gradient can be expressed in terms of that of the free atom:

$$\frac{eq_{mol}}{eq_{atom}} = \frac{(eq^2 Q/h)_{mol}}{(eq^2 Qq/h)_{atom}} = (1 - s)(2 - b) \qquad (3.23)$$

This follows from Equation (3.22) with $c = 2$. This is the Townes–Dailey equation generally applied to halogen n.q.r. when the p_x and p_y orbitals are not involved in any π bonding. Since this expression includes a ratio of the field gradients it is implicit that the contributions from the inner electronic orbitals will be common to both terms and will therefore cancel. The field gradient from the inner orbitals is not zero since they are susceptible to polarization; the lower levels by the quadrupolar moment of the nucleus itself and the valence shell electrons have a similar effect on the upper levels of the inner orbitals.

Equation (3.23) is sometimes expressed in terms of the ionic character of the bond, i:

$$\frac{(e^2 Qq/h \ _{\text{mol}}}{(e^2 Qq/h)_{\text{atom}}} = (1-s)(1-i) \tag{3.24}$$

since the net ψ_b population, $b = 1 + i$. For chlorine the amount of sp-hybridization, s, was previously thought to be significant (up to 15 per cent) but the more modern view is that s is normally quite small [3.11], and the $(1-s)$ term is often neglected in this approximate treatment.

When both $3p_x$ and $3p_y$ orbitals are involved in bonding so that their populations fall to $(2-\pi)$ electrons, Equation (3.24) has the form:

$$\frac{(e^2 Qq/h)_{\text{mol}}}{(e^2 Qq/h)_{\text{atom}}} = (1-s)(1-i) - \pi \tag{3.25}$$

where π is the amount of π character of the bond. This is alternatively written:

$$\frac{(e^2 Qq/h)_{\text{mol}}}{(e^2 Qq/h)_{\text{atom}}} = -\left(\frac{N_x + N_y}{2} - N_z\right) \tag{3.26}$$

where N_x, N_y are the populations of the p_x and p_y orbitals, previously $(2-\pi)$, and N_z is the population of the p_z orbital, equal to $(1 + i + s)$. The small term (is) is neglected. The quantity in parentheses on the right in Equation (3.26) is sometimes called the 'p-electron defect' since N_z is usually less than N_x or N_y (i.e. when halogen is bonded to a more electropositive atom). When the halogen is bonded to a more electronegative atom, the term is 'p-electron excess' and Equation (3.25) becomes:

$$\frac{(e^2 Qq/h)_{\text{mol}}}{(e^2 Qq/h)_{\text{atom}}} = (1-s)(1+i) - \pi \tag{3.27}$$

If the molecule is planar then often only one of the p_π orbitals can be involved in multiple bonding (e.g. chlorine $3p_x$ orbitals in chlorobenzenes). In this instance the value of the asymmetry parameter η will be closely related to the degree of double-bond character. The expression is:

$$\pi = \frac{2}{3}\eta \frac{(e^2 Qq/h)_{\text{mol}}}{(e^2 Qq/h)_{\text{atom}}} \tag{3.28}$$

It is possible for this equation to overestimate π since some contribution to the asymmetry parameter may arise from intermolecular sources.

In the above equations the value of the ratio of molecular coupling constant to that of the free atom is calculated from the observed transition frequency. For ^{35}Cl, for example, $(e^2 Qq/h)_{\text{atom}}$ is equal to 109.746 MHz, and assuming small η values, the n.q.r. frequency, $\nu(\text{Cl})$ is just half of the molecular quadrupole coupling constant [Equation (3.10)]:

$$\frac{\nu(\text{Cl})}{54.873} = \frac{(e^2 Qq/h)_{\text{mol}}}{(e^2 Qq/h)_{\text{atom}}}$$

Now some examples will be considered. The ^{35}Cl n.q.r. frequency in the chlorine molecule is 54.5 MHz at 20 K. Therefore, if s is zero, Equation (3.24) indicates that bonding in Cl_2 is practically 100 per cent covalent, as might be expected. On the other hand, the chloride ion has all its 3p orbitals fully occupied so its n.q.r. frequency should be close to zero, since $i = 1$ (100 per cent ionic). There is in fact an approximately linear relationship between the ionic character as derived from Equation (3.24) (assuming $s = 0$) and the difference in electronegativity of the two bonded atoms, In the case of ^{35}Cl, the linearity falls away for bonds which are greater than 85 per cent ionic.

Substituent effects

Apart from the electronegativity differences, the molecular n.q.r. frequency will be influenced by the inductive or conjugative effects of substituents in the molecule. Figure 3.10 shows the ^{35}Cl n.q.r. transition frequencies for two series of molecules, $CH_{4-n}Cl_n$ and $(CH_3)_{n-1}CH_{4-n}Cl$. In the first series from $CHCl_3$ to CCl_4, the frequency increases as the number of chlorine atoms (n) increases.

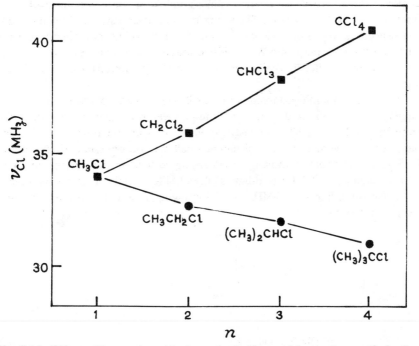

Fig. 3.10 Effect of increasing chlorine substitution and increasing methyl substitution on the ^{35}Cl frequency in two series, $CH_{4-n}Cl_n$ and $(CH_3)_{n-1}CH_{4-n}Cl$.

This can be rationalized with the electron-withdrawing power of chlorine which should cause the covalency of the C–Cl bond to be increased. Equation (3.24) shows that the n.q.r. frequency should rise accordingly. The opposite effect is apparent in the second series. Here the electron-releasing effect of the increasing number of methyl groups $(n-1)$ pushes more charge on to the chlorine atom, which results in a more ionic C–Cl bond; i.e. according to Equation (3.24) the frequency drops. Both effects appear to have a linear relationship with the n.q.r. frequency, and a good correlation has been found (for carbon–halogen compounds) with the Taft parameter, σ_i^*, a measure of the inductive effect.† For substituted methyl chlorides, $R_1 R_2 R_3 CCl$, the relationship is found to be experimentally [3.12]:

$$\nu(^{35}Cl \text{ at } 77\,K) = 32.05(1 + 0.0317 \sum_i \sigma_i^*) \pm 0.35\,MHz \qquad (3.29)$$

Not all substituent effects are inductive, however, When the substituents R have conjugative as well as inductive effects a more general relationship has been proposed for tetrahedral molecules of group IV, $R_1 R_2 R_3 MCl$ [3.13].

$$\frac{\nu - \nu_0}{\nu_0} = \alpha \sum \alpha_i + \beta \sum \alpha_c \qquad (3.30)$$

ν_0, α, and β are constants for a particular group IV atom; α_i and α_c are the inductive and conjugative parameters as measured from ^{19}F n.m.r. shifts of fluorine-substituted benzenes. The values of the transmission coefficients α and β reflect the relative ease with which inductive and conjugative effects are transmitted to the chlorine atom. Where M has available d-orbitals, for example, the n.q.r. frequency is in some cases reduced by participation of the chlorine $3p_\pi$ electrons.

The effects of conjugation are more easily seen when a halogen is bonded directly to an unsaturated atom. The lone-pair electrons then participate in the bonding and Equation (3.25) must be used to interpret the data. π-bonding has the effect of lowering the n.q.r. frequency below that expected entirely on the basis of the electron withdrawing (or releasing) effects of the substituents. The ^{35}Cl frequency for 2-chloropyridine is at 34.19 MHz, which is lower than that for 3-chloropyridine (35.23 MHz) probably owing to a contribution from the resonance structure

† σ_i^* (R) $= \log \dfrac{[K_a(RCH_2 COOH)]}{[K_a(CH_3 CH_2 COOH]}$; K_a is the acid dissociation constant.

The value of the asymmetry parameter is often significant in compounds of this type. In phosgene, $COCl_2$, the asymmetry parameter has a value of 25 per cent, which suggests that the structure

is important in the ground state of the molecule. In the case of *p*-chloroaniline the value of η is less than that of *p*-dichlorobenzene owing to a contribution from the resonance form

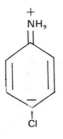

Inorganic compounds

In the field of inorganic chemistry, n.q.r. studies have in some cases been directed at the question of the involvement of d-orbitals in bonding. Such an example is that of the Group IV tetrahalides. Table 3.2. gives the observed ^{35}Cl n.q.r. frequencies at 77 K., and the electronegativity difference $(\chi_{Cl} - \chi_M)$. In relation to the difference in electronegativity between M and Cl the observed frequencies for $SiCl_4$ appear to be anomalously low. The reason suggested for this is that 3d silicon orbitals are available for bond formation with the lone-pair electrons of chlorine. This π bonding will result in the $3p_\pi$ populations [N_x and N_y in Equation (3.26)] falling below 2, which leads to a decrease in the resonance frequency. The hypothesis of π bonding in $SiCl_4$ is consistent with the short Si–Cl bond length observed in the compound. Four separate ^{35}Cl lines are observed for $SiCl_4$, $GeCl_4$, and $SnCl_4$, which means that, in certain types of crystal, the tetrahedra would be distorted. A crystal structure study of $SnBr_4$ (which also exhibits four ^{81}Br n.q.r. lines) in fact does show some distortion.

For the trihalide ions, X_3^- or X_2Y^-, the charge distribution of each atom in the ion can be calculated using Equation (3.23). These ions are all practically linear with the central atom usually unsymmetrically placed. Two possible bonding shcemes can be considered; a four-electron three-centre bond, or a d-orbital for the central atom may be involved, to yield two dp hybrid orbitals. In the latter case there should be a deficiency in the number of p-electrons as calculated using Equation (3.23). Table 3.3 shows the n.q.r. data and the calculated p_σ populations.

197

Table 3.2 ^{35}Cl n.q.r. frequencies and electronegativity differences for group IV tetrachlorides

	^{35}Cl n.q.r. frequency[†] (V/MHz at 77 K)	$\chi_{Cl} - \chi_{M}$ (Pauling)
CCl$_4$	40.629[‡]	0.5
SiCl$_4$	20.273	1.2
	20.408	
	20.415	
	20.464	
GeCl$_4$	25.450	1.2
	25.715	
	25.735	
	25.745	
SnCl$_4$	23.720	1.2
	24.140	
	24.226	
	24.296	

[†] Data from ref. [3.20].
[‡] Average of 15 lines at 77 K

Table 3.3 Nuclear quadrupole coupling constants and the calculated bonding orbital populations for I$_3^-$ and ICl$_2^-$ [3.9]

Molecule	$h^{-1}e^2Qq(^{127}I)/$ MHz	$h^-e^2Qq(^{35}I)/$ MHz	p_{σ} populations	Total
NH$_4$I$_3$	466.8		1.80	4.00
	1725.0		1.25	
	2458.7		0.95	
RbI$_3$	744.2		1.66	3.97
	1449.8		1.37	
	2465.7		0.94	
CsI$_3$	819.0		1.64	3.95
	1436.6		1.37	
	2477.5		0.97	
KICl$_2$		36.81, 37.53	1.65 (×2)	3.95
		38.00, 39.25		
	3081		0.65	
	3129			
CsICl$_2$		39.72	1.64 (×2)	3.93
	3099		0.65	

The total p_{σ} populations in each case are approximately equal to four, which is evidence for the three-centre four-electron bond theory, and d-orbital involvement is not expected to any great extent.

In inorganic chloro-complexes the magnitude of the chlorine resonance frequency is very sensitive to the *cis*- and *trans*-influence of ligands, and this has been used to compare the relative abilities of ligands to donate σ electrons. Complexes of the form *trans*-$L_2 PdCl_2$ show a gradual decrease in ^{35}Cl resonance frequency as the ligand L increases in its σ donor properties [3.14]. As L donates more charge to the metal, some of this is transferred to Cl which causes the resonance frequency to be reduced. The predicted order of ligand σ donor ability from the n.q.r. measurements is:

$$piperidine > pyridine > R_3 As > R_3 P > EtCN > PhCN$$

This order correlates well with the chlorine reactivity in the corresponding *trans*-$L_2 PtCl_2$ complexes where the rate of replacement of chlorine by weak nucleophiles increases in a similar order.

Interpretation of ^{14}N quadrupole coupling constants

N.q.r. data from quadrupolar nuclei situated in multi-coordinated atoms may also be interpreted by the semi-empirical (Townes–Dailey type) treatment. This is illustrated using the ^{14}N and ^{15}Cl n.q.r. data from *N*-chlorodimethylamine, $(CH_3)_2 NCl$. The presence of two n.q.r. nuclei in the molecule provides an opportunity for testing the self-consistency of the Townes–Dailey method. Schempp [3.15] observed three ^{14}N resonance frequencies ν_+ at 5.997 MHz, ν_- at 3,501 MHz, and ν_0 at 2.496 MHz (77 K). Equation (3.11) leads to a value for $e^2 Qq/h$ equal to 6.333 MHz and η has a very high value of 78.83 per cent. The latter follow from a large difference in ionic characters of the N–Cl and N–C bonds. The bond angles are approximately tetrahedral and the nitrogen bonding orbitals are assumed to be well represented by sp^3 hybrids. If the z axis is chosen as the lone-pair direction, the orbitals can be written:

$$\psi_{N-Cl} = \tfrac{1}{2}s - (2/3)^{\frac{1}{2}}p_y + (1/12)^{\frac{1}{2}}p_z$$
$$\psi_{N-C_1} = \tfrac{1}{2}s + (1/6)^{\frac{1}{2}}p_y + (1/2)^{\frac{1}{2}}p_z + (1/12)^{\frac{1}{2}}p_z$$
$$\psi_{N-C_2} = \tfrac{1}{2}s + (1/6)^{\frac{1}{2}}p_y - (1/2)^{\frac{1}{2}}p_x + (1/12)^{\frac{1}{2}}p_z$$
$$\psi_{lone\ pair} = \tfrac{1}{2}s - (3/4)^{\frac{1}{2}}p_z \tag{3.31}$$

The y axis is taken as the molecular plane of symmetry. The field gradient in the z direction can then be assessed, using q_x, q_y, q_z to represent field gradients due to an electron in a $2p_x, 2p_y,$ or $2p_z$ orbital respectively. The populations of $\psi_{N-Cl} = b_1, \psi_{N-C} = b_2; \psi_{lone\ pair}$ is assumed to be 2. The total z direction field gradient, q_{mol}, is:

$$(\tfrac{2}{3}q_y + \tfrac{1}{12}q_z)b_1 + (\tfrac{1}{6}q_y + \tfrac{1}{2}q_x + \tfrac{1}{12}q_z + \tfrac{1}{6}q_y + \tfrac{1}{2}q_x + \tfrac{1}{12}q_z)b_2 + \tfrac{3}{2}q_z$$

Using the expression $q_x = q_y = -\tfrac{1}{2}q_z$ [see Equation (3.19)]:

199

$$-\frac{q_{mol}}{q_z} = \frac{(e^2 Qq/h)_{mol}}{(e^2 Qq/h)_{atom}} = (\tfrac{3}{2} - \tfrac{1}{4}b_1 - \tfrac{1}{2}b_2) \qquad (3.32)$$

Similar calculations for the field gradient in the x and y directions and Equation (3.5) yield:

$$\frac{(e^2 Qq/h)_{mol}}{(e^2 Qq/h)_{atom}} \; \eta = b_2 - b_1 \qquad (3.33)$$

Substitution of the experimental data then gives the N–Cl σ bond population b_1 equal to 0.80e and the N–C σ bond population b_2 equal to 1.31e using a value of $(e^2 Qq/h)_{atom}$ of 9.8 MHz. The value for the latter quantity is not known exactly for ^{14}N, but estimates range between 8 and 10 MHz. Theoretically a value of 9.65 MHz has been predicted [3.16].

The observed chlorine n.q.r. frequency of 43.70 MHz gives a value of 1.20e for the population of the chlorine $3p_\sigma$ orbital using Equation (3.23). Taken together, a net transfer of 0.2 electrons from the nitrogen to the chlorine atom is indicated. The fact that the nitrogen and chlorine σ bond populations sum to 2.0e agrees very well with the usual two-electrons chemical bond theory.

Lucken has given a systematic review of the general application of the Townes–Dailey method [3.9].

Cotton–Harris M.O. method

Cotton and Harris [3.17] have put forward an approximate method of calculating $e^2 Qq/h$ which takes into account orbital overlap and does not attempt to separate the effects of hybridization and ionic character. A set of N orthogonal molecular orbitals, Φ_p is assumed, each of which is found from a linear combination of n orthogonal atomic orbitals, φ_1:

$$\Phi_p = c_{1p}\varphi_1 + c_{2p}\varphi_2 + \ldots + c_{np}\varphi_n = \sum_i c_{ip}\varphi_i \qquad (3.34)$$

i indicates an atomic orbital centred on the quadrupolar nucleus, while j is an orbital centred on another nucleus. The expression for calculating chlorine quadrupole coupling constants, corrected by Kaplansky and Whitehead [3.18] is:

$$(e^2 Qq/h)_{mol} = 109.7 \sum_i \sum_p (2 - [N_p(c_{ip}^2 + \sum_{j>i} c_{ip}c_{jp} S_{ij})]) \qquad (3.35)$$

where S_{ij} is the overlap integral $\int \varphi_i{}^* \varphi_j d\tau$. Since the input parameters for this expression are typically those obtained from M.O. calculations, this method is used in the opposite sense to that of Townes and Dailey, namely to predict from eigenvalues and overlap integrals a value for $e^2 Qq/h$, which can then be compared with an experimental value. Cotton and Harris have used the expression (3.35) to calculate ^{35}Cl coupling constants for $K_2 PtCl_4$, and $K_2 MCl_6$ (where M = Re,

Os, Ir, Pt), and the values agree with the experimental data within 10 per cent
[3.19].

3.4.2 Molecular structure

In many instances the n.q.r. spectrum itself can provide direct structural information. Halogen quadrupole resonance frequencies respond predominantly to
the ionic character of the M—Cl bond, and on this basis frequency shift tables
can be drawn up. Tables for M—Cl and C—Cl group frequencies which have been

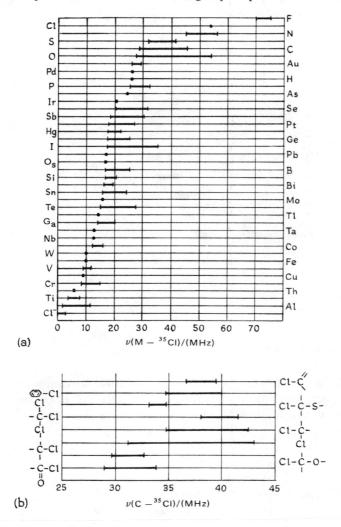

Fig. 3.11 ^{35}Cl n.q.r. frequency shift tables for (a) M − Cl compounds and (b)
C − Cl compounds (after Semin [3.20]).

published by Semin [3.20] are reproduced in Fig. 3.11. In both tables there is a considerable spread of frequencies in a particular band, which is not surprising in view of the sensitivity of n.q.r. frequencies to substituent effects (Section 3.4.1) and lattice effects (Section 3.5). As an example of the use of the group frequency table Fig. 3.11(b), consider the n.q.r. spectrum of trichloroacetyl chloride, CCl_3COCl. Four ^{35}Cl resonances are observed at 33.72, 40.13, 40.47, and 40.6 MHz [3.21]. Following Fig. 3.11(b) the lower frequency line is ascribed to the COCl group and the upper three lines to the CCl_3 group. This type of analysis can be used for many halogen-containing organic compounds. Fitzsky has published a similar table for ^{14}N n.q.r. frequencies [3.22] (Fig. 3.12).

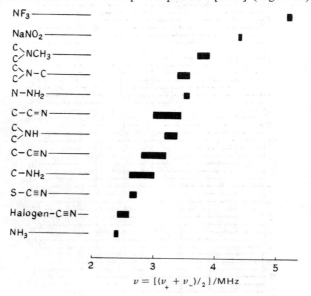

Fig. 3.12 ^{14}N n.q.r. frequency shift table (after Fitzsky [3.22]. Courtesy *Angewandte Chemie*).

In ionic compounds the n.q.r. frequencies of halide-containing ions are often a useful 'fingerprint' for their identification in compounds of unknown structure. Such ions as BCl_4^-, $AlCl_4^-$, PCl_4^+, PCl_6^-, and $SbCl_6^-$ have been identified in this way [3.23].

Atoms at different chemical sites in a molecule will also exhibit a difference in their n.q.r. frequency. For example, the n.q.r. spectrum of tetrachloro(phenyl) arsenic(V), $PhAsCl_4$, shows that four ^{35}Cl resonances in the form of two doublets, separated by almost 10 MHz (see Fig. 3.13). This spectrum is consistent with a trigonal bipyramidal structure, in which the phenyl group is equatorially located. As expected chemically the two axial chlorines are considerably more ionic than those in equatorial positions and therefore their resonances occur at a lower frequency.

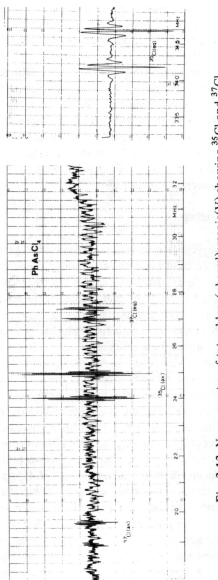

Fig. 3.13 N.q.r. spectra of tetrachloro(phenyl)arsenic(V) showing ^{35}Cl and ^{37}Cl resonances from axially and equatorially located chlorine atoms.

The two axial chlorine atoms are not crystallographically equivalent and consequently have slightly different resonance frequencies. The same is true for the equatorial chlorine atoms. ^{37}Cl resonances are also shown in Fig. 3.13. These occur at a fixed ratio of the ^{35}Cl resonance frequency, equal to the ratio of their quadrupole moments $[eQ(^{35}Cl)/eQ(^{37}Cl) = 1.2688]$. ^{37}Cl resonances are always less intense than those of ^{35}Cl because of the lower natural abundance, which is approximately 1/4 that of ^{35}Cl.

Bridged and terminal halogen atoms can also be distinguished by n.q.r. The resonance frequencies for a bridged halogen are usually significantly less than those for the terminal atoms. Some examples of n.q.r. frequencies from bridged compounds (dimers) are given in Table 3.4 (data from ref. [3.20]). The asymmetry parameter of the bridged halogen is usually non-zero and is related to the bridge inter-bond angle. This has been calculated for the Group III triiodides and the results are close to $90°$ showing that chlorine s-orbital participation in the bonding orbital is small [3.9].

3.4.3 Hydrogen bonding

The quadrupole coupling constants of nuclei in many hydrogen-containing molecules is often observed to be greater in the vapour state (as obtained by microwave measurements) than in the solid (n.q.r.). For example, the ^{35}Cl quadrupole coupling constant for HCl is 21 per cent smaller in the solid state than in the vapour; the ^{14}N q.c.c. in NH_3 is also smaller by 22.5 per cent in the solid state. A reduction in quadrupole coupling constant on going from the gas to the solid state is often observed, but these shifts are very much larger than the average and appear to be due to a reduction in the field gradient at the nucleus caused by strong intermolecular hydrogen bonding in the solid state.

Hydrogen bonds are also detectable by solid-state measurements alone. For instance, the ^{35}Cl n.q.r. spectrum of sodium tetrachloroaurate dihydrate observed by Fryer and Smith [3.14] shows four resonances, one of which is about 2.5 MHz lower at 77 K (see Fig. 3.14). The lower frequency signal is attributed to a chlorine atom which is involved in two hydrogen bonds to adjacent water molecules. This resonance shows a shift in frequency upon deuteriation of the sample whereas the upper resonance frequencies are unaffected. The anomalous temperature dependence of the lower frequency resonance is caused by the gradual breakdown of the hydrogen bonding because of the larger thermal motion of the H_2O molecule as the temperature is increased. The n.q.r. deuteriation shift is negative because the substitution of D for H is expected to reduce the motional amplitude of the water molecule, and increase the hydrogen bonding interaction.

The ^{35}Cl n.q.r. spectrum of 4-chloropyridinium hexachlorostannate(*IV*) also shows an anomalously low frequency line due to the $SnCl_6^{2-}$ ion. Brill and Welsh observed four lines at 14.97, 17.32, 17.52, and 35.92 MHz (298 K) [3.24]. (The latter high frequency line is due to C—^{35}Cl from the cation). An

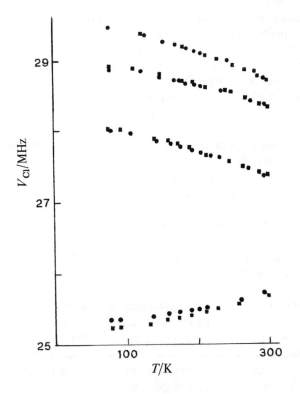

Fig. 3.14 Temperature dependence of ^{35}Cl resonance frequencies in $NaAuCl_4 \cdot 2H_2O$ (circles) and $NaAuCl_4 \cdot 2D_2O$ (squares). (After Fryer and Smith [3.14]. Courtesy of the Chemical Society).

X-ray study in fact showed a tetragonally distorted anion, the two long Sn–Cl bonds of which were involved in N–H. . .Cl hydrogen bonding with the cation [3.25].

Table 3.4 N.q.r. frequencies observed for molecules containing halogen bridges

Molecule	Nucleus	Terminal halogen frequency (MHz)	Bridging halogen frequency (MHz)	Temp(K)
$GaCl_3$	^{35}Cl	19.084, 20.225 (η = 8, 9, 3.4%)	14.667 (η = 47.3%)	306
$AuCl_3$	^{35}Cl	36.116, 33.340	23.285	77
ICl_3	^{35}Cl	33.413, 34.918	13.740	297
$AlBr_3$	^{79}Br	113.790, 115.450	97.945	77

The simplest hydrogen bond which is open to investigation by n.q.r. is that in the hydrogen dichloride ion, HCl_2^-. Both asymmetrical (Cl–H. . .Cl, Type I) and symmetrical (Cl–H–Cl, Type II) formsof the ion exist and their different structures are reflected in a large shift (8 MHz) in their ^{35}Cl resonance frequencies. Type I ions were found to have ^{35}Cl frequencies of about 20 MHz (assigned to the more covalently bonded chlorine atom) while Type II ions exhibited a lower frequency near 12 MHz [3.26, 3.27] . Type I ions showed a positive deuteriation shift of about 0.7 MHz, which is of opposite sign to that observed in $NaAuCl_4$.2 H_2O. This is largely because the hydrogen bond is no longer intermolecular but is the major source of field gradient in the ion. Type II ions showed hardly any deuteriation shift but this is to be expected if the proton or deuteron is situated in a symmetrical potential well.

3.5 SOLID STATE APPLICATIONS

3.5.1 Line splittings due to crystallo-graphic inequivalence

In the solid state a molecule will be subjected to slight perturbations due to its environment by means of many intermolecular forces, and this often has the effect of causing a distortion from ideal symmetry. Thus a molecule like $SiCl_4$ or $GeCl_4$, which has tetrahedral symmetry in the vapour phase, loses this in the solid state and the four chlorine atoms become crystallographically inequivalent. This results in a four-line n.q.r. spectrum (Table 3.2). Here the solid-state splitting is of the order of 200 kHz but values up to 500 kHz are typically found. Atoms which are rather different chemically usually give rise to splittings greater than this. The number of lines which are observed for each transition in an n.q.r. spectrum is equal to the number of crystallographically inequivalent nuclear sites in the asymmetric unit of the unit cell. This means that a full n.q.r. spectrum can decide the point symmetry of the molecule if the space group and the number of molecules in the unit cell are known. The latter can be found from the unit cell dimensions and the density.

3.5.2 Phase transitions

The ability of n.q.r. to detect crystallographic inequivalences can be used as a means for studying phase transitions in the solid state. A phase transition is indicated when there is a change in the number of resonance lines or when there is an abrupt change in the curve of n.q.r. frequency versus temperature. These effects occur because of the sensitivity of the frequency to intermolecular (as well as intramolecular) forces and to low frequency modes, so any change in these will be detected. The phase transitions in many molecules have been studied by this method, including those of the general form R_2MX_6 where M = Re, Se, Te [3.28] .

3.5.3 Zeeman studies

The resonance line splitting caused by the application of a magnetic field to a single-crystal specimen enables η to be determined and the orientation of the principal axes of the e.f.g. tensor to be found with respect to those of the crystal. If the molecular symmetry enables these axes to be related to the molecular axes, the orientation of the molecules in the crystal lattice can be found. However, large single crystals are not always available so the technique is not easy to apply. The splitting of the zero-field lines is dependent upon the orientation of the crystal with respect to the magnetic field direction. By studying the nature of this dependence the orientation of the field gradient axes can be found. In the case of *p*-chlorophenol a single-crystal ^{35}Cl n.q.r. study was able to give almost a full crystal structure determination [3.8].

When a powder sample is used the crystallites are oriented in a random fashion, so the application of a magnetic field causes the lines to be smeared out. However, if the zero-field resonances are of sufficient intensity, the application of a weak magnetic field allow some residual splitting to be observed. The magnitude of this can be related to the value of η if the field strength is known. This technique is useful for spin 3/2 nuclei (e.g. ^{35}Cl) where η is not obtainable directly from the n.q.r. spectrum.

3.5.4 Molecular motion

The normally observed reduction in n.q.r. frequency with increase in temperature is caused in the majority of cases by the increasing amplitudes of molecular thermal motions which have a partial averaging effect on the electric field gradient. Angular motions or librations are often responsible for the temperature dependence, and the frequency of these can be calculated using an expression by Bayer [3.8]. Some correlation with observed vibration frequencies is possible using infrared, Raman, and neutron inelastic scattering spectroscopy. The temperature dependence of the spin–lattice relaxation time, T_1, is also a powerful technique for the evaluation of energy barriers associated with solid state molecular motion.

Nuclear quadrupole resonance spectroscopy is the subject of books by Das and Hahn [3.8], Lucken [3.9], Semin, Babushkina, and Jacobson [3.20] (in Russian), and a periodical edited by Smith [3.29]. There are also review articles by Kubo and Nakamura [3.28], Van Bronswyk [3.30], Jeffrey and Sakurai [3.31], and Smith [3.32]. Recent tables of n.q.r. frequencies have been published by Semin [3.20], and Biryukov [3.33].

REFERENCES

3.1 Dehmelt, H.G. and Kruger, H., *Naturwissenschaften*, **37**, 111 (1950).

3.2   Pound, R.V. and Knight, W.D., *Rev. Sci. Instrum.,* 21, 219 (1950).

3.3   Robinson, F.N.H., *J. Sci. Instrum.,* 36, 481 (1959).

3.4   Slusher, R.E. and Hahn, E.L. *Phys. Rev.,* 166, 332 (1968).

3.5   Edmonds, D.T. and Speight, P.A., *Phys. Letters,* A34, 325 (1971).

3.6   Hsieh, Y., Koo, J.C., and Hahn, E.L., *Chem. Phys. Lett.* 13, 563 (1972).

3.7   Herzog, B. and Hahn, E.L. *Phys. Rev.* 103, 148 (1956).

3.8   Das, T.P. and Hahn, E.L., *Nuclear Quadrupole Resonance Spectroscopy,* Academic Press, New York and London (1958).

3.9   Lucken, E.A.C. *Nuclear Quadrupole Coupling Constants,* Academic Press, London and New York (1969).

3.10   Townes, C.H. and Dailey, B.P., *J. Chem. Phys.,* 17, 782 (1949).

3.11   Schoemaker, D., *Phys. Rev.,* 149, 693 (1961).

3.12   Biryukov, I.P. and Voronkov, M.G. *Coll. Czech. Chem. Comm.,* 32, 830 (1967).

3.13   Semin, G.K. and Bryuchova, E.V., *Chem. Comm.,* 605, (1968).

3.14   Fryer, C.W. and Smith, J.A.S., *J. Chem. Soc. (A),* 1029 (1970.

3.15   Schempp, E., *Chem. Phys. Lett.,* 8, 562 (1971).

3.16   Bonaccorsi, R., Scrocco, E., and Tomasi, J. *J. Chem. Phys.,* 50, 2940 (1969).

3.17   Cotton, F.A. and Harris, C.B. *Proc. Nat. Acad. Sci. (US),* 56, 12 (1966).

3.18   Kaplansky, M., and Whitehead, M.A., *Trans. Faraday Soc.,* 65, 641 (1969).

3.19   Cotton, F.A. and Harris, C.B. *Inorg. Chem.* 6, 369, 376 (1967).

3.20   Semin, G.K., Babushkina, T.A. and Jacobson, G.G., *Applications of Nuclear Quadrupole Resonance in Chemistry,* Izdavatelstvo "Khimiya" Leningrad (1972).

3.21   Bray, P.J., *J. Chem. Phys.,* 23, 703 (1955).

3.22   Fitzsky, H.G., Wendisch, D. and Holm, R. *Angew. Chem.,* 11, 979 (1972).

3.23   Lynch, R.J. and Waddington, T.C., *Advances in Nuclear Quadrupole Resonance,* 1, 37, Heyden, London (1974).

3.24   Brill, T.B., and Welsh, W.A. *J. Chem. Soc. Dalton,* 357 (1973).

3.25   Gearhart, R.C., Brill, T.B., Welsh, W.A. and Wood, R.H., *J. Chem. Soc. Dalton,* 359 (1973).

3.26   Evans, J.C. and Lo G.Y–S., *J. Phys. Chem.,* 71, 3697 (1967).

3.27   Ludman, C.J., Waddington, T.C., Salthouse, J.A., Lynch, R.J., and Smith J.A.S., *Chem. Comm.* 6, 405 (1970).

3.28   Kubo, M. and Nakamura, D., *Adv. Inorg. Chem. Radiochem.,* 8, 257 (1966).

3.29   Smith, J.A.S. (Editor) *Advances in N.Q.R.,* Heyden, London (1974).

3.30   Van Bronswyk, W. *Structure and Bonding,* 7, 87 (1970).

3.31   Jeffrey, G.A. and Sakurai, T. *Progress in Solid State Chemistry,* 1, 380, Pergamon, London (1964).

3.32   Smith, J.A.S., *J. Chem. Education,* 48, 39, A77, A147, A243 (1971).

3.33   Biryukov, I.P., Voronkov, M.G., and Safin, I.A. *Tables of NQR Frequencies,* Jerusalem: Israel Program for Scientific Translation (1969).

4 Electron spin resonance (e.s.r.) spectroscopy

4.1 INTRODUCTION

Electron spin resonance (e.s.r.) is a spectroscopic technique confined to the study of those species having one or more unpaired electrons. The method takes advantage of the spin of the electron and its magnetic moment to reveal a wealth of information. It has been possible in many cases, for example, to obtain detailed information on the unpaired electron (spin) density distribution. Obviously the necessity for at least one unpaired electron limits the number of applications of the technique; for this reason it is unlikely that e.s.r. spectroscopy will achieve the analytical importance to the organic chemist of n.m.r. spectroscopy. Despite this restriction, however, the technique has been applied to the study of a large range of paramagnetic systems, among the most important of which have been free radicals, triplet states, and transition-metal ions.

4.1.1 The e.s.r. experiment

The origin of e.s.r. spectroscopy lies in the spin of the electron and its associated magnetic moment, and the technique is perhaps best introduced by considering the alignment of magnetic dipoles in the presence of an applied magnetic field. It should be noted at this stage that magnetic dipoles associated with electrons may arise from both 'spin' and 'orbital' angular momenta (or a combination of the two). Orbital angular momentum has its origins in the 'motion' of an electron in an orbital, while spin angular momentum has its origins in the rotation of an electron about its own axis. In the majority of cases of immediate interest the magnetic dipole arises from spin angular momentum only, but cases where there is coupling between the two angular momenta will also be considered.

The energy E of a magnetic dipole of moment μ in the presence of magnetic induction B^* is given by

$$E = -\mu \cdot B \tag{4.1}$$

* Please see Page vi for an explanation of the use in this book of the symbol B

209

From the point of view of classical mechanics any orientation of the magnetic dipole with respect to the magnetic field would be possible. One orientation, that in which the magnetic dipole is aligned parallel to the applied field, corresponds to E_{min}, and one orientation, that in which the magnetic dipole is aligned antiparallel to the applied field, corresponds to E_{max}. If the energy of the interaction is large compared to kT, the alignment of the great majority of dipoles would correspond to E_{min}. However, as electrons obey quantum and not classical mechanics, only two orientations of the electron magnetic moment are allowed and, in addition, the energy of interaction in the e.s.r. experiment is considerably smaller than kT.

The spin and charge of the electron confer upon it a magnetic moment μ, the component of which along the direction of an applied magnetic induction is:

$$\mu_z = -g\mu_B M_s \tag{4.2}$$

where g is a dimensionless proportionality constant (often referred to as the g-factor or spectroscopic splitting factor), μ_B is the Bohr magneton ($=eh/4\pi m_e$, where e and m_e are the charge and mass of the electron respectively, and h is Planck's constant), and M_s is the electron spin quantum number. A combination of Equations (4.1) and (4.2) gives:

$$E = g\mu_B B M_s$$

The electron spin quantum number has two allowed values, $+\frac{1}{2}$ and $-\frac{1}{2}$, and consequently there are two energy states:

$$E_1 = +\frac{1}{2}g\mu_B B$$

$$E_2 = -\frac{1}{2}g\mu_B B$$

The difference in energy between these two states (see Fig. 4.1) is $g\beta B$ and the application of electromagnetic radiation of a suitable frequency ν will induce transitions between the two states when the energy of the radiation is equal to ΔE, i.e. when:

$$h\nu = g\mu_B B \tag{4.3}$$

It is the transition between these two energy states which is studied by e.s.r.

Before discussing this important relationship in more detail it is useful to consider the nature of electromagnetic radiation. Electromagnetic radiation can be regarded as coupled electric and magnetic fields, oscillating perpendicular both to the direction of propagation and to each other. It is the magnetic component of this electromagnetic radiation which interacts with the magnetic dipole associated with the unpaired electron. Consequently transitions between the two energy levels in Fig. 4.1 occur when the application of the electromagnetic radiation results in a re-orientation of the electron magnetic moment. If the electromagnetic radiation is applied such that the oscillating magnetic field is *parallel* to the applied magnetic induction field (B), an oscillation of the energy levels occurs but no re-orientation of the electron magnetic moment. It follows that, if the e.s.r. absorption

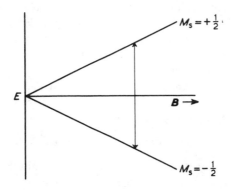

Fig. 4.1 Energy levels for a single unpaired electron as a function of magnetic field strength.

is to be observed, the oscillating magnetic field must be applied *perpendicular* to the applied magnetic field.

The successful observation of the e.s.r. spectrum requires suitable values for the frequency (ν) and the magnetic induction field (B) to satisfy the condition given in Equation (4.3). The choice of frequency is governed by the magnitude of the magnetic field, which is itself determined by instrumental and theoretical considerations. The proportionality constant g in Equation (4.3) has a value of 2.0023 for the free electron. Since magnetic fields in the range 0–1.2 T are used in many commercial spectrometers, a frequency in the microwave region is required. Magnetic fields of 0.35 T are often selected for commercial spectrometers with a corresponding frequency requirement of ~9.5 GHz. Such spectrometers are referred to as 'X-band' spectrometers. Some commercial spectrometers ('Q-band') operate at higher magnetic fields (1.25 T) with a correspondingly higher microwave frequency (35 GHz).

4.1.2 Thermal equilibrium and relaxation

It is instructive at this stage to consider why a magnetic induction of ~0.35 T is commonly selected. The condition for e.s.r. absorption would be equally well satisfied by a magnetic induction of (say) 1 mT and a frequency of 28 MHz. The smaller magnetic induction would certainly be more convenient from some points of view. In order to appreciate the reason for this choice it is necessary to consider the difference between the populations of the two energy states which exists at thermal equilibrium. If the number of spins occupying the upper and lower energy states are n_1 and n_2 respectively, then:

$$n_1/n_2 = e^{-\Delta E/kT}$$

where k is Boltzmann's constant and T is absolute temperature. Substitution of $g\mu_B B$ for ΔE in this expression reveals that the population difference between the

211

two states is extremely small for a sample having the free electron g-factor. Even with an applied magnetic field of 0.35 T the population difference is only about 0.0014 for thermal equilibrium at room temperature, and it follows that magnetic inductions in this order of magnitude are required if the sensitivity of the technique is to be reasonable.

A further complication appears to be present when an oscillating (magnetic) field is applied and transitions are stimulated between the two energy states. The rate of change in the population of the upper level is given by:

$$dn_1/dt = -p_1 n_1 + p_2 n_2$$

where p_1 is the transition probability of a transition from the upper to the lower level, and p_2 is the reverse transition probability. Under conditions of stimulated transitions $p = p_1 = p_2$ where p is the stimulated transition probability, and therefore:

$$dn_1/dt = P(n_2 - n_1) = P \cdot n$$

where n is the population difference $(n_2 - n_1)$. Now if a total of N spins are present:

$$N = n_1 + n_2$$

and therefore

$$2n_1 = N - n$$

and

$$dn_1/dt = -\tfrac{1}{2}\, dn/dt$$

By substitution therefore

$$dn/dt = -2Pn$$

and

$$n_t = n_0 \cdot e^{-2Pt}$$

where n_t is the population difference at time t, and n_0 is the initial population difference at $t = 0$. According to this relationship therefore the population difference between the two states decreases with time.

By analogy with the above arguments, the rate at which energy is absorbed by the sample is given by:

$$dE/dt = -n_1 P_1(\Delta E) + n_2 P_2(\Delta E)$$
$$= P \cdot \Delta E \cdot n$$

It follows that, if the population difference n is decreasing with time, the magnitude of the electromagnetic energy absorbed by the sample must also decrease with time, ultimately to zero.

212

Fortunately, the intensity of an e.s.r. spectrum is normally independent of time, and consequently there must be some process operating which maintains a population difference. That is, there must be some mechanism operating which allows spins occupying the upper level to lose energy and return to the lower level other than by means of a radiative transition. The non-radiative process occurring is the transfer of energy from the spin system to other degrees of freedom within the environment (the lattice) and is termed 'relaxation'.

It is usual to refer to the time scale for such processes as the spin–lattice relaxation time. A short relaxation time (i.e. rapid loss of energy to the lattice) allows the population difference to be maintained at relatively high temperatures. However, if the relaxation time is long the population difference may not be maintained, leading to a condition referred to as 'saturation'.

4.2   EXPERIMENTAL METHODS

The condition to be satisfied in order to observe the e.s.r. absorption is:

$$h\nu = g\mu_B B$$

There are, therefore, two quite distinct ways in which a spectrometer could be designed in order to observe the e.s.r. absorption. The frequency (ν) could be varied at a fixed d.c. magnetic induction (B); alternatively, the magnetic induction could be varied at a fixed frequency. However, in the microwave region of the spectrum there are a number of experimental difficulties involved in the former approach. Consequently, e.s.r. spectrometers operate with a fixed frequency (of ~9 GHz), and a magnetic induction which can be varied linearly over the range 0.35 ± (say) 0.30 T. Two further factors support the choice of 9 GHz as a suitable frequency. First, this frequency corresponds to a sample cavity with reasonably large dimensions (approx. 2 cm × 1 cm × 4 cm), allowing a convenient sample size. (Higher frequencies necessitate a smaller sample cavity with consequent restrictions on sample size.) Second, at higher frequencies larger magnetic inductions are required, with an increasing risk that the magnetic induction experienced by the sample may not be homogeneous to the required degree. A broadening of the absorption curve would result.

4.2.1 A simple e.s.r. spectrometer

It is now possible to summarize a few basic requirements necessary to construct an X-band e.s.r. spectrometer and to observe an e.s.r. spectrum:

(a) a suitable electromagnet capable of supplying a homogeneous magnetic induction field which can be varied linearly either side of 0.35 T;

(b) a source of microwave radiation in the region of 9 GHz;

(c) a suitable sample cavity;

(d) a crystal detector to measure variations in microwave power;

(e) a suitable recorder to display the output signal from the detector.

On this basis a simple spectrometer such as that illustrated in the block diagram in Fig. 4.2 could be constructed. A suitable source of electromagnetic radiation is a klystron oscillator which produces monochromatic radiation of the required frequency. (Fine adjustment of the frequency is achieved by variation of the d.c. voltage applied to one of the klystron electrodes known as the 'reflector'.) Typically the klystron is required to give approximately 30 mW of power, attenuation of which allows a power as low as 0.25 mW to be used.

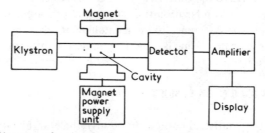

Fig. 4.2 Block diagram of a simple electron spin resonance spectrometer.

The monochromatic output from the klystron is transmitted to the sample cavity by means of a waveguide consisting of a length of gold-plated brass tubing, rectangular in cross section, with internal dimensions appropriate to the klystron frequency (2.29×1.02 cm for an X-band spectrometer). The 'transmission' type of sample cavity consists of a section of blanked off waveguide with two small holes in the end walls through which the microwave power is transmitted. The distance between the end walls is carefully chosen so as to concentrate the microwave power within the cavity by means of multiple reflections off the end walls. The cavity is situated between the poles of the electromagnet such that the sample is at the centre of the applied magnetic induction. The sample itself is usually contained in a thin-walled quartz tube, free of paramagnetic impurities. The outside diameter of the tubing varies from X-band to Q-band; in the former an outside diameter of between 4 and 5 mm would be fairly typical. The microwave power transmitted through the cavity is then detected (a semiconducting crystal detector is usually used for this purpose), amplified, and fed to a suitable recorder.

The output from a spectrometer such as that described would be in the form of a decrease in detector current at that magnetic induction where the absorption condition given in Equation (4.3) is satisfied (see Fig. 4.3). However, the spectrometer would suffer from a number of disadvantages, perhaps the most serious of which is that its performance would be limited by the 'noise' associated with a crystal detector operating at high currents. This difficulty could be overcome by reducing the power reaching the crystal detector, but this would not be compatible with a 'transmission' cavity. However, it is possible to reduce the power reaching the crystal detector, while still maintaining a high microwave power in the sample cavity, if a microwave bridge is employed in conjunction with a 'reflection' cavity.

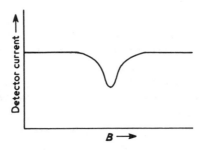

Fig. 4.3 Variation of detector current with magnetic induction obtained using a transmission cavity.

4.2.2. The 'reflection' cavity and micro-wave bridge

The 'reflection' cavity is usually constructed from a section of the waveguide with a length corresponding to two wavelengths (approx. 4.5 cm for an X-band spectrometer) with a single aperture at the junction between the cavity and the waveguide. The internal dimensions of the cavity are such that multiple reflections of the microwave radiation from the end walls create a stationary wave such as that shown in Fig. 4.4. It will be noted that the position corresponding to the maximum intensity of the oscillating magnetic field lies on a plane running through the centre of the cavity, and this is therefore the correct position to locate the sample tube. It should also be noted that this position corresponds to the node in the oscillating electric field. Interaction between the oscillating electric field and solvents are therefore at a minimum at this position. However, for solvents with a high dielectric constant a flat quartz sample tube of small path length (~ 1 mm) may be used. This design of the 'reflection' cavity is by no means unique. For example, cylindrical cavities have found application particularly in the study of gaseous samples.

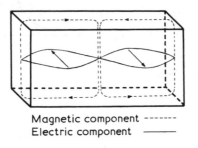

Magnetic component - - - - - - -
Electric component ————

Fig. 4.4 A reflection cavity showing the microwave magnetic and electric field pattern.

A schematic representation of a microwave bridge used in conjunction with the 'reflection' cavity is shown in Fig. 4.5 and incorporates a device known as a magic-T. At the magic-T microwave power is prevented from passing directly from the klystron to the crystal detector. Instead the power is equally divided between two arms, one leading to the sample cavity and the other to the 'matched load'. The impedance of the load can be varied so as to balance exactly that of the cavity, so that no power reaches the detector. However, when the applied magnetic field is such that Equation (4.3) is satisfied, the sample absorbs some of the microwave power, so unbalancing the microwave bridge and allowing power to reach the detector. This procedure has the advantage that the conditions can be selected such that the power reaching the crystal detector does so at its minimum noise level. This is achieved by permanently offsetting the load on the microwave bridge to achieve the optimum conditions. The output would now have the form shown in Fig. 4.6(a).

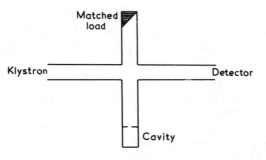

Fig. 4.5 A magic-T microwave bridge.

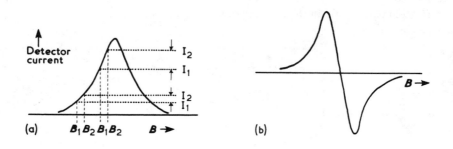

Fig. 4.6 The output obtained from an e.s.r. spectrometer when employing (a) a reflection cavity, and (b) when also employing magnetic field modulation. The effect of applying 100 KHz magnetic field modulation of small amplitude is illustrated in (a) where magnetic field modulation varying between B_1 and B_2 produces an output signal varying between I_1 and I_2.

216

4.2.3 Magnetic field modulation

Such a spectrometer would have a good signal-to-noise ratio but sensitivity and resolution would still be rather poor. This final problem is overcome by modulation of the d.c. magnetic induction. In practice the most common modulation frequency selected for this purpose is 100 KHz, and it is applied by means of small coils mounted within the walls of the sample cavity. The effect of magnetic induction modulation is perhaps best understood through a discussion of the features shown in Fig. 4.6.

Consider the d.c. magnetic induction sweep to be moving steadily through the absorption, and the amplitude of the 100 KHz magnetic induction modulation to be relatively small, it is clear that the microwave power reaching the crystal detector is also modulated at the same frequency but that the amplitude of this latter modulation (the 'output signal') varies with the position of the magnetic field sweep. A phase-sensitive detector is now employed which compares the output signal from the microwave bridge with a reference signal of the same frequency (100 kHz) and is sensitive to both the amplitude and phase of the output signal. As the d.c. magnetic field is swept linearly through the absorption, the detector outputs a sine wave the amplitude of which, starting from zero, increases to a maximum, returns to zero, and then with a 180° change in phase increases to a maximum again before returning finally to zero. The output from the detector is amplified, rectified, and then integrated to give a d.c. output to the recorder equal in intensity to the amplitude of the modulated output signal. Provided that the amplitude of the 100 kHz magnetic induction modulation is small compared with the width of the e.s.r. absorption, the final output to the recorder is proportional to the slope of the absorption curve and can be regarded as its first derivative.

The improved spectrometer now has considerable advantage over the simple spectrometer described previously. In particular, the spectrometer is now far more sensitive to small changes in the shape of the absorption curve. This improvement in resolution is of considerable importance as we shall see in Section 4.4. In addition, a further improvement in the signal-to-noise ratio is now obtained since coherent signals are obtained only from those input signals having the same frequency as the reference.

4.3   THE E.S.R. SPECTRUM

4.3.1 Characteristics of the *g*-factor

The value of the *g*-factor will be independent of magnetic field direction only in isotropic systems such as organic free radicals in low-viscosity solutions. In a large number of other systems the *g*-factor is anisotropic, varying markedly with the orientation of the sample.

It is possible, by means of classical electrodynamics, to show that the orbital magnetic moment μ_L, produced by a charge $-e$ of mass m_e moving in a circular path, has the value:

$$\mu_L = \frac{-e}{2m_e} \cdot P_L$$

where P_L is the orbital angular momentum. The orbital angular momentum of the electron is quantized and has values which are integral multiples of $h/2\pi$. Consequently:

$$\mu_L = \frac{-eh}{4\pi m_e}$$

and the quantity $eh/4\pi m_e$ is the Bohr magneton. The magnetic moment associated with orbital angular momentum is therefore that expected from a simple classical treatment. It is necessary to assume, however, that the magnetic moment associated with spin angular momentum has twice the value expected from a classical treatment [4.1] i.e.

$$\mu_s = \frac{-eh}{2\pi m_e}$$

The g-factor for systems where the magnetic moment arises from spin angular momentum only will be 2, but for a system where the magnetic moment arises from orbital angular momentum only, g will be 1. (Actually it is necessary to apply a small correction so that the value of g for a spin-only system is 2.0023.)

In a large number of organic free radicals the unpaired electron occupies a widely delocalized molecular orbital, and g-factors close to 2 are observed. A similar value will be expected for any other system in which the unpaired electron occupies an orbital having little or no orbital contribution. In contrast in those systems where the unpaired electron is located in an atomic orbital associated with a single atom, it may have significant orbital angular momentum and the g-factor can differ significantly from the free-spin value.

It has already been mentioned that isotropic g-factors are only to be expected for metal ions in regular octahedral crystal fields. However, for systems with lower symmetry the g-factor varies with the relative orientation of the crystal axes with respect to the applied magnetic field, and in order to describe such systems it is usual to introduce a subscript. For example, if the magnetic field is applied along the x-axis of the crystal the observed g-factor will be g_{xx}. Similarly, g_{yy} and g_{zz} can be defined as the g-factors obtained when the field is applied along the y and z axes respectively.

4.3.2 Absorption intensity and concentration measurements

The magnitude of the microwave radiation absorbed is proportional to the population difference between the two energy states. At a particular microwave frequency (and at constant temperature) the population difference between the energy states is a constant proportion of the total number of unpaired electrons present. It is therefore possible to relate the integrated intensity of the e.s.r. absorption to the number of unpaired electrons in the system. The applications of this relationship to kinetic studies of, for example, the rate of decay of relatively unstable organic free radicals, are obvious. In practice the peak-to-peak amplitude of the derivative curve is often used for this purpose with reasonable accuracy.

The absolute number of unpaired electrons present in a particular sample is difficult to determine and involves the spectrometer sensitivity. For a particular spectrometer the minimum number of unpaired electrons which can be detected, N_{min}, is given by the expression [4.2] :

$$N_{min} \propto \frac{T_s}{g^2 S(S+1)} \cdot \frac{\Delta\omega}{\eta Q} \cdot \left(\frac{\Delta\nu \cdot F}{P_0}\right)^{\frac{1}{2}} \qquad (4.4)$$

where T_s, g, and S are the temperature, g-factor, and spin associated with the sample, $\Delta\omega$ is the linewidth parameter, Q is the 'Q-factor' of the cavity and η is the filling factor of the cavity, P_0 is the microwave power, $\Delta\nu$ the system bandwidth, and F represents the detector noise level.

The above equation therefore represents the relationship between the minimum number of unpaired electrons that can be detected and those parameters which may vary for a particular sample in a particular spectrometer system. The effect of each of these parameters is discussed briefly below.

The values of g and S are fixed for a particular sample; the equation indicates the way in which changes in these quantities affect N_{min}. T_s can be decreased with advantage. The linewidth of many samples also decreases with temperature, resulting therefore in an improvement in $\Delta\omega$. Apart from the possible influence of temperature, $\Delta\omega$ is also determined by conditions associated within the sample itself, which in most cases cannot be readily optimized. The filling factor η varies with the precision with which the sample tube is placed in the magnetic component of the microwave field. Interaction with the electric component of the microwave field results in losses which can markedly affect η and also reduces the Q-factor. Consequently the geometry of the sample tube is a further factor which governs spectrometer sensitivity. Equation (4.4) also shows that N_{min} increases with the square-root of the microwave power P_0, which can therefore be increased as required provided that saturation is avoided.

Provided that the e.s.r. spectrometer is operating at its most favourable signal-to-noise ratio, and that all instrumental and sample variables have been optimized, the value of N_{min} for a typical commercial X-band spectrometer is in the range $10^{-8} - 10^{-9}$ M (for a single having $g = 2$, $S = \frac{1}{2}$). There are two other factors which affect the absorption intensity. The first of these is the modulation amplitude. If the modulation amplitude is too large, distortion of the e.s.r. line-shape can result. For magnetic field modulation techniques to be employed successfully the modulation amplitude B_m should be smaller than the peak-to-peak derivative linewidth ΔB_{pp}. If the derivative line-shape is to be strictly the first derivative of the absorption curve, the portion of the absorption curve scanned by the magnetic field modulation must be effectively linear. It can be concluded therefore that as B_m increases with respect to ΔB_{pp} a point will be reached where this linear relationship no longer holds, and as B_m is increased further the peak-to-peak derivative amplitude can actually decrease. In the absolute determination of concentration this is therefore a further factor which has to be considered although it may be necessary to allow some deviation from true first derivative line-shape in order to obtain greater sensitivity.

The second factor is the concentration of the sample which can itself lead to a line broadening. This broadening is a result of electron spin—spin interaction discussed in Section 4.3.3. It is possible in some samples to achieve a situation where the peak-to-peak amplitude of the derivative absorption curve actually decreases with concentration as the broadening of the absorption curve becomes serious. Consequently, care should be taken to work with samples of fairly low concentration ($<10^{-4}$ M) when making measurements of absolute concentration. Naturally the above discussion assumes that the derivative curve rather than the absorption curve is being used in the determination. Measurements made from the absorption curve are more satisfactory in most respects provided that a satisfactory integration of the derivative absorption curve can be obtained.

The most reliable method of concentration determination is by comparison of the sample with a standard of known concentration. Provided that the following precautions are taken, reasonable results may be obtained. (a) Both the sample and the standard should be in the same solvent, and the same sample tube should be used for both measurements. (b) The same microwave power level should be used and hence the microwave power level selected should be such that saturation is avoided in both the sample and the standard. (c) Both measurements should be made at the same temperature. (d) A standard should be selected with g and S similar to that of the sample, and more reliable comparisons will also be found if the two linewidths are fairly similar. Provided that these precautions are observed, the ratio of the two concentrations, determined from the absorption curves, is given by:

$$\frac{\{X\}}{\{Y\}} = \frac{A_x}{A_y} \cdot \frac{B_x}{B_y} \cdot \left(\frac{g_y}{g_x}\right)^2 \cdot \frac{G_y}{G_x} \cdot \frac{B_{my}}{B_{mx}} \cdot \frac{[S(S+1)]_y}{[S(S+1)]_x}$$

where A is the area beneath the absorption curve, B is the magnetic field range scanned, and G is the spectrometer receiver gain.

4.3.3 Factors influencing the absorption line-shape

The width of an e.s.r. absorption depends upon two instrumental factors, the homogeneity of the magnetic field and the frequency of the magnetic induction modulation. Additional factors affecting the width of the absorption are the so-called spin–lattice and spin–spin interactions.

The spin–lattice interaction has already been encountered in Section 4.1.3. Readers will recall that if this interaction is weak the population of the two energy states can equalize, leading to saturation. The quantity $1/(P_1 + P_2)$ has the dimension of time and is termed the spin–lattice relaxation time. The width of the absorption is related to the spin–lattice relaxation time T_1 $[= 1/(P_1 + P_2)]$ from a consideration of the Heisenberg uncertainty principle:

$$\Delta E \cdot \Delta t = h/2\pi$$

Clearly, if T_1 is small, Δt is small, and ΔE must be large, leading to a spread in the energy states between which transitions are occurring. In effect the absorption condition is satisfied over a small range of magnetic field values. Conversely, if the spin–lattice interaction is weak, the mean spin lifetime is relatively long and narrow absorption lines ($\sim 7\mu T$) usually result. This latter situation is usually found for organic free-radicals in solution. The former solution, of a strong spin–lattice interaction, is often encountered in transition-metal ions (where the spin–lattice interaction is often the dominating factor governing the linewidth), and linewidths of 10 mT are not unusual.

The second major interaction, the spin–spin interaction, results from the interaction between spin dipoles. The magnetic field experienced by any individual spin will be that due to the applied magnetic field plus any small magnetic fields produced by neighbouring unpaired electrons. This slight variation in magnetic field experienced by each individual spin again results in a small variation in the value of the magnetic field at which absorption occurs. In crystals this additional magnetic field can be significant; however, in dilute solution the nearest neighbours are sufficiently well removed for the effect to be considerably reduced.

Two other factors which can lead to line-broadening are worth mentioning at this point. The first of these results when there is a small unresolved component to the hyperfine structure when the absorption line recorded consists of a small envelope of many lines. The second results from anisotropy in either g or the hyperfine interaction in solutions of moderate or high viscosity.

Some chemical reactions also give rise to line broadening, notably electron spin exchange, and electron transfer reactions. These will be discussed in a later section.

4.4   HYPERFINE STRUCTURE

There can be little doubt that the feature which makes e.s.r. spectroscopy so fascinating is the appearance of hyperfine structure in the spectrum. This structure arises from the interaction of the unpaired electron with the magnetic moments of nuclei within its orbital. Clearly, the only nuclei where such an interaction would not be expected would be those having a nuclear spin quantum number of zero (i.e. ^{12}C).

4.4.1 Origins of hyperfine interaction

There are two quite distant ways in which an unpaired electron may interact with magnetic nuclei to produce hyperfine structure. The first of these is the dipole—dipole interaction between the electron and nuclear magnetic moments. This interaction is directional since it depends upon the angle made between the magnetic field and a line joining the two dipoles. Its magnitude decreases rapidly ($\propto r^{-3}$) as the distance between the dipoles increases. Since this interaction is directional it is referred to as the anisotropic interaction. For a system such as an organic free-radical in solution, the orientation of the radical with respect to the magnetic field changes rapidly and the interaction averages to zero. The hyperfine structure observed in solution must therefore result from an alternative mechanism.

  The second mechanism by which interaction can occur is the 'Fermi' or 'contact' interaction which is a result of a finite unpaired electron density at the nucleus. This condition is satisfied when the unpaired electron occupies an s-orbital, but not when it occupies a *p-*, *d-*, or *f*-orbital where there will be an orbital node at the nucleus. This interaction is independent of orientation with respect to the magnetic induction and is therefore isotropic.

4.4.2 Energy levels for a radical with
$S = \frac{1}{2}$ and $I = \frac{1}{2}$

As already noted, the hyperfine structure observed in the e.s.r. spectrum arises from interaction of the unpaired electron with any nuclear magnetic dipole in its vicinity. In other words, the total magnetic field experienced by an unpaired electron is that due to the applied magnetic field plus that arising from any small local fields.

  Consider the simplest possible free radical, the hydrogen atom, which has one unpaired electron ($S = \frac{1}{2}$) and one proton ($I = \frac{1}{2}$). There are therefore two orientations of the nuclear magnetic moment for each orientation of the electron magnetic moment, leading to the four energy levels illustrated in Fig. 4.7. One selection rule for an e.s.r. transition, $\Delta M_s = \pm 1$, has been inferred in Section 4.1.1. Since the total angular momentum must remain unchanged, a second selection rule applies when interactions with nuclear magnetic moments are present, i.e. $\Delta M_I = 0$.

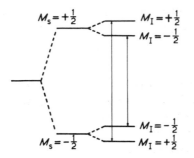

Fig. 4.7 Energy levels (at constant magnetic induction) and allowed transitions for the hydrogen atom.

For the hydrogen atom two transitions are therefore allowed (Fig. 4.7) and are indeed observed [4.3]. The spacing between the two absorptions corresponding to these two transitions is termed the hyperfine 'splitting' constant (a). The fields at which the absorptions occur are therefore:

$$B = B' \pm a/2 \qquad (4.5)$$

where B' is the field where the absorption would occur if $a = 0$.

One further feature of the spectrum of the hydrogen atom is the relative intensity of the two absorptions. The population difference of the two nuclear spin states for a particular electron spin state is only 0.0000025 at room temperature, and n.m.r. spectroscopy, of course, depends on this small difference. However, from the point of view of e.s.r. spectroscopy, the two states are effectively equally populated, and the two absorptions will be of equal intensity within experimental accuracy. The spectrum of the hydrogen atom therefore consists of two equally intense absorptions.

It would be helpful at this stage to introduce a term which is useful when dealing with more complicated e.s.r. spectra, i.e. the spin Hamiltonian operator. An operator represents an instruction to carry out a defined mathematical operation on a particular function. In e.s.r. spectroscopy the operator of interest, the spin Hamiltonian operator, operates on the spin part of the wave function describing a particular state to give the energies of the system. In the absence of any nuclear interactions, combination of the magnetic moment operator $\hat{\mu}_z (= -\mu_N \hat{S}_z)$ with Equation (4.1) gives the spin Hamiltonian operator $\hat{\mathcal{H}}$:

$$\hat{\mathcal{H}} = g\mu_B B \hat{S}_z \qquad (4.6)$$

from which the energy states $\pm g\mu_N B/2$ are obtained.

This form of the Hamiltonian is obviously inadequate in those situations where interaction with nuclear magnetic moments occurs. The additional term required, the Hamiltonian operator for isotropic hyperfine interaction, can be shown to have the form:

$$\hat{\mathcal{H}}_{iso} = hA\hat{S}_z \hat{I}_z$$

223

where A is the isotropic hyperfine 'coupling' constant, and has the unit of frequency. The complete spin Hamiltonian for the hydrogen atom is therefore:

$$\hat{\mathcal{H}} = g\mu_N B \hat{S}_z + hA \, \hat{S}_z \hat{I}_z \qquad (4.7)$$

[Actually both Equations (4.6) and (4.7) are incomplete since they do not include the nuclear Zeeman term $(=-g_N \beta_N B \hat{I}_z)$. However, this latter term can be omitted from the Hamiltonian for the purpose of the present discussion since it does not affect the transition energies.]

There are four spin states for the hydrogen atom, and application of the spin Hamiltonian gives the energies of these states:

$$E_1 = +\tfrac{1}{2} g\mu_N B + \tfrac{1}{4} hA$$

$$E_2 = +\tfrac{1}{2} g\mu_N B - \tfrac{1}{4} hA$$

$$E_3 = -\tfrac{1}{2} g\mu_N B + \tfrac{1}{4} hA$$

$$E_4 = -\tfrac{1}{2} g\mu_N B - \tfrac{1}{4} hA$$

Applying the e.s.r. selection rules the allowed transitions are:

$$\Delta E_{1,4} = g\mu_N B + \tfrac{1}{2} hA$$

$$\Delta E_{2,3} = -g\mu_N B - \tfrac{1}{2} hA$$

These equations are analogous to Equation (4.5), i.e.

$$B_{1,4} = B' - hA/2g\mu_N$$

$$B_{2,3} = B' + hA/2g\mu_N$$

and the relationship between the hyperfine splitting constant (a) and the isotropic hyperfine coupling constant (A) is therefore:

$$a = hA/g\mu_N$$

It can be seen from this equation that a cannot be converted to A unless the g-factor is known. If g is known then:

$$a \text{ (mT)} = 0.0357 \left(\frac{g_e}{g} \right) A \text{ (MHz)}$$

where g is the g-factor for (say) the hydrogen atom and g_e is the free-electron g-factor (2.0023).

It is possible to represent complicated hyperfine splitting patterns in a very compact form by means of the spin Hamiltonian. In much of the immediate discussion that follows, full advantage will not always be taken of this simplification although the advantages of doing so will rapidly become apparent.

4.4.3 Energy levels for a radical with a single set of equivalent protons

Obviously most radicals encountered experimentally are considerably more complex than the hydrogen atom, and the above discussion can be extended to a consideration of the hyperfine splitting patterns to be expected when an unpaired electron interacts with a single set of equivalent protons.

Consider a free radical in which an unpaired electron is able to interact with two equivalent protons. For each arrangement of the electron magnetic moment there are four arrangements of the nuclear magnetic moments, i.e.

	Proton 'A'	Proton 'B'
I	$+\frac{1}{2}$	$+\frac{1}{2}$
II	$+\frac{1}{2}$	$-\frac{1}{2}$
III	$-\frac{1}{2}$	$+\frac{1}{2}$
IV	$-\frac{1}{2}$	$-\frac{1}{2}$

Arrangements I and IV have a net nuclear spin of $+1$ and -1 respectively. On the other hand, arrangements II and III have a set nuclear spin of zero and are clearly indistinguishable. This situation is summarized in Fig. 4.8(a). It should be noted that three transitions are now allowed, but since the $\Sigma M_I = 0$ state is doubly degenerate this transition will have double the intensity of the other two transitions. The energy levels for the $\Sigma M_I = +1$ and $\Sigma M_I = -1$ states are equally spaced from the $\Sigma M_I = 0$ state and the e.s.r. spectrum consists of three equally spaced absorptions with $1 : 2 : 1$ relative intensities.

The situation for the interaction of the unpaired electron with three equivalent protons may be deduced by following the same arguments. For each arrangement of the nuclear spins given in the text above there will be two arrangements of the nuclear spin of the third proton (proton 'C'), There will now be one arrangement where all three nuclei have their magnetic moments parallel to the applied induction ($\Sigma M_1 = +3/2$), and one arrangement where all magnetic moments are antiparallel to the applied induction ($\Sigma M_I = -3/2$). There will be three arrangements where one magnetic moment is parallel and two are antiparallel to the applied induction ($\Sigma M_1 = -\frac{1}{2}$), and finally three arrangements where one magnetic moment is antiparallel and two are parallel to the applied induction ($\Sigma M_I = +\frac{1}{2}$). Four transitions are now allowed [Fig. 4.8(b)] leading to four equally spaced absorptions in the e.s.r. spectrum; However, the two central absorptions are now triply degenerate and have an intensity three times that of the outer absorptions.

The corresponding energy-level diagram for four equivalent protons can now be deduced following an extension of the above argument, and leads to five equally space absorptions with relative intensities $1 : 4 : 6 : 4 : 1$. It is now apparent that this information may be summarized in the form of two general rules:

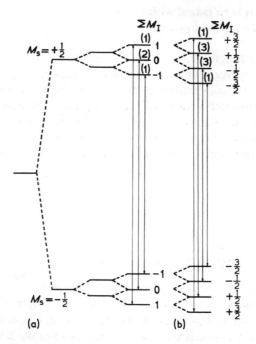

Fig. 4.8 Energy levels and allowed transitions for the interaction of an unpaired electron with (a) two equivalent nuclei with $I = \frac{1}{2}$ and (b) three equivalent nuclei with $I = \frac{1}{2}$. (The relative intensities of the transitions are given in brackets on the upper set of energy levels).

(1) n equivalent protons give $(n + 1)$ equally spaced absorptions in the e.s.r. spectrum;

(2) the relative intensities of these absorptions will be proportional to the coefficients of a binomial expansion of order n.

The relative intensities are readily found from the appropriate Pascal's triangle, i.e.

n	Relative intensity of e.s.r. absorptions							Number of energy levels for each value of M_s
0				1				1
1			1		1			2
2			1	2	1			4
3		1	3		3	1		8
4	1	4		6		4	1	16
5	1	5	10		10	5	1	32
6	1	6	15	20	15	6	1	64

4.4.4 Energy levels for a radical with multiple sets of equivalent protons

The structure of most organic free radicals is such that there will be multiple sets of equivalent protons. The construction of the appropriate energy level diagrams for such systems becomes more complicated but the general procedure can be readily illustrated with two examples.

The simplest example is that in which the unpaired electron interacts with two magnetically non-equivalent protons. The construction of the appropriate energy level diagram can be followed by considering the process in two stages. First consider interaction with the proton having the largest coupling constant. This gives a diagram similar in general appearance to that for the hydrogen atom. Now consider interaction with the second proton (having the smaller coupling constant). Each of the energy levels obtained at the end of the first stage will now be split into two further levels, although the spacing between the levels introduced for this latter interaction will be correspondingly smaller. The complete energy level diagram is shown in Fig. 4.9. Four equally intense transitions are now allowed, the spacing between component pairs being a_1 and a_2.

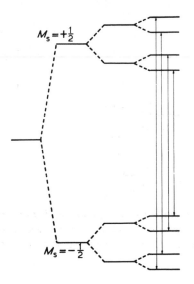

$M_s = +\frac{1}{2}$

$M_s = -\frac{1}{2}$

Fig. 4.9 Energy levels and allowed transitions for the interaction of an unpaired electron with two non-equivalent nuclei with $I = \frac{1}{2}$.

The energy level diagram in Fig. 4.9 differs from that for two equivalent protons [Fig. 4.8(a)] only in that the spacing $\frac{1}{2}hA_1$ differs from $\frac{1}{2}hA_2$. As the magnitude of the spacing $\frac{1}{2}hA_2$ increases the central pair of transitions will approach one another until, when $a_1 = a_2$, a degenerate pair of energy levels is obtained.

As a second example consider the interaction of the unpaired electron with one pair of equivalent protons (splitting constant a_1) and with a single proton (splitting constant a_2). Assuming $a_2 > a_1$ consider first the interaction a_2, this will lead, as before, to an energy level diagram analogous to that of the hydrogen atom. The interaction of the unpaired electron with two equivalent protons leads to three equally spaced energy levels, the central energy level being doubly degenerate. Each energy level obtained at the end of the first stage of the process is now split into further energy levels leading to the final diagram shown in Fig. 4.10(a). [This situation is found experimentally in the pyrogallol radical dianion; see Fig. 4.10(b).]

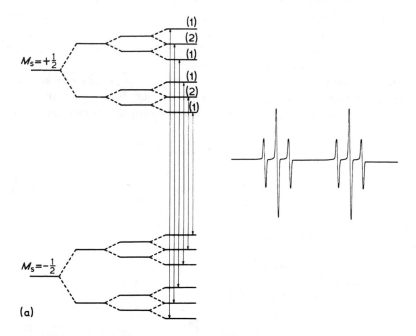

Fig. **4.10** (a) Energy levels and allowed transitions for the interaction of an unpaired electron with a single nucleus with $I = \frac{1}{2}$ and also with two equivalent nuclei with $I = \frac{1}{2}$. (b) The e.s.r. spectrum of the pyrogallol radical dianion (computer simulation).

The energy level diagrams for the two samples described are simple to construct. However, the construction of energy level diagrams for more complicated examples rapidly becomes more involved. Consider, for example, the naphthalene radical anion in which there are two sets of four equivalent protons. From the above discussion an energy level diagram consisting of 50 levels (25 transitions) would be expected. (In fact there are a total of 512 energy levels, but many of them are degenerate.) In the naphthalene radical anion the two splitting constants are 0.490 and 0.183 mT [4.4], and the interested reader may wish to try the above process, using these values and taking care to keep the correct spacings between the levels.

The correct interpretation of the hyperfine patterns arising from complex organic free radicals in solution is obviously of prime importance, and the whole of the next section has been devoted to this. However, it is necessary before discussing spectra interpretation to consider interactions that can arise from nuclei other than ^{1}H. The only other nuclei with $I = \frac{1}{2}$ likely to be encountered in organic free radicals are ^{13}C, ^{19}F, and ^{31}P. Of these the latter two have a natural abundance of 100% and the hyperfine patterns arising from the interaction of the unpaired electron with these nuclei will be indistinguishable from ^{1}H patterns. However, ^{13}C has a natural abundance of only 1.1% (for ^{12}C, $I = 0$) and hyperfine splitting arising from interaction with this nucleus is usually difficult to observe. Consider as an example the methyl radical. Only 1.1 in every 100 methyl radicals will be ^{13}CH$_3$. In each of these latter radicals the interaction of the unpaired electron with the ^{13}C nucleus will produce two equally intense absorptions. Each of the lines arising from this interaction will therefore have an intensity of only 0.55% of that of the lines arising from interaction with ^{1}H. The intensity of these satellite lines increases slightly in larger radicals. For example, in the cyclopentadienyl radical 5.5% of the radicals present will have a ^{13}C nucleus. In those spectra where the hyperfine splitting is relatively simple, splitting arising from interaction with ^{13}C can be observed at high spectrum amplifications.

4.4.5 Hyperfine interactions from nuclei with $I > \frac{1}{2}$

Nuclei with spin $I = 1$ are often present in organic free radicals, e.g. ^{14}N and ^{1}D. The natural abundance of the latter nucleus is again very small but the study of spectra of free radicals which have been deuterated at one or more carbon atoms can be informative.

The hyperfine splitting patterns obtained from nuclei with spin $I = 1$ differ significantly from those discussed so far. Three orientations of the nuclear spin (corresponding to $M_I = +1, 0$, and -1) are now allowed for each orientation of the electron spin. Figure 4.11(a) gives the energy level diagram for a single nucleus of spin $I = 1$. The allowed transitions are as before, so three equally spaced, equally intense, absorptions are expected. The spectrum of the nitrogen atom shows this pattern [Fig. 4.11(b)]. The energy level diagram for two equivalent nuclei of spin $I = 1$ can be readily constructed, predicting five equally spaced absorptions with relative intensities $1 : 2 : 3 : 2 : 1$.

A number of nuclei with spin $I = 3/2$ may be encountered in organic free radicals, viz. ^{7}Li, ^{23}Na, ^{33}S, ^{35}Cl, ^{37}Cl, and ^{39}K. The three alkali metals are included here because one very common means of preparing radical anions of aromatic molecules is by reaction with an alkali metal. The transfer of an electron from the metal to the lowest unoccupied π molecular orbital of the molecule produces the radical anion together with an alkali metal counterion. Interaction of the unpaired electron with the counterion is often observed in spectra of radical anions prepared by this technique. Four orientations of the nuclear spin are now allowed for each orientation of the electron spin, and consequently the interaction of the unpaired electron with one nucleus of spin $I = 3/2$ produces four equally spaced, equally intense, lines.

The spin and natural abundance of some of the common nuclei encountered in organic free radicals are summarized in Table 4.1.

Table 4.1 The spin and natural abundance of some common nuclei (compiled from the 'Handbook of Chemistry and Physics', The Rubber Co., Ohio, U.S.A., 48th Edition, 1968).

Nucleus	Spin	Natural Abundance (%)
^{1}H	1/2	99.98
^{2}D	1	0.02
^{6}Li	1	7.42
^{7}Li	3/2	92.58
^{13}C	1/2	1.11
^{14}N	1	99.63
^{15}N	1/2	0.37
^{17}O	5/2	0.04
^{19}F	1/2	100
^{23}Na	3/2	100
^{29}Si	1/2	4.70
^{31}P	1/2	100
^{33}S	3/2	0.76
^{35}Cl	3/2	75.53
^{37}Cl	3/2	24.47
^{39}K	3/2	93.10

4.5   THE INTERPRETATION OF E.S.R. SPECTRA IN SOLUTION

4.5.1 Interpretation of spectra

The interpretation of many e.s.r. spectra is quite straightforward; for example, the spectrum of the pyrogallol dianion radical [Fig. 4.10(b)] can be readily interpreted in terms of a doublet of triplets. The interpretation of many other spectra may be just as simple provided that there are relatively few groups (say two or three) of equivalent protons and the magnitudes of the splitting constants for each group are markedly different. Such spectra should present no problems, although a reconstruction of the spectrum should always be made (as described later); this procedure provides a check that the interpretation is not in error.

In radicals where the magnitudes of the splitting constants are fairly similar, overlapping, and often superimposition, of hyperfine lines will occur. An example of this is found in the spectrum of the naphthalene radical anion. However, in cases such as this, where the structure of the radical is known, the symmetry of the radical is helpful. Symmetry considerations will immediately reveal the number of groups of equivalent protons to be found and the relative intensities of the lines within each group. If the structure of the radical is not known interpretation can be more difficult.

A number of different approaches may be made to the interpretation of e.s.r. spectra. For example, the 'wings' of the spectrum will suffer least from the problems of overlapping, and a careful study of this region of the spectrum is usually rewarding and can often reveal one or perhaps two of the smaller splitting constants. A second procedure which is often helpful is to start with the central line and search towards the wings for other intense lines. These are usually associated with the larger splitting constants although they may result from the coincidence of two (or more) less intense lines. This can sometimes lead to an interpretation error, but this should be spotted in the spectrum reconstruction. If no central intense line is present, this procedure may still be adopted if a start is made with one of the pair of most intense lines.

The above points, and several others which may be helpful, are summarized below.

(1) Check that the spectrum is symmetric about the centre. If the spectrum is asymmetric, two radicals (with slightly different g-factors) may be present.

(2) Check the wings of the spectrum. The spacing between the outermost two lines gives the smallest splitting constant, and a second splitting constant can often be found from a study of this region.

(3) Check for the most intense lines, working from the centre towards the wings. This should reveal further splitting constants.

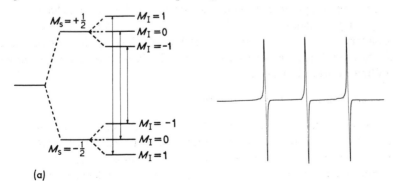

$M_s = +\frac{1}{2}$ $M_I = 1$ $M_I = 0$ $M_I = -1$

$M_I = -1$ $M_I = 0$ $M_I = 1$ $M_s = -\frac{1}{2}$

(a)

Fig. 4.11 (a) Energy levels and allowed transitions for the interaction of an unpaired electron with a single nucleus with $I = 1$. (b) The e.s.r. spectrum of the peroxylamine disulphonate ion.

(4) Check for groups of equally spaced lines which conform to the expected intensity patterns. A pair of calipers is invaluable in this aspect of the work.

(5) Where the structure of the radical is known, check that all the nuclei have been accounted for in the interpretation.

(6) The sum of the splitting constants (for nuclei with spin $I = \frac{1}{2}$) should, allowing for equivalence, equal the separation between the centre of the spectrum and the outermost line. For example, in the naphthalene radical ion the two splitting constants, each corresponding to four equivalent protons, are 0.490 and 0.183 mT. The separation should therefore be $2 \times 0.49 + 2 \times 0.183 = 1.346$ mT.

(7) Perform a reconstruction of the spectrum based on the measured splitting constants as described below. This should account for all the lines present in the spectrum and also, within experimental error, their relative intensities. (This assumes that all the lines in the spectrum have the same width, this may not always be so.)

The method of spectra reconstruction follows the procedure:

(a) Starting with a single line on a piece of graph paper, construct immediately below this line the intensity pattern expected from the interaction of the unpaired electron with the group of equivalent protons having the largest splitting constant, each line being (equally) spaced by a distance proportional to the splitting constant.

(b) Take each of the lines obtained in Stage (a) in turn, and construct immediately below each line the intensity pattern expected from the interaction of the unpaired electron with the group of equivalent protons having the next largest splitting constant. Take care to space these latter groups of lines the correct distance apart proportional to their splitting constant, and preserve carefully the relative intensities of the lines.

(c) Repeat this procedure until all of the splitting constants have been accounted for.

This procedure is illustrated for the naphthalene radical anion in Fig. 4.12. A suitable computer program is helpful, allowing a number of possible values for the splitting constants to be tried simply and quickly. Many commercial spectrometers have a built-in interface enabling a small computer to reconstruct the spectrum for display on the spectrometer. This latter procedure also allows a range of linewidths to be tested.

4.5.2 Origin of proton hyperfine coupling

A large number of studies have been made of the e.s.r. spectra of radical anions formed from the reaction of an alkali metal with an aromatic hydrocarbon in solvents such as tetrahydrofuran. A characteristic of these aromatic hydrocarbons and other conjugated organic molecules is the π-molecular orbitals which are formed by the overlap of the $2p_z$ orbitals on adjacent atoms. The unpaired electron, transferred from the alkali metal atom, enters the lower previously unoccupied π-molecular orbital which has a node in the molecular plane. The Fermi contact interaction (Section 4.4.1) requires a finite unpaired

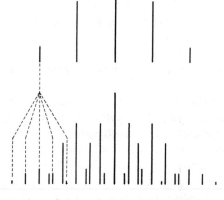

Fig. 4.12 A reconstruction of the e.s.r. spectrum of the naphthalene radical anion.

electron density at the nucleus, and this condition is not satisfied in radical anions derived from aromatic hydrocarbons. On the other hand, the e.s.r. spectrum of the naphthalene radical anion, for example, shows hyperfine structure which can only arise from interaction of the unpaired electron with ring protons. How then does this interaction arise?

The answer to this problem is best understood by considering a $>\dot{C}-H$ fragment of an aromatic molecule. There are two possible arrangments [(A) and (B)] of the spins of the electrons forming the C $-$ H σ-bond for a particular spin of the electron in the $2p_\pi$ orbital. If the pairing of spins were perfect,

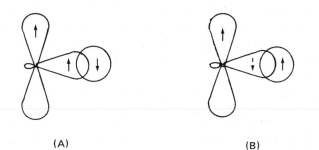

(A)   (B)

the two arrangements would be of equal importance. Fortunately, from our point of view, when one allows for the interaction between the π and σ systems, there is a slight preference for arrangement (A). This is due to a slightly more favourable exchange interaction between the π electron and the carbon σ electron when their spins are parallel. As a result there is a small but significant unpairing of spins in the $C - H$ bond leading to a finite unpaired electron density at the hydrogen nucleus. It should be noted that the spin at the proton is opposite to that of the spin in the $2p_\pi$ orbital.

Obviously one would expect the magnitude of such an effect to be small; this is found experimentally. Compared with the hydrogen atom, where the hyperfine splitting constant is approximately 50 mT, the hyperfine splitting constant for unit unpaired electron density in a $2p_\pi$ orbital in such a $> \overset{\cdot}{C}H$ fragment would be about 2.5 mT.

4.5.3 Use of molecular orbital calculations and negative spin densities

From the above discussion one would expect intuitively that the magnitude of the hyperfine splitting constant, for a particular proton, would depend upon the magnitude of the unpaired spin density in the $2p_\pi$ orbital at the carbon atom to which the proton is attached. Consequently hyperfine splitting constants provide a sensitive test for theories of electron distribution. The simple Hückel molecular orbital approach is particularly interesting in this respect. It allows the unpaired π electron density $\rho_n(=c_n^2$, where c_n is the coefficient of the molecular orbital on carbon atom n) on each carbon in the π system to be calculated. A comparison with the experimental splitting constants can then be made. Such a comparison is shown in Fig. 4.13(a) for selected hydrocarbons. The agreement between a_H and ρ is encouraging and leads to the relationship (due to McConnell [4.5]):

$$a_H = Q \cdot \rho$$

The value of Q that satisfies this relationship varies from one aromatic hydrocarbon to another but averages at about -2.8 mT. One might expect to find similar relationships for ^{14}N, ^{17}O, ^{19}F, and ^{33}S, although Q would have a different value for each nucleus.

Obviously McConnell's relationship would be of greater value if the agreement between experiment and theory could be improved. One of the biggest difficulties with the molecular orbital approach is that zero (or close to zero) unpaired spin densities are often predicted at carbon atoms where substantial hyperfine splitting constants are found. Two examples are the pyrene radical anion and the allyl radical.

The e.s.r. spectrum of the pyrene radical anion can be readily interpreted in terms of two quintets [$a_H(4) = 0.475$ and 0.208 mT respectively] and a triplet [$a_H(2) = 0.109$ mT] [4.6]. The molecular orbital calculation predicts a zero (unpaired) spin density at positions 2 and 7, but the triplet splitting observed

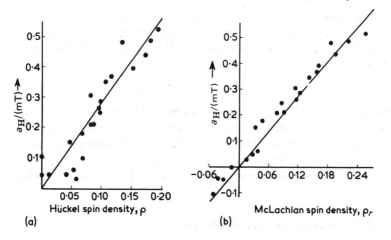

Fig. 4.13 Proton hyperfine splitting constants for a selection of radical anions
(a) vs. HMO unpaired electron densities and (b) vs. McLachlan spin densities

in the spectrum can only arise from interaction with the protons at these
positions.

Consider the simpler allyl radical. The molecular orbital calculation gives
three molecular orbitals, ψ_1, ψ_2, and ψ_3:

$$\psi_3 = \tfrac{1}{2}(\phi_1 - \sqrt{2}\phi_2 + \phi_3)$$
$$\psi_2 = (1/\sqrt{2})(\phi_1 - \phi_3)$$
$$\psi_1 = \tfrac{1}{2}(\phi_1 + \sqrt{2}\phi_2 + \phi_3)$$

Two electrons occupy the ψ_1 orbital and are paired. The remaining electron
occupies the ψ_2 orbital and is unpaired. The prediction is, clearly, for the
unpaired electron density to be $\tfrac{1}{2}$ on both of the end carbon atoms and zero
on the central carbon atom. On the other hand, the e.s.r. spectrum of the
allyl radical shows a doublet splitting of 0.406 mT [4.7] which can only arise
from the interaction of the unpaired electron with the proton bonded to the
central carbon atom.

To allow for this deficiency it is necessary to consider electron correlation
effects. If the two electron wave functions overlap, the Coulombic repulsion
between the two electrons will be reduced if they have the same spin. The
orbital occupied by the unpaired electron, ψ_2, is concentrated on the two end
carbon atoms, forcing, by virtue of the favourable exchange energy, the spin-
positive bonding electron in orbital ψ_1 on to the end carbon atom also, thus
forcing the ψ_1 negative spin towards the central carbon atom. Since the spin
density determined by e.s.r. reveals the magnitude but not the sign of the
spin, the conclusion is that the splitting constant of 0.406 mT in the allyl
radical corresponds to a negative spin density in the ψ_2 orbital.

235

Negative spin densities are often predicted in radical anions of aromatic hydrocarbons, and a most useful self-consistent field approach taking electron correlation effects into account has been made by McLachlan [4.8]. The simple formula:

$$\rho_r = c_{or}^2 - \lambda \sum_s \pi_{rs} c_{os}^2$$

has been obtained (for alternant hydrocarbons), where c_{or} is the Hückel coefficient of the odd orbital on atom r, π_{rs} is the mutual polarizability of atoms r and s; and λ is a constant of value ~ 1.2. The formula is simple to apply and gives rewarding results. Figure 4.13(b) shows a plot of ρ_r against a_H for the same hydrocarbons as selected for the plot in Fig. 4.13(a). The value of Q which best satisfies the plot is -2.4 mT.

4.5.4 The absolute assignment of splitting constants

The application of molecular orbital theory to the interpretation of e.s.r. spectra, through McConnell's relationship, is invaluable when the structure of the radical is known. Not only does the theory permit prediction of the experimental hyperfine splitting constants, but it also gives valuable information on the relative order of magnitudes of the splitting constants. In many spectra absolute assignment of splitting constants to particular protons in the radical can be made in this way. However, difficulties may arise when theory predicts two groups of equivalent protons to have approximately the same splitting constants. If both groups contain the same number of protons absolute assignment may not be possible.

Deuterium substitution is most useful in distinguishing between carbon atoms with a similar unpaired spin density. Not only is the interaction with a nucleus of spin $I = \frac{1}{2}$ replaced by interaction with a nucleus of spin $I = 1$, but the magnitude of the splitting constant is reduced. The study of the [2,2] paracyclophane radical anion (I) is an example where the technique has proved very successful [4.9]. The spectrum may be interpreted in terms of two sets of eight equivalent protons with splitting constants 0.297 and 0.103 mT. However,

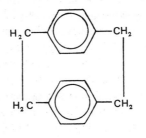

(I)

it is not immediately obvious which of these splitting constants is to be associated with which set of protons. The problem was solved by deuterium substitution at the 4 and 12 positions. The spectrum of the 4,12-dideuterio[2,2]-
paracyclophane radical anion may be interpreted in terms of splitting constants
of 0.297 (six equivalent protons), 0.103 (eight equivalent protons), and 0.046
mT (two equivalent deuterons). Consequently the splitting constant of 0.103 mT
is that associated with the methylene protons.

4.5.5 Linewidth variation

Two relaxation processes (the spin–spin and spin–lattice interactions) which
determine the width of the e.s.r. absorption were discussed in Section 4.3.3. In
addition to these two basic relaxation processes, however, a number of chemical
reactions can lead to linewidth variations. Many of these are of special interest,
revealing information on the rates of chemical exchange reactions which
would be difficult to obtain by any other technique.

One of the first exchange reactions to be studied in detail was that of the
electron transfer reaction between the naphthalene radical anion and neutral
naphthalene:

$$N_1 + (N_2 \cdot)^- \rightleftharpoons (N_1 \cdot)^- + N_2$$

Consider the interaction of an unpaired electron with four equivalent protons
only. In the absence of an electron transfer reaction the e.s.r. spectrum would
consist of five equally spaced lines of relative intensity 1 : 4 : 6 : 4, 1. However, this will not be the case when we consider the effects of an electron
transfer reaction. When the electron jumps there will be 16 (i.e. $1+4+6+4+1$)
possible arrangements of the nuclear spins on the new molecule. Consequently
for $\Sigma M_I = \pm 2$ a possibility of only 1 in 16 electron transfers will leave the
unpaired electron interacting with a set of nuclei in the same configuration as
those on the molecule it has just left (i.e. only 1 in 16 transfers will leave the
position of absorption in the magnetic induction unchanged). For
$\Sigma M_I = \pm 1$, 4 in 16 transfers will result in the same nuclear configuration, and
for $\Sigma M_I = 0$, 6 in the 16 transfers will result in the same configuration. The
outer lines in the spectrum will therefore be broadened to a greater extent
than the inner lines.

The broadening observed depends upon the concentration of the neutral
molecules present and on the velocity constant (k_e) for the electron transfer
reaction. Generally values of k_e are in the range 10^6-10^7 l mol^{-1} s^{-1}, from
which it can be inferred that only a small fraction of the collisions between a
radical anion and a neutral molecule lead to electron transfer.

A second exchange reaction, often with a much larger velocity constant, is
electron spin exchange. This arises when two unpaired electrons exchange their
spins at collisions involving two radicals. Much the same effect as that described
above is often observed with the spectrum sometimes reducing to a single broad

line. However, in contrast to electron transfer reactions which can only occur if the neutral molecule is also present, electron spin exchange can occur in any system and needs to be reduced by dilution if high resolution, narrow line, spectra are to be obtained.

In the effects so far discussed all the lines in the spectrum are broadened, although not always to the same extent. However, by far the most striking of all the linewidth effects is the alternating linewidth effect. Consider a simple hypothetical example:

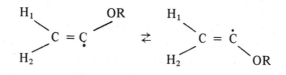

In such a radical protons H_1 and H_2 are in different environments and will have different splitting constants in the e.s.r. spectrum if the rate of interconversion between the two isomers is slow. An e.s.r. spectrum consisting of a doublet of doublets would therefore be expected [Fig. 4.14(a)]. If, on the other hand, the rate of interconversion was very rapid, protons H_1 and H_2 would become equivalent and a 1 : 2 : 1 triplet would be expected [Fig. 4.14(c)]. The alternating linewidth effect will be observed when the rate of interchange is intermediate between these two extremes, when the outer two lines remain sharp whilst the central line(s) appear broad [Fig. 4.14(b)]. The reason for this effect is that the position of the absorption due to those radicals contributing to the outer lines remains unchanged during an interconversion whilst that due to radicals contributing to the central line(s) does not.

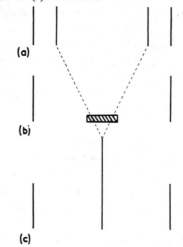

Fig. 4.14 A representation of the e.s.r. spectrum to be expected for a free radical with two protons under conditions of (a) very slow interconversion, (b) and intermediate interconversion rate, and (c) very rapid interconversion.

At intermediate rates of interconversion the above situation can lead to the e.s.r. spectrum consisting of a 1 : 1 doublet (with $a = a_1 + a_2$) when the central line is too broad to be easily detected. The hyperfine splitting constant under conditions of rapid interconversion (a_{av}) will be given by:

$$a_{av} = \tfrac{1}{2}(a_1 + a_2)$$

The situation observed for four equivalent protons (i.e. two pairs) is more complicated but the same arguments hold. At intermediate interconversion rates the e.s.r. spectrum can consist of a 1 : 4 : 1 equally spaced triplet.

4.6 ANISOTROPIC SYSTEMS

It has been realized since the early days of e.s.r. spectroscopy that the examination of free radicals in the solid state formed a fruitful area for study. First, there are a variety of methods available for preparing such free radicals and, second, once formed the radicals are generally far more stable than free radicals formed as reaction intermediates in solution. It was also appreciated at an early stage that irradiation of a single crystal produced radicals aligned in specific orientations and the resulting spectra were therefore orientation dependent.

4.6.1 Anisotropic g-factors

In order to simplify the discussion in the early stages it is helpful to assume that there is no hyperfine interaction in the spectrum and that the Hamiltonian:

$$\hat{\mathcal{H}} = \mu_N g \cdot B \cdot \hat{S}$$

can therefore be taken to represent such a system. The simplest case to examine is that in which the unpaired electron occupies a site about which there is cubic or spherical symmetry. In such a case g remains isotropic and the Hamiltonian has the form:

$$\hat{\mathcal{H}} = g\mu_N(B_x\hat{S}_x + B_y\hat{S}_y + B_z\hat{S}_z)$$

This isotropy is lost in crystals with axial, tetragonal, or tetrahedral symmetry. There is now one axis which is unique, and it is usual to label this as the z axis. The terms $g_\parallel$ and $g_\perp$ are introduced to represent the values of g measured when the magnetic induction is applied parallel and perpendicular respectively to the z axis. Consequently $g_x = g_y = g_\perp$ and $g_z = g_\parallel$ and the spin Hamiltonian is now:

$$\hat{\mathcal{H}} = \mu_N[g_\parallel B_z\hat{S}_z + g_\perp(B_x\hat{S}_x + B_y\hat{S}_y)]$$

In the case of lower symmetry each axis is unique and the spin Hamiltonian becomes:

$$\hat{\mathcal{H}} = \mu_N\{g_{xx}B_x\hat{S}_x + g_{yy}B_y\hat{S}_y + g_{zz}B_z\hat{S}_z\}$$

239

Suppose that in a particular crystal the symmetry axis of the paramagnetic unit coincides with the crystal axes. If such a crystal is mounted in the spectrometer cavity such that the symmetry (c) axis is horizontal, it is possible to determine $g_\parallel$ and $g_\perp$ by rotation of the crystal about a vertical $[a \text{ (or } b)]$ axis. As the crystal is rotated the value of g varies between $g_\parallel$ and $g_\perp$. Rotation of the crystal about the b (or a) axis will result in the same behaviour, while rotation about the c axis will give a constant value of $g_\perp$.

If, however, the symmetry axis of the paramagnetic unit does not coincide with the crystal axes, it is necessary to resort to the methods of matrix algebra in order to determine the g-factors. Suppose the crystal is mounted such that it can be rotated in turn about three mutually perpendicular axes, x, y, and z. It can be shown that for rotation about the x axis the measured value of g^2 varies

$$g^2 = g_{yy}^2 \cos^2\theta + 2g_{yz}^2 \cos\theta \sin\theta + g_{zz}^2 \sin^2\theta$$

For rotation about the y axis g^2 varies:

$$g^2 = g_{zz}^2 \cos^2\theta + 2g_{zx}^2 \cos\theta \sin\theta + g_{xx}^2 \sin^2\theta$$

and for rotation about the z axis g^2 varies:

$$g^2 = g_{xx}^2 \cos^2\theta + 2g_{xy}^2 \cos\theta \sin\theta + g_{yy}^2 \sin^2\theta$$

It is apparent that rotation about each axis leads to the determination of three values of g^2. For example, rotation about the x axis gives the value of g_{yy}^2 when $\theta = 0°$, g_{zz}^2 when $\theta = 90°$, and g_{yz}^2 when $\theta = 45°$ or $135°$. Six independent components of the g^2 tensor can be found in this way. The tensor is symmetric and hence the remaining three components are also known. Diagonalization of the tensor using standard procedures should reveal the three principal values of g^2 from which the principal g-factors g_{xx}, g_{yy}, and g_{zz} are readily found.

4.6.2 Anisotropy of hyperfine coupling

In a large number of orientated systems anisotropy of the nuclear hyperfine coupling constant is found and is often quite large. The appearance of the e.s.r. spectrum can therefore change rapidly as the crystal is rotated.

The origins of the anisotropic hyperfine interaction have already been discussed, arising as a result of the interaction between electron and nuclear dipoles. (The reader will recall that in solution this interaction averages to zero.) Suppose the hyperfine coupling constant for a particular nucleus is measured by rotation of the crystal about the x, y, and z axes as before. The measured value of A^2 is related to θ by the equations:

$$A^2 = A_{yy}^2 \cos^2\theta + 2A_{yz}^2 \cos\theta \sin\theta + A_{zz}^2 \sin^2\theta$$

$$A^2 = A_{zz}^2 \cos^2\theta + 2A_{xz}^2 \cos\theta \sin\theta + A_{xx}^2 \sin^2\theta$$

$$A^2 = A_{xx}^2 \cos^2\theta + 2A_{xy}^2 \cos\theta \sin\theta + A_{yy}^2 \sin^2\theta$$

for rotation about the x, y, and z axes respectively. Again the tensor is symmetric and hence the six independent components determined from the above equations are sufficient to establish the tensor. Diagonalization of the resulting tensor should then establish the principal values of A^2 and hence A_{xx}, A_{yy}, and A_{zz} as before.

4.6.3 Interpretation of spectra in the solid state

Aliphaţic free radicals formed by irradiation of single crystals are trapped in sites related to the original crystal. The first step is to establish the principal values of the g and A tensors. It may also be possible in some cases to determine H–C–H bond angles directly from such measurements [4.10].

Among the most common free radicals formed by irradiation of an aliphatic molecule are those in which a hydrogen atom is removed leaving an unpaired electron localized in the $2p_z$-orbital of a trigonal carbon atom. In such a free radical the p-orbital is perpendicular to the radical plane and the coupling arising from the α-hydrogen will be highly anisotropic. It is possible to predict qualitatively the magnitude and sign of the anisotropic hyperfine interaction.

The magnitude of the electron-nuclear dipole–dipole interaction varies with θ according to the function $1 - 3 \cos^2 \theta$. Taking into account the negative value of the electron magnetic moment there will be boundaries where the term $(3 \cos^2 \theta - 1)$ changes sign. (In fact $3 \cos^2 \theta - 1$ will be zero when $\cos^2 \theta = 1/3$, when $\theta \sim 55°$. The space around the nucleus can therefore be divided into four regions, two in which $3 \cos^2 \theta - 1$ will be positive and two in which it will be negative.) Consider the applied magnetic induction to be along the C − H bond, i.e. $B \| Y$. This situation is represented in Fig. 4.15(a). It can be seen that the p-orbital is located predominantly in a region where $3 \cos^2 \theta - 1$ is positive. Thus anisotropic hyperfine coupling with $B \| Y$ should be large and positive. The situation for B perpendicular to the C − H bond and to the p-orbital $(B \| X)$ is represented in Fig. 4.15(b). Now the p-orbital lies in a region where

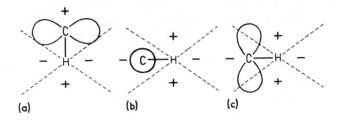

Fig. 4.15 The anisotropic hyperfine interaction for an α proton showing areas of positive and negative electron-nuclear dipolar interaction for (a) $B \| Y$, (b) $B \| X$, and (c) $B \| Z$.

241

$3 \cos^2 \theta - 1$ is large and negative. Finally when B is perpendicular to the $C - H$ bond but parallel to the p-orbital $(B\|Z)$ [Fig. 4.15(c)], the p-orbital extends into both regions almost equally. A small anisotropic hyperfine coupling is therefore expected for this orientation.

Both anisotropic and isotropic interactions contribute to the experimental spectrum. The hyperfine coupling constant determined experimentally will be the sum of both contributions from which the purely isotropic component is $(1/3)(A_{xx} + A_{yy} + A_{zz})$. Suppose, for example, that the experimental hyperfine coupling constants measured for orientations of the magnetic field along the molecular x, y, and z axes are $-92, -31$, and -60 MHz respectively. The isotropic component is thus -61 MHz and hence the anisotropic component may be determined:

$$\begin{vmatrix} -92 & 0 & 0 \\ 0 & -31 & 0 \\ 0 & 0 & -60 \end{vmatrix} = \begin{vmatrix} -61 & 0 & 0 \\ 0 & -61 & 0 \\ 0 & 0 & -61 \end{vmatrix} + \begin{vmatrix} -31 & 0 & 0 \\ 0 & +30 & 0 \\ 0 & 0 & +1 \end{vmatrix}$$

The situation for hyperfine coupling constants arising from the interaction of the unpaired electron with β-protons is rather different. Since the magnitude of the electron-nuclear dipole–dipole interaction decreases rapidly with distance $(\propto r^{-3})$ the anisotropic contribution is much reduced compared with α-proton couplings. Experimental results indicate that the value of the anisotropic component lies between 10 and 20% of that for an α-proton. On the other hand, the isotropic component for a β-proton remains similar in magnitude. The relationship:

$$A_\beta = B + B_0 \cos^2 \theta \tag{4.8}$$

has been proposed for β-proton coupling constants [4.11], where B is usually less than 10 MHz and B_0 is approximately $+140$ MHz.

An interesting and often quoted radical is that derived from alanine by γ-irradiation, i.e. $CH_3 \overset{\cdot}{C}HCOOH$ [4.12]. Here the β-protons are associated with a methyl group which, at room temperature, is freely rotating. Hence $\cos^2 \theta = \frac{1}{2}$ and the isotropic component should be in the region of 75 MHz [Equation (4.8)]. In fact the isotropic component is $+70$ MHz, and the anisotropic components are $-3.0, -2.5$, and $+6.5$ MHz. At 77 K, however, the anisotropic components of the hyperfine coupling constants for the three β-protons are $+120, +77$, and $+14$ MHz. These values strongly suggest that the methyl group is no longer rotating freely but is 'locked' into a particular conformation. The average of the three coupling constants, 71 MHz, is virtually the same as the single room-temperature value, and from Equation (4.8) the appropriate values for θ in the locked conformation are $18°$, $138°$, and $258°$.

4.7   THE TRIPLET STATE

4.7.1 Energy levels for a system with $S = 1$.

In the systems so far discussed the spin S has always been $\frac{1}{2}$. However, in the triplet state there are two unpaired electrons and the spin $S = 1$. For the triplet state the total molecular Hamiltonian is the sum of a spin-independent part and a spin-dependent part, $\hat{\mathcal{H}}_s$, such that

$$\hat{\mathcal{H}} = \mu_N B \cdot g \cdot \hat{S} + \hat{\mathcal{H}}_s$$

in the absence of any hyperfine splitting. The term $\hat{\mathcal{H}}_s$ is itself the sum of two parts, the spin–orbit interaction $\hat{\mathcal{H}}_{so}$ and the spin–spin interaction $\hat{\mathcal{H}}_{ss}$. The first of these two terms, $\hat{\mathcal{H}}_{so}$, is of less importance from the e.s.r. point of view and is not usually considered in discussions of the e.s.r. study of triplet states. It is, however, important in those transition metal ions with $S = 1$, where it can be at least comparable to, and sometimes greater than, spin–spin effects.

For an atom with spherical symmetry the interaction energy between two electron magnetic moments averages to zero and hence the triplet degeneracy remains. However, in the case of a molecule this is no longer true and in the absence of an external magnetic induction the interaction energy between the spins differs in different directions. The spectrum is then highly anisotropic and in a sample consisting of randomly orientated crystals resonance occurs over a range of magnetic inductions often covering several tens of mT. This explains why the search for the e.s.r. absorption of a triplet state was not successful until 1958 [4.13]. The study of triplet states induced in single crystals has obvious advantages.

In planar aromatic molecules the axis perpendicular to the plane (the Z axis) will always be a symmetry axis. For molecules with axial symmetry the X and Y axes remain equivalent and two of the three energy levels remain degenerate. However, for molecules of lower symmetry the triplet degeneracy will be completely removed.

The spin–spin Hamiltonian $\hat{\mathcal{H}}_{ss}$ required by the electron spin–spin interaction is given by:

$$\hat{\mathcal{H}}_{ss} = \hat{S} \cdot D \cdot \hat{S}$$

and the full molecular Hamiltonian becomes:

$$\hat{\mathcal{H}} = \mu_N B \cdot g \cdot \hat{S} + \hat{S} \cdot D \cdot \hat{S}$$

D is again a tensor, which may be diagonalized to give the principal values D_{xx}, D_{yy}, and D_{zz}. It is also traceless, so the sum of the principal values is zero. Strictly g should also be a tensor as in Section 4.6.1. However, for most triplet state systems of interest g is nearly always isotropic and has been assumed to be so in the discussion that follows. Hence:

$$\hat{\mathcal{H}}_{ss} = D_{xx}\hat{S}_x^2 + D_{yy}\hat{S}_y^2 + D_{xx}\hat{S}_z^2$$

and

$$\hat{\mathcal{H}} = g\mu_N \cdot B \cdot \hat{S} + D_{xx}\hat{S}_x^2 + D_{yy}\hat{S}_y^2 + D_{zz}\hat{S}_z^2$$

Since the D tensor has a trace of zero, it is necessary to have only two independent parameters. These are usually designated D and E such that:

$$D = (3/2)D_{zz} \text{ and } E = \tfrac{1}{2}(D_{xx} - D_{yy})$$

and hence the Hamiltonian can be rewritten:

$$\hat{\mathcal{H}} = g\mu_N B \cdot \hat{S} + D(\hat{S}_z^2 - 2/3) + E(\hat{S}_x^2 - \hat{S}_y^2)$$

The D and E terms introduced here are referred to as the zero-field splitting
parameters. For those molecules with axial symmetry the values of the triplet
energy levels, in the absence of a magnetic field, are:

$$E_x = E_y = (1/3)D; E_z = -(2/3)D$$

Application of a magnetic field removes the remaining degeneracy to produce,
typically, an energy level diagram such as that in Fig. 4.16(a). (If D is negative
the states at zero field are reversed.) For molecules with lower symmetry the
triplet degeneracy will be completely removed in zero field and the three energy
levels are:

$$E_x = (1/3)D-E; E_y = (1/3)D+E; E_z = -(2/3)D$$

The values of both D and E can be either positive or negative, and may be of
opposite sign for a particular triplet state. Application of a magnetic field
results in the energy level diagrams in Fig. 4.16(b). It can be seen from these
diagrams that transitions occur at widely different magnetic inductions.
Furthermore, these fields differ for each of the three orientations of the molecular axes, and consequently the spectrum of the triplet state is, as mentioned
before, significantly anisotropic. However, if a single crystal is mounted about
each of three mutually perpendicular axes in turn, and measurements are
made as the crystal is rotated in the magnetic field, it should be possible to
establish the D tensor and hence to find the values of D and E.

In a glass, however, the crystals are orientated at random with respect to the
applied magnetic field, and owing to the anisotropic nature of the spectrum it
is difficult to observe the e.s.r. absorption. Under conditions of high sensitivity,
sufficient measurements can often be made to establish D and E to a reasonable
degree of accuracy. Consider a single crystal mounted in the cavity of the
spectrometer and rotated with respect to the magnetic induction. If the magnetic
induction is applied along one of the crystal axes, a pair of absorptions will be
observed both of which will vary in position periodically as the crystal is
rotated. This situation is illustrated for a typical absorption in Fig. 4.17. Now
consider a random distribution of crystals; as in a glass, a large number of triplets are in resonance at the 'turning points' compared with fields in between
these values. This should result in two distinct magnetic fields where it should
be possible to observe e.s.r. absorptions. (The theory of line-shapes for such a
situation is not really within the scope of this discussion but the absorptions
will not have the usual derivative shape. The interested reader is referred to

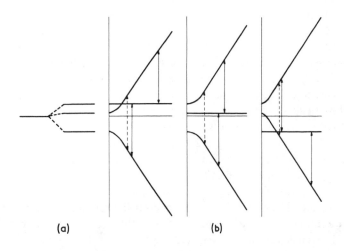

(a)   (b)

Fig. 4.16 (a) Zero-field splitting of a triplet state. (b) the influence of a magnetic induction applied parallel to each of the three molecular axes ($\Delta M_s = 1$ transitions are shown as solid lines, $\Delta M_s = 2$ transitions as dotted lines).

those texts that deal with this topic more fully.) The inductions at which these 'absorptions' occur are given by the expressions [4.14]

$$B_x^2 = [(B_0 \pm D^* \mp E^*)(B_0 \mp 2E^*)]$$

$$B_y^2 = [(B_0 \mp D^* \mp E^*)(B_0 \mp 2E^*)]$$

$$B_z^2 = [(B_0 \mp D^*)^2 - E^{*2}]$$

where $E^* = E/g\mu_N$, $D^* = D/g\mu_N$, and $B_0 = h\nu/g\mu_N$.

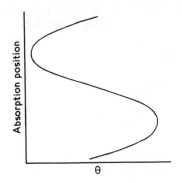

Fig. 4.17 The angular dependence of the position of an anisotropic e.s.r. absorption as obtained from the rotation of a single crystal in a magnetic inductance.

245

4.7.2 '$\Delta M_s = 2$' transitions

Undoubtedly the biggest advance in the study of triplet states in glasses has been made by a study of the so-called '$\Delta M_s = 2$' transitions. Such a transition is usually forbidden in e.s.r. spectroscopy but can nevertheless be observed in triplet state spectra. The prime reason for the violation of the usual selection rule ($\Delta M_s = \pm 1$) is that the energy of interaction between the electron dipoles is comparable to $g\mu_N B$. Under such conditions the usual selection rules break down at low fields since the two spin angular momentum vectors are not independently quantized, and it is not possible to assign M_s values to the energy states. (The use of the term $\Delta M_s = 2$ is therefore misleading but it is commonly used to refer to transitions between the two outermost energy levels.)

The most important feature of the $\Delta M_s = 2$ transition is that it is virtually isotropic. This can be clearly seen from Fig. 4.16, where it is shown as the dotted transition. The absorption is enhanced if the microwave field is applied parallel and not (as is normal) perpendicular to the external magnetic field. However, in a single crystal the $\Delta M_s = 2$ transition still has about 2% of the intensity of the $\Delta M_s = 1$ transition when observed with the normal cavity arrangement. In a glass this is sufficient to make the $\Delta M_s = 2$ absorption readily detectable compared with the $\Delta M_s = 1$ absorptions owing to the isotropic nature of the former compared with the anisotropic nature of the latter. The $\Delta M_s = 2$ transition is therefore of great value in the study of triplet states in the glass phase. In addition it is still possible to determine the values of D and E from this transition (from the minimum field at which resonance occurs) although with a reduced degree of accuracy, i.e.

$$B_{min} = \frac{1}{g\mu_N} [(h\nu)^2/4 - (D^2 + 3E^2)/3]^{1/2}$$

4.8   TRANSITION METAL IONS

The e.s.r. study of the transition metal, rare earth, and actinide ions would be worthy of a text to itself. It is the intention in this section to give only a brief introduction to these studies, and one group, the 3d group, has been selected for a more detailed approach.

4.8.1. Orbital degeneracy and crystal fields

Consider, by way of an example, the first transition series. These metals have n 3d electrons in the valence shell outside an argon core. The metal atom will possess one or two 4s electrons but in the metal ion, as a result of the positive charges on the ion, the d-orbitals become relatively more stable and the metal ions adopt a $3d^n$ configuration.

In the free ion the five d-orbitals have identical energy, and the electronic energy levels are then determined by three interactions. These are (a) Coulombic repulsion between the d-electrons, (b) exchange energies between d-electrons of

the same spin, and (c) spin—orbit coupling. In fact (a) and (b) are usually significantly greater than (c), and the ground-state electron configuration of the
free ion will be 'high spin'.

In a transition metal ion complex a new factor, the 'ligand-field splitting', is
introduced. Consider an octahedral complex in which there are six negatively
charged ligands located along the x, y, and z axes and equidistant from the metal
ion. From an inspection of the orbitals (illustrated in Fig. 4.18) it can be seen
that those orbitals which have a high electron density near to the ligands will be
repelled to a greater extent than those orbitals which avoid the ligands. The
d_{xy}, d_{yz}, and d_{zx} orbitals clearly remain degenerate and have a lower energy than
the d_{z^2} and $d_{x^2-y^2}$ orbitals which also remain degenerate. In a tetrahedral field
this arrangement is reversed.

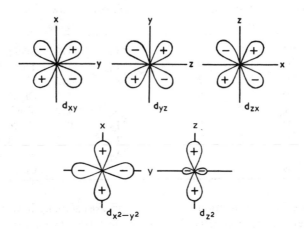

Fig. 4.18 A schematic representation of d-orbitals.

The main task in interpreting the e.s.r. spectra of transition metal ions is to
understand how the various interactions affect the energy levels. The simplest
approach is to consider each interaction in turn, starting first with the strongest
interaction, then the next strongest, and so on. (The influence of a magnetic
field is the last of the interactions to be considered in the 3d series.)

The influence of the crystal field in this context depends upon its strength
compared with the other energy terms, and three typical cases may be distinguished
The first of these, the weak crystal field, is found in the rare earths where the
4f electrons are well shielded by the closed shells of 5s and 5p electrons and the
spin—orbit interaction is considered before that of the crystal field. The second
case, that of moderate crystal fields, is encountered in the first transition series
where the magnitude of the crystal field energy exceeds the spin—orbit interaction energy. The splitting of the d-orbital energies that results for this transition series has already been discussed. The final case, that of a strong crystal
field, is found in the 4d and 5d transition series. In the strong field case it is

necessary to construct the ground state by first considering the effect of the crystal field on each of the d-orbitals. The d-orbitals are split into the lower triplet and upper doublet already described. The d-electrons then fill the lower triplet before the upper doublet to give the spin of the ions along the series as $1/2$, 1, $3/2$, 1, $1/2$, 0, $1/2$, 1, $1/2$ respectively (see Table 4.2). In conclusion it should be remembered that the discussion has been concerned with the effect of octahedral fields; other arrangements will lead to different results.

Table 4.2 Ground states and degeneracies of the first transition series in octahedral fields

d config- uration	Ground term in free ion	Total spin S (high spin)	Orbital degeneracy of free ion	Ground term in octahedral complex	Total spin S (low spin)	Examples
1	^2D	$1/2$	5	$^2T_2$	$1/2$	Sc(2), Ti(3), VO(2)
2	^3F	1	7	$^3T_1$	1	Ti(2), V(3), Cr(4)
3	^4F	$3/2$	7	$^4A_2$	$3/2$	V(2), Cr(3), Mn(4)
4	^5D	2	5	^5E	1	Cr(2), Mn(3)
5	^6S	$5/2$	1	$^6A_1$	$1/2$	Mn(2), Fe(3), Co(4)
6	^5D	2	5	$^5T_2$	0	Fe(2)
7	^4F	$3/2$	7	$^4T_1$	$1/2$	Fe(1), Co(2), Ni(3)
8	^3F	1	7	$^3A_2$	1	Ni(2), Cu(3)
9	^2D	$1/2$	5	^2E	$1/2$	Cu(2)

In fact not many transition metal ion complexes actually have regular octahedral or tetrahedral symmetry. The commonest distortion encountered in the 3d series is a tetragonal distortion which, in the case of an octahedral complex, usually results in four short and two long metal—ligand bonds. A trigonal distortion is also found fairly frequently.

Before considering the individual ions in the 3d transition series it is necessary to introduce two important theorems which determine whether an e.s.r. spectrum is likely to be observed for a particular transition metal ion. First, the *Jahn—Teller* theorem, which states that in any *orbitally* degenerate ground state there will be a distortion to remove the degeneracy. The exceptions to this theorem are linear molecules and those systems having Kramer's doublets. The second theorem, *Kramer's* theorem, rules that in any system having an odd number of electrons, the zero-field ground state will be at least two-fold degenerate. This electronic degeneracy is only removed by a magnetic induction and an e.s.r. spectrum should be observable.

4.8.2 Interpretation of spectra for 3d transition metal ion complexes

d^1, $S = \frac{1}{2}$. In a regular octahedral crystal field the ^2D state of the free ion is split leaving a $^2T_2$ ground state [Fig. 4.19(a)]. Spin-orbit coupling splits this ground state but with $S = \frac{1}{2}$ there should be a Kramer's doublet. If interaction with the E_g states is not included, then:

$$g_\parallel = \frac{3(2\delta + \lambda)}{[(2\delta + \lambda)^2 + 8\lambda^2]^{1/2}} - 1 \quad \text{and} \quad g_\perp = \frac{2\delta - 3\lambda}{[(2\delta + \lambda)^2 + 8\lambda^2]^{1/2}} + 1$$

where λ is the free ion spin-orbit coupling constant. For regular octahedral symmetry $\delta = 0$ and therefore $g_\parallel = g_\perp = 0$ and the e.s.r. spectrum is not observable.

A tetragonal distortion splits the $^2T_2$ ground state leaving a $^2B_1$ ground state and a ^2E excited state. Spin–orbit coupling removes the remaining degeneracy leaving Kramer's doublets and hence e.s.r. spectra should be observed. If the distortion is small, excited states lie close to the ground state leading to a short spin–lattice relaxation time and broad line spectra. It is often necessary, therefore, to resort to very low temperatures in order to observe the spectra.

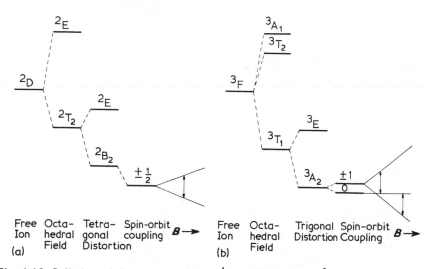

Fig. 4.19 Splitting of the states of (a) a d^1 ion and (b) and d^2 ion.

d^2, $S = 1$. In a regular octahedral crystal field the ^3F state of the free ion is split leaving a $^3T_1$ ground state [Fig. 4.19(b)]. In the case of a $3d^2$ ion the distortion is usually trigonal and this results in the splitting of the $^3T_1$ state leaving a $^3A_2$ ground state. Spin–orbit coupling leaves a non-degenerate lowest level ($M_s = 0$) and a doubly degenerate upper level ($M_s = \pm 1$). If the zero field splitting is large the spectrum will not be observed at the usual magnetic induction. A $\Delta M_s = 2$ transition may be observable.

d^3, $S = 3/2$. In a regular octahedral crystal field the ^4F state is split leaving a $^4A_2$ ground state [Fig. 4.20(a)]. The Jahn–Teller theorem does not apply leaving the Kramer's doublets. The excited states are well removed from the ground state, resulting in relatively long spin–lattice relaxation times and narrow line spectra, which should therefore be readily observable.

Tetragonal distortion leads to zero field splitting. Spectra should be readily obtained but if the zero field splitting is large some of the transitions may not be observed. However, the transitions between the $M_s = \pm\frac{1}{2}$ states should always be observable.

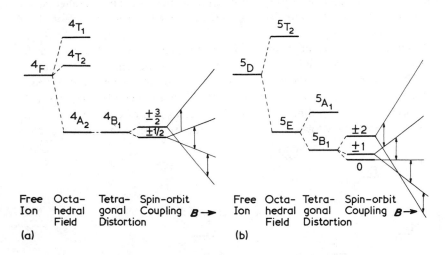

Fig. 4.20 Splitting of the states of (a) a d^3 ion and (b) a d^4 ion.

d^4, $S = 2$. In a regular octahedral crystal field the ^5D state is split leaving a ^5E ground state. The Jahn–Teller theorem is applicable and a tetragonal distortion results in the further splitting of the ^5E state into a $^5B_1$ ground state [Fig. 4.20(b)]. Zero field splitting is often large in $3d^4$ complexes, and since the $M_s = 0$ state lies lowest, spectra are not readily obtainable.

d^5, $S = 5/2$. The ^6S state is not split by a regular octahedral crystal field, and generally there are no nearby excited states. Consequently there is no spin–orbit coupling and a virtually isotropic narrow line absorption is expected at $g \sim 2$ which should be readily observed at room temperature.

Zero field splitting occurs if there is a tetragonal distortion (Fig. 4.21) leaving three Kramer's doublets. If the zero field splitting is large some of the transitions will not be observable.

d^5, $S = \frac{1}{2}$. In a strong octahedral crystal field the d-electrons are paired leaving a $S = \frac{1}{2}$ ground state. A tetragonal distortion is usually present in these complexes and the states are connected by spin–orbit coupling. Low temperatures are often required, therefore, in order to detect the absorption.

d^6, $S = 2$. In a regular octahedral crystal field the ^5D state splits leaving a $^5T_2$ ground state, connected to excited states by spin–orbit coupling. Short relaxation times result and the absorption is difficult to observe. Tetragonal

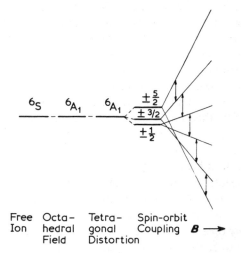

Fig. 4.21 Splitting of the states of a d^5 ion.

distortion splits the $^5T_2$ state leaving a $^5B_2$ ground state [Fig. 4.22(a)] . Spin–orbit coupling is again operative leaving the $M_s = 0$ level lowest, and owing to a large zero field splitting a spectrum is not usually obtainable.

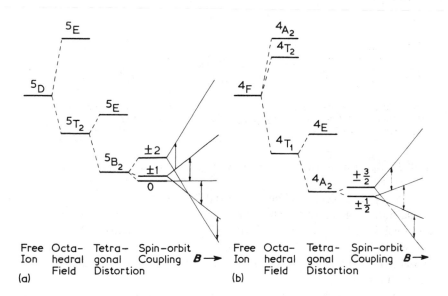

Fig. 4.22 Splitting of the states of (a) a d^6 ion and (b) a d^7 ion.

d^7, $S = 3/2$. In a regular octahedral crystal field the ^4F state is split leaving a $^4T_1$ triplet ground state [Fig. 4.22(b)] . Spin–orbit coupling connects the ^4T

state to the excited states leaving a Kramer's doublet as the lowest level. An e.s.r. spectrum should therefore be obtainable, but relaxation times are short and low temperatures are required. The lowest lying Kramer's doublet is usually the only level occupied at the low temperatures required.

Tetragonal distortion splits the ^4T state leaving a $^4A_2$ ground state, and spin–orbit coupling is again operative leading to similar problems as above.

d^7, $S = \frac{1}{2}$. In strong crystal fields a doublet $S = \frac{1}{2}$ becomes the lowest energy state. The Jahn–Teller theorum is operative leading to distortion. In the case of tetragonal distortion the system behaves very much like the d^1 case already discussed, and with excited states now further removed from the ground state, spectra are observable at higher temperatures.

d^8, $S = 1$. In a regular octahedral crystal field the ^3F state is split leaving a $^3A_2$ ground state [Fig. 4.23(a)]. Spin–orbit coupling leaves the ground state triply degenerate in spin, and since there are no nearby excited states e.s.r. spectra should be readily observed. A tetragonal distortion leads to a splitting of the spin states, and since the lowest lying state is not a Kramer's doublet it may not always be possible to observe a spectrum.

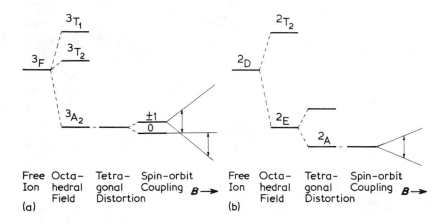

Fig. 4.23 Splitting of the states of (a) a d^8 ion and (b) a d^9 ion.

d^9, $S = \frac{1}{2}$. In a regular octahedral crystal field the ^2D state is split into a ^2E ground state [Fig. 4.23(b)]. The Jahn–Teller theorem is applicable, and a tetragonal distortion leads to a splitting of the ^2E state leaving a ^2A ground state which is a Kramer's doublet. When the distortion is large excited states are well removed and narrow line spectra are obtainable at quite high temperatures.

A number of conclusions may be drawn from the above discussion of individual ions. In a regular octahedral crystal field the ground state will still be orbitally degenerate except in the d^3, d^5, and d^8 configurations. Some of the degeneracy remaining in the other configurations may be removed by spin–orbit interaction or by the Jahn–Teller theorem. The latter theorem allows distortion to remove not only orbital but also spin degeneracy except in those systems with an odd number of electrons (Kramer's theorem). Excited states often remain nearby, leading to short spin–lattice relaxation times. The prediction is, therefore, that for octahedral complexes only the d^3, d^5, and d^8 configurations lead to narrow line spectra at room temperature. The other configurations either require very low temperatures or complexes with symmetry that produces orbitally non-degenerate ground states. In many other configurations, notably d^2, d^4, and d^6, zero field splittings can be large and with a $M_s = 0$ ground state the $\Delta M_s = \pm 1$ transition may not be observable at normal magnetic inductions.

It should also be noted that many of the 3d transition metal atoms have nuclei with non-zero nuclear spin (Table 4.3) and metal hyperfine structure is often obtained in the spectrum. The hyperfine structure will have both isotropic and anisotropic components as before. In dilute solution an isotropic spectrum is usually obtained with average values for both g and A. However, when the solution is frozen an anisotropic spectrum is obtained which should be the same as that obtained from a 'dilute' powder of the sample. The resolution of a powder spectrum is usually quite poor, although the g and A values obtained from such studies should be in agreement with those from single crystals.

Table 4.3 The spin and natural abundance of some 3d transition metal nuclei (compiled from the 'Handbook of Chemistry and Physics', The Rubber Co., Ohio, U.S.A., 48th Edition, 1968).

Nucleus	Spin	Natural abundance (%)
^{45}Sc	7/2	100
^{47}Ti	5/2	7.28
^{49}Ti	7/2	5.51
^{51}V	7/2	99.76
^{53}Cr	3/2	9.55
^{55}Mn	5/2	100
^{57}Fe	1/2	2.19
^{59}Co	7/2	100
^{61}Ni	3/2	1.19
^{63}Cu	3/2	69.09
^{65}Cu	3/2	30.91

REFERENCES

4.1   Goudsmit, S., and Uhlenbeck, G.E., *Nature,* **117**, 264 (1926).
4.2   Poole, C.P., *Electron Spin Resonance'*, Interscience, New York (1967).
4.3   Kroh, J., Green, B.C., and Spinks, J.W.T., *Canad. J. Chem.,* **40**, 413 (1962).

4.4   Carrington, A., Dravnieks, F., and Symons, M.C.R., *J. Chem. Soc.*, 947 (1959).

4.5   McConnell, H.M., *J. Chem. Phys.*, **24**, 632 (1956).

4.6   Hoijtink, G.J., Townsend, J. and Weissman, S.I., *J. Chem. Phys.*, **34**, 507 (1960).

4.7   Fessenden, R.W., and Schuler, R.H., *J. Chem. Phys.*, **39**, 2147 (1963).

4.8   McLachlan, A.D., *Mol. Phys.*, **3**, 233 (1960).

4.9   Gerson, F., and Martin, W.B., *J. Amer. Chem. Soc.*, **91**, 1883 (1969).

4.10   Horsfield, A., Morton, J.R., and Whiffen, D.H. *Mol. Phys.*, **4**, 327 (1961).

4.11   Heller, C., and McConnell, H.M., *J. Chem. Phys.*, **32**, 1535 (1960).

4.12   Horsfield, A., Morton, J.R. and Whiffen, D.H., *Mol. Phys.*, **5**, 115 (1962).

4.13   Hutchison, C.A. Jr., and Mangum, B.W., *J. Chem. Phys.*, **29**, 952 (1958).

4.14   Wasserman, E., Snyder, L.C., and Yager, W.A., *J. Chem. Phys.*, **41**, 1763 (1964).

5 Mössbauer spectroscopy

5.1 THE MÖSSBAUER EFFECT

Mössbauer spectroscopy is concerned with the phenomena of nuclear resonant absorption and fluoresence. Most modes of radioactive decay produce a daughter nucleus in a highly excited state, which then decays to the ground state by emitting a series of γ-ray photons. This is clearly analogous to electronic de-excitation, the main differences being in the much higher energies involved in nuclear transitions. It was recognized in the 1920s that it should be possible to observe a parallel phenomenon to the already well known atomic resonant fluorescence. A γ-ray emitted during a nuclear transition from an excited state to the ground state should be capable of exciting a second ground state nucleus of the same isotope, thereby giving rise to nuclear resonant absorption and fluoresence. Unfortunately the high energy and momentum of the photon cause the nucleus to have a high recoil energy, and reduce the probability of detecting this resonant absorption to almost negligible proportions.

A solution to this problem was found by R.L. Mössbauer in 1958 [5.1]. Under appropriate experimental conditions *in the solid state* it is possible to produce *recoilless* emission and absorption events, so that resonant absorption and fluoresence can be detected quite easily. His classic experiments formed the basis for a completely new branch of spectroscopy which is now popularly referred to as Mössbauer spectroscopy or nuclear gamma resonance (n.g.r.). The technique can only be applied to the solid state.

That there is a link between recoilless nuclear resonant absorption and chemistry is at first sight surprising. The nuclear transitions with an energy of $> 10^6$ kJ mol^{-1} are far more energetic than chemical bonds (~ 200 kJ mol^{-1}) or molecular vibrations (~ 10 kJ mol^{-1}). However, the γ-rays emitted in recoilless events are an unusually good source of monochromatic radiation, and have a spread in energy of $< 10^{-6}$ kJ mol^{-1}. In consequence it becomes possible to observe the weak interactions of the nucleus with its chemical environment.

Some of these interactions have already been described in a different context in
the chapters on nuclear magnetic resonance and nuclear quadrupole resonance.
They have energies of $< 10^{-3}$ kJ mol^{-1}. One of the unusual features of
Mössbauer spectroscopy is the way in which energy differences of the order of
1 part in 10^{12} are measured. Such differences can be produced by the Doppler
effect of relative motion. Thus if a γ-ray of energy E_γ is emitted from a source
moving with a relative velocity of $+v$, the 'apparent' energy is $(1 + v/c)E_\gamma$
where c is the velocity of light. A velocity of 0.3 mm s^{-1} corresponds to an
energy shift of 1 in 10^{12}. The emitting nuclei can therefore be moved in or out
of resonance with the absorbing nuclei merely be a relative mechanical motion.

The first observations of hyperfine interactions with the extranuclear electrons (magnetic dipole interactions in 1959 by Pound and Rebka [5.2], and
electric monopole and quadrupole interactions in 1960 by Kistner and Sunyar
[5.3]) resulted in the rapid development of Mössbauer spectroscopy. It has
proved to be a versatile technique with many diverse applications in solid-state
physics and chemistry.

5.1.1 Recoilless emission and absorption

Let us examine in detail the energetics of γ-ray emission and absorption. Consider an isolated atom which has a nuclear excited state at an energy E above
the ground state. We need only consider the translational motion of the atom
along the direction in which the γ-photon is to be emitted, because the components of motion perpendicular to this axis remain unaffected by the decay.
If the excited atom is initially moving with velocity V along this axis and has
mass M, the total energy above the ground state *at rest* is $E + \frac{1}{2}MV^2$. If a
γ-ray of energy E_γ is emitted, the nucleus recoils so that its new velocity is
$(V + v)$ (this is a vector sum because the directions may be opposed), and the
kinetic energy of the nucleus is now $\frac{1}{2}M(V + v)^2$. By conservation of energy:

$$E + \tfrac{1}{2}MV^2 = E_\gamma + \tfrac{1}{2}M(V + v)^2 \qquad (5.1)$$

so the actual energy of the photon as emitted is deficient in energy and is
given by:

$$E_\gamma = E - \tfrac{1}{2}Mv^2 - MvV$$
$$= E - E_R - E_D \qquad (5.2)$$

The term $E_R = \frac{1}{2}Mv^2$ is independent of the initial velocity V and is termed the
recoil energy, while the term $E_D = MvV$ is proportional to V and is therefore
called a Doppler-effect or thermal energy.

In this emission process we must also have conservation of momentum. The
momentum of the emitted γ-ray photon is E_γ/c so that:

$$MV = M(V + v) + E_\gamma/c \qquad (5.3)$$

Thus $E_\gamma/c = -Mv$, from whence:

$$E_R = E_\gamma^2/(2Mc^2) \tag{5.4}$$

The mean value of E_D can be related to the mean kinetic energy per translational degree of freedom $\bar{E}_K \approx \frac{1}{2} kT$ by:

$$\bar{E}_D = 2\sqrt{(\bar{E}_K E_R)} = E_\gamma\sqrt{[2\bar{E}_K/(Mc^2)]} \tag{5.5}$$

where k is Boltzmann's constant and T is the absolute temperature. The statistical distribution in energy of emitted γ-rays is therefore displaced from the true excited-state energy by $-E_R$ and broadened by E_D into a Gaussian distribution of width $2\bar{E}_D$.

It is now easy to show that in the reverse process of absorption, the γ-ray distribution is similar but displaced by an energy $+E_R$. For values of E of $\sim 10^4$ eV, E_R and E_D are typically $\sim 10^{-2}$ eV. [γ-Ray energies have formerly been expressed in units of electron-volts. Since $e = 1.6022 \times 10^{-19}$ C, 10^4 eV $\equiv$ 1.6022×10^{-15} J.] Nuclear resonant absorption can only be observed as a successive emission (to generate the γ-photon) and absorption, and will only have a significant probability if the emission and absorption profiles overlap strongly. As can be seen in Fig. 5.1, the overlap may be quite small and therefore difficult to detect.

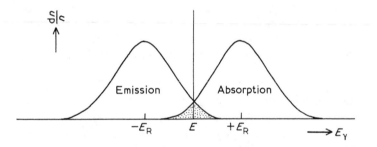

Fig. 5.1 The statistical distributions of the γ-ray energy for emission and absorption. The resonance overlap which is shown shaded is small.

It is interesting to note that these equations are not peculiar to nuclear resonant absorption, and could equally be applied to the resonant absorption of ultraviolet radiation by atoms. However, in the latter case the recoil energy is very much less than the Doppler broadening, so the emission and absorption profiles overlap strongly. The recoil is only a significant problem for very energetic transitions.

Initial attempts to increase the resonant absorption included the use of high temperatures (to broaden the distribution via $\bar{E}_D$) or large velocities of about 7×10^5 mm s^{-1} (to compensate for E_R), but the increases were only marginal. The Mössbauer effect is significantly different in that the recoil and Doppler energy are not merely compensated but entirely eliminated. This results in the width of the energy distribution becoming effectively that required by the

Heisenberg uncertainty principle. The excited state of the nucleus has a mean life τ less than 10^{-6} s. From the uncertainty principle, the width of the γ-ray distribution at half-height Γ_s is given by:

$$\Gamma_s \tau \geqslant \hbar \tag{5.6}$$

The mean-life τ and the half-life $t_{1/2}$ are related by $\tau = \ln 2 \times t_{1/2}$ so, if Γ_s is in eV and $t_{1/2}$ in seconds:

$$\Gamma_s = 4.562 \times 10^{-16}/t_{1/2} \tag{5.7}$$

For example, for the first excited state of ^{57}Fe at 14.4 keV, $t_{1/2} = 97.7$ ns and $\Gamma_s = 4.67 \times 10^{-9}$ eV. Thus, if recoil and thermal broadening are eliminated, a monochromaticity of 1 part in 10^{12} can be achieved, thereby giving a potential resolution greater by orders of magnitude than that in any other type of measurement extent.

The key to the problem lies in the behaviour of a recoiling nucleus in a crystal lattice. The recoil energy is less than the chemical binding energy of the lattice, but similar in magnitude to the lattice vibration phonon energies. The recoil energy can be transferred to vibrational energy in the crystal, in which case the energy of the photon is still degraded. From a naive viewpoint, we can consider the vibrational energy of the crystal to be quantised so that only discrete amounts of energy 0, $\pm \hbar\nu$, $\pm 2\hbar\nu$, etc. can be transferred where ν is the frequency of the phonon. Considering a large number of emission or absorption processes, there is a probability that a zero-phonon transition (one in which there is no energy transfer to the lattice) takes place given by:

$$f = 1 - E_R/(\hbar\nu) \tag{5.8}$$

and this quantity f is finite if $E_R \neq \hbar\nu$. However, the atom is unable to recoil freely because it is chemically bound to its site, so the crystal mass as a whole must move. Equations (5.4) for E_R and (5.5) for $\bar{E}_D$ contain the reciprocal mass $1/M$, which has now become that of a crystal containing perhaps 10^{15} atoms. Thus both E_R and $\bar{E}_D$ become very small and are much less than Γ_s.

This model is a simplification, because in any real solid there exists a wide range of lattice frequencies. Fortunately, it is difficult to excite the low frequencies, and there is still a significant fraction f of nuclear events in which the recoil momentum is taken up by the crystal as a whole and the γ-ray energy is not degraded.

In summary, there is a finite probability *in a solid matrix* that a γ-ray emission or absorption can take place without recoil or thermal broadening, and that the width of the energy distribution is defined by the Heisenberg uncertainty principle. For this reason f is usually referred to as the recoilless or recoil-free fraction, and can be quantitatively related to the vibrational properties of the lattice. In general:

$$f = \exp\left(-\frac{E^2\langle x^2\rangle}{(\hbar c)^2}\right) \tag{5.9}$$

where $\langle x^2 \rangle$ is the mean-square vibrational amplitude of the nucleus in the direction of the γ-ray. Thus f will only be large for low-energy γ-rays and for atoms tightly bound to the lattice so that their mean-square displacement is small (the highest energy for which a Mössbauer effect is known is 187 keV in ^{190}Os). If the lattice vibrations are assumed to follow the Debye model, then:

$$f = \exp\left[-\frac{6E_R}{k\theta_D}\left\{\frac{1}{4}+\left(\frac{T}{\theta_D}\right)^2\int_0^{\theta_D/T}\frac{x\,dx}{e^x-1}\right\}\right] \tag{5.10}$$

where θ_D is the Debye temperature of the lattice. From the form of this equation, f is large in a strong lattice (large θ_D) at low temperature (small T).

5.1.2 The Mössbauer spectrum

In the knowledge of the preceding discussion it is possible to design an experiment to demonstrate the Mössbauer effect of recoilless resonant absorption. A solid matrix containing the excited nuclei of a given isotope (the source) is placed next to a second matrix containing the same isotope in the ground state (the absorber). The intensity of the γ-ray beam transmitted through the absorber can be measured as a function of the temperature of the matrix. In the absence of resonant absorption the transmission rate would be independent of temperature, but instead we will now expect to see a decrease in the counting rate (i.e. an increase in the resonant absorption) as the temperature is lowered and the recoilless fraction increases.

This experiment is essentially that first demonstrated by Mössbauer, but he subsequently developed a more subtle technique which is now used universally. We have already seen that the strength of resonant absorption is determined by the overlap of the energy profiles for the source and absorber. Because of this a Mössbauer resonance is specific to atoms of a particular isotope of an element, and is therefore a highly selective form of spectroscopy. A movement of the source and absorber relative to each other with velocity v will alter the effective value of E_γ 'seen' by the absorber by a small Doppler shift energy of $\epsilon = (v/c)E_\gamma$. If the effective resonance energies for source and absorber exactly match at a particular velocity v, resonance absorption will be at a maximum. Any increase or decrease in v can only decrease the absorption. In the extreme of large velocities, such that there is no overlap of the two energy distributions, there will be no resonant absorption. Thus the basic principle of obtaining a Mössbauer spectrum is to record the transmission of γ-rays from a source through an absorber as a function of the Doppler velocity v. This is illustrated schematically in Fig. 5.2.

The line-shape of the absorption derives from the mathematics of the source and absorber energy distributions. The source has a recoilless fraction f_s and a Heisenberg width Γ_s. The number of transitions $N(E)$ between $(E_\gamma - E)$ and $(E_\gamma - E + dE)$ is given by the Lorentzian distribution:

259

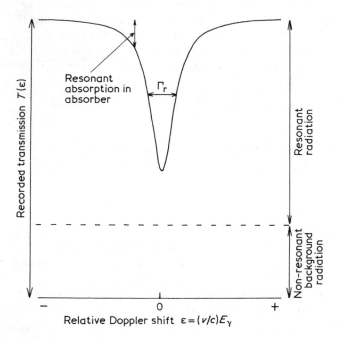

Fig. 5.2 A Mössbauer transmission spectrum recorded by Doppler scanning.

$$N(E)\,\mathrm{d}E \;=\; \frac{f_s\Gamma_s}{2\pi}\;\frac{\mathrm{d}E}{(E-E_\gamma)^2+(\Gamma_s/2)^2} \tag{5.11}$$

which has a maximum at $E = E_\gamma$. The cross-section for resonant absorption $\sigma(E)$ is similarly expressed as:

$$\sigma(E) \;=\; \sigma_0\,\frac{(\Gamma_a/2)^2}{(E-E_\gamma)^2+(\Gamma_a/2)^2} \tag{5.12}$$

where Γ_a is the Heisenberg width of the absorber energy distribution. The cross-section σ_0 is a constant for a particular transition.

The resonance line-shape as measured experimentally can be related to $N(E)$ and $\sigma(E)$ and the thickness of the absorber, but is most usefully considered in the limit of a very thin source and absorber for which the line-shape is Lorentzian in shape but with a width of $\Gamma_r = 2\Gamma_s$. An increase in absorber thickness causes a gradual increase in the magnitude of Γ_r but with minimal distortion from the Lorentzian shape.

The intensity of the absorption maximum can be related to $\sigma(E)$, the absorber thickness, and the recoilless fraction in the absorber f_a. However, as well as resonant absorption there are also other competitive non-resonant processes such as Compton scattering which can cause an increase in the radiation background. These place an effective limit on the absorber thickness that can be used, and in particular make it impossible to determine f_a or f_s by direct measurement of the intensity of absorption.

5.2 EXPERIMENTAL METHODS

A Mössbauer spectrum is a record of the transmission rate of the resonant γ-ray as a function of the Doppler velocity between the source and the absorber, i.e. transmission as a function of energy. In this respect there is no difference from, for example, infrared spectroscopy. The major problem arises from the health hazard in using very strong radioactive sources. The relatively low photon flux-densities that can be used safely necessitate much longer counting times to achieve a given resolution than in optical spectroscopy. The statistics of γ-ray counting result in the standard deviation in N registered γ-counts being $\sqrt{N}$; thus the standard error in 10 000 counts is 1%, and in 1 000 000 counts is 0.1%. Although increasing the counting time increases spectrum definition, this must be balanced against the experimental time required and the long-term reproducibility of the measuring equipment.

There are two approaches to measuring the γ-ray transmission as a function of Doppler velocity. The total number of counts can be registered in a fixed time at constant velocity, and subsequently re-measured at several different velocities. This is most easily achieved by mounting the source on a mechanical constant-velocity device, but has many disadvantages. Alternatively, one can scan rapidly over the whole velocity range on a cyclical basis so that the whole spectrum is accumulated simultaneously. This method has many advantages in that it is much easier to automate and can be continuously monitored.

A typical modern Mössbauer spectrometer is shown schematically in Fig. 5.3. It incorporates a device known as a multichannel analyser, which can store an accumulated total of γ-counts in any one of several hundred registers (or channel addresses) in a similar way to a small computer. Each channel is held open in turn for a short fixed time interval governed by a constant-frequency clock, and any counts registered during that time interval are added to the accumulated total stored in the channel. The scan through the channel addresses is completed in about 1/20 of a second, and is then repeated *ad infinitum*. The timing pulses from the clock can also be converted electronically into a voltage signal which increases or decreases linearly with time. This signal is used as a reference signal to control an electromechanical servo-drive system. The physical velocity of the drive shaft which imparts the Doppler shift is sensed by the voltage induced in a monitor-coil, and is compared with the reference command signal by the servo-amplifier. A voltage increasing linearly with time corresponds to a motion with constant acceleration in which the drive shaft spends an equal time interval at each velocity increment. The multichannel analyser and the drive are coupled so that during a single scan the velocity changes linearly from $-v$ to $+v$ with increasing channel address. There is then a short wait while the drive returns to its starting point, and then the scan is repeated. Alternatively, the velocity changes from $-v$ to $+v$ in the first half of the scan, and then reverts to $-v$ during the second half, thereby accumulating a mirror-image spectrum which may be folded on to the first half.

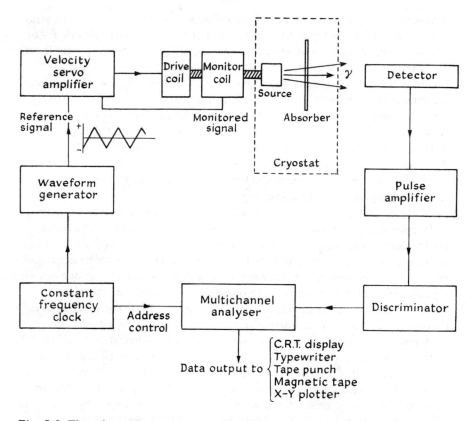

Fig. 5.3 The schematic arrangement of a Mössbauer spectrometer incorporating a servo-drive mechanism.

The γ-rays transmitted through the absorber are detected by a scintillation counter, gas proportional counter, or a lithium-drifted germanium detector; all three instruments may be used in the 10–100 keV region, and each has its advantage and disadvantages. The pulses from the detector are amplified, and are then passed through a discriminator which rejects any pulse outside a chosen voltage range. In this way some of the X-ray or γ-ray events which are un-connected with the resonant transition can be discarded. The remaining pulses are then fed to the open channel address.

In this way the Mössbauer spectrum of total accumulated count against velocity is built up. It can be continuously monitored by an oscilloscope display, and final numerical data read-out can be obtained by typewriter, tape punch, or other peripheral device. It is the digital nature of the data which results in the rather characteristic appearance of the Mössbauer spectrum.

The majority of useful Mössbauer isotopes require a range of Doppler velocities of less than $\pm 10\,\mathrm{mm\,s^{-1}}$, and the overall displacement of the source during a 1/20 second scan is barely visible to the naked eye. This causes considerable difficulty in calibrating a spectrometer accurately. It is possible to achieve an absolute calibration by mounting a diffraction grating or optical interferometer on the drive shaft, but precision optical equipment is very expensive. The usual alternative is to utilise an absorber which has several resonance lines for which exact velocities have been measured on a calibrated instrument. In this way both the velocity amplitude and the linearity can be checked with acceptable accuracy. Thus it is common practice to use a single-line ^{57}Co source with a magnetic α-iron foil which has six resonance lines (see Section 5.3) as a calibrant. It is not easy to establish the exact channel address corresponding to zero Doppler velocity, but this can also be avoided by quoting all measurements relative to the centroid of a reference spectrum, which in ^{57}Fe Mössbauer spectroscopy is again the iron foil.

It was observed from the form of Equation (5.10) that the Mössbauer absorption will be greater at 'low' temperatures. Indeed, although it is possible to obtain strong resonances in ^{57}Fe and ^{119}Sn at room temperature, in many instances one must resort to cooling at least the absorber, and sometimes the source also, in a cryostat. Since the source is moving, this presents some difficulty. Furthermore, there must be no parasitic vibrations in the cryostat, and there must be an optically transparent path for the γ-rays through the walls, e.g. using beryllium or aluminised mylar windows. Refrigerants commonly used are liquid nitrogen (down to 78 K) and liquid helium (4.2 K).

Low temperatures are not merely used to increase the resonant effect. Several of the hyperfine interactions described in the next section are temperature dependent, and for this reason cryostats and ovens with the absorber at a variable controlled temperature are now in common use. Another type of experiment inspired by the hyperfine interactions involves placing the absorber in a large magnetic field. This is usually done in a superconducting magnet installation with which fields of up to 10 T can now be obtained. Pressure-dependence studies require specially designed equipment which is less commonly available.

In view of the problem of counting statistics already mentioned, it is highly desirable to make the resonant absorption as strong as possible. This can in principle be done by increasing the absorber thickness, but the mathematics of the process reveal a saturating effect on the resonance as well as an increase in Compton scattering. Having optimised the absorber thickness, it is possible to increase the recoilless fractions f_s and f_a by cooling. It is very important to choose a source which gives a single line unbroadened by hyperfine interactions and with a large value for f_s. The best choice of matrix for the radioisotope is usually a high-melting metal or refractory oxide with a high Debye temperature. In the case of ^{57}Fe, the parent isotope is ^{57}Co which is thermally diffused into a metal matrix such as palladium, platinum, or rhodium.

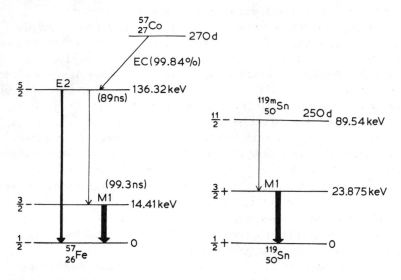

Fig. 5.4 The decay schemes for ^{57}Co and ^{119}Sn showing the Mössbauer transitions.

The decay schemes of ^{57}Co and ^{119m}Sn are illustrated in Fig. 5.4. Although about 85% of the decays from the 136.32 keV level in ^{57}Fe populate the Mössbauer level at 14.41 keV, only about 10% of the decays of this state generate a 14.41 keV γ-ray, the remainder producing internal conversion electrons and an X-ray at 7 keV. Good discrimination in the detection system is therefore necessary. Although a Mössbauer resonance involving the second excited state of ^{57}Fe is possible, this is extremely weak because of the very small recoilless fraction associated with the high-energy γ-ray. The abundance of ^{57}Fe in natural iron is only 2.17%, but the high resonant cross-section σ_0 and the large recoilless-fraction obtained for this low-energy γ-transition result in satisfactory results from most iron compounds at room temperature. In special cases such as compounds with a very low iron content (e.g. dilute iron alloys or biological iron-proteins) it is possible to obtain an adequate cross-section by using iron enriched to 90% in ^{57}Fe, but this material is expensive.

Although a large number of Mössbauer resonances have been detected, only a comparatively small number of these have been used for chemical applications. Some of the more important isotopes are listed in Table 5.1, together with some of the relevant nuclear parameters.

5.3   HYPERFINE INTERACTIONS

We have seen how the Mössbauer effect allows the observation of nuclear resonance as an absorption line with Lorentzian shape and width Γ_r. This would have a very limited usefulness if it were not for the fact that the energy of the

Table 5.1 Nuclear parameters for selected Mössbauer transitions

Isotope	E_γ(keV)	Γ_r(mm s^{-1})	I_g	I_e	Natural abundance (%)	Nuclear decay
^{57}Fe	14.41	0.192	1/2−	3/2−	2.17	^{57}Co(EC 270d)
^{99}Ru	90	0.147	5/2+	3/2+	12.63	^{99}Rh(EC 16d)
^{119}Sn	23.87	0.626	1/2+	3/2+	8.58	^{119m}Sn(IT 250d)
^{121}Sb	37.15	2.1	5/2+	7/2+	57.25	^{121m}Sn(β^-76y)
^{125}Te	35.48	5.02	1/2+	3/2+	6.99	^{125}I(EC 60d)
^{127}I	57.60	2.54	5/2+	7/2+	100	^{127m}Te(β^- 109d)
^{129}I	27.72	0.59	7/2+	5/2+	Nil	^{129m}Te(β^-33d)
^{129}Xe	39.58	6.85	1/2+	3/2+	26.44	^{129}I(β^- 1.7 × 10^{7}y)
^{151}Eu	21.6	1.44	5/2+	7/2+	47.8	^{151}Gd(EC 120d)
^{161}Dy	25.65	0.37	5/2+	5/2−	18.88	^{161}Tb(β^- 6.9d)
^{169}Tm	8.40	9.3	1/2+	3/2+	100	^{169}Er(β^- 9.4d)
^{182}W	100.10	2.00	0+	2+	26.4	^{182}Ta(β^- 115d)
^{193}Ir	73.0	0.60	3/2+	1/2+	61.5	^{193}Os(β^- 31h)
^{197}Au	77.34	1.87	3/2+	1/2+	100	^{197}Pt(β^- 18h)
^{237}Np	59.54	0.073	5/2+	5/2−	Nil	^{241}Am(α 458y)

nucleus is slightly influenced by its chemical environment. The high definition of the resonance (about 1 part in 10^{12}) is similar in magnitude to the energy differences that occur, and it therefore becomes possible to detect them.

The Hamiltonian describing the energy of the nucleus may be written as:

$$\mathcal{H} = \mathcal{H}_0 + E_0 + M_1 + E_2 \qquad (5.13)$$

where $\mathcal{H}_0$ contains all terms other than the hyperfine interactions with the environment. E_0 is a Coulombic interaction with the electrons which alters the separation of the ground and excited states of the nucleus, and thus can cause a shift of the resonance line. It is therefore known as the chemical isomer shift (often referred to as the isomer shift or centre shift). The M_1 term refers to an interaction of the nuclear spin with a magnetic field. The latter may be intrinsic or extrinsic to the atom, and the resultant multiplet line structure is known as magnetic hyperfine splitting. The E_2 term describes the interaction of the nuclear quadrupole moment with the local electric field gradient produced by the surrounding charge distribution. This also results in line splitting, and is known as an electric quadrupole interaction.

All three terms may occur together, but only the magnetic hyperfine and electric quadrupole interaction have a complex inter-relation because they are both directional. It is convenient to discuss them individually.

5.3.1 Chemical isomer shift

It is normally adequate to consider an electron as being influenced by Coulombic interaction with a positively charged *point* nucleus. In this model, the decay of a nuclear excited state to a ground state causes no change in the Coulombic interaction energy. However, in the current context it must be recognized that the nucleus has a finite size. Using the appropriate relativistic equations for the electrons, it may be seen that an s-electron has a finite probability of being at the origin; i.e. it may be *inside* the nucleus. The integrated electrostatic energy for an electron of charge $-e$ moving in the field of a point nucleus of charge $+Ze$ is given by:

$$E_0 = \frac{-Ze^2}{4\pi\epsilon_0} \int_0^\infty \psi^2 \frac{d\tau}{r} \qquad (5.14)$$

where ϵ_0 is the permittivity of a vacuum, r is the radial distance, and $d\tau$ is a volume element. This equation overestimates the interaction within the nuclear radius R. It also fails to allow for the fact that the nucleus expands (or contracts) as it changes its energy state. The change in nuclear radius δR results in a difference in energy of:

$$\Delta W = \frac{6(\rho + 1)|\psi_s(0)|^2 Ze^2}{\epsilon_0(2\rho + 1)(2\rho + 3)\Gamma^2(2\rho + 1)} \left(\frac{2Z}{a_H}\right)^{2\rho-2} R^{2\rho} \frac{\delta R}{R} \qquad (5.15)$$

where $\rho = \sqrt{[1 - Z^2(e^2/\hbar c)^2]}$, a_H is the first Bohr radius, and $\psi_s(0)$ is the non-relativistic Schrödinger wave-function at $r = 0$. This equation is rather intractable, but if one compares *two* chemical environments A and B and assumes that $\rho \approx 1$, one can predict an energy difference of:

$$\delta = \frac{Ze^2}{5\epsilon_0} R^2 \frac{\delta R}{R} \{|\psi_s(0)_A|^2 - |\psi_s(0)_B|^2\} \qquad (5.16)$$

This equation contains the product of a nuclear term, which is a constant for a given Mössbauer transition, and a chemical term $\{|\psi_s(0)_A|^2 - |\psi_s(0)_B|^2\}$ which represents the difference in s-electron density *at the nucleus*. δ is called the chemical isomer shift.

If $\delta R/R$ is positive, a positive value of δ implies that the s-electron density at the nucleus in matrix A is greater than in matrix B. A and B can be the absorber and source in a Mössbauer experiment, or B can be a reference absorber to which A is compared.

$|\psi_s(0)|$ contains contributions from all the occupied s-orbitals, 1s, 2s, 3s, ... , but is obviously most sensitive to change through the outer shells participating in chemical bonding. Other electrons with p-, d-, and f-character do not have a direct interaction within the nucleus, but nevertheless have a significant indirect effect on δ by shielding of the outer s-electrons. We shall see examples of this in subsequent sections.

The effect of the chemical isomer shift on the Mössbauer spectrum is to shift the resonance line to higher or lower velocity according to the sign of δ. An excellent example of this is given by comparing the ^{127}I and ^{129}I spectra of $Na_3H_2IO_6$ in Fig. 5.5. The sources used [5.4] were the appropriate tellurium precursors ^{129m}Te and ^{127m}Te in zinc telluride matrices, and were therefore chemically equivalent. The resonances occur on opposite sides of zero velocity because $\delta R/R$ is negative in ^{127}I and positive in ^{129}I. Note the different line-widths for ^{127}I ($\Gamma_r = 2.54$ mm s^{-1}) and ^{129}I ($\Gamma_r = 0.59$ mm s^{-1}).

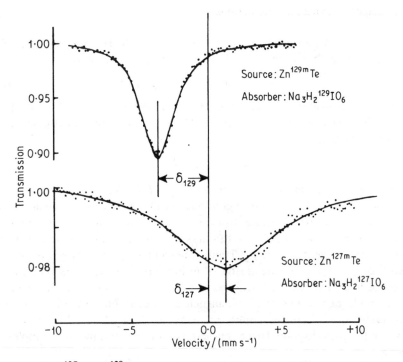

Fig. 5.5 The ^{127}I and ^{129}I spectra for $Na_3H_2IO_6$ using ZnTe sources showing the different line-widths, and the larger chemical isomer shift of opposite sign in ^{129}I. [5.4].

However, the observed shift also contains a second contribution in addition to the true chemical isomer shift which results from the vibration of an atom on its lattice site. The relativistic equation for the Doppler effect on the emitted γ-ray predicts a frequency shift from that of a stationary atom of:

$$\nu = \nu_0(1 + \langle v^2 \rangle/2c^2) \tag{5.17}$$

where $\langle v^2 \rangle$ is the mean-square velocity of the atom in the direction of emission. Thus there is a change in the energy of the Mössbauer line by:

$$\delta E_\gamma/E_\gamma = -\langle v^2 \rangle/2c^2 \tag{5.18}$$

Vibrations increase with rising temperature, so δE_γ is temperature dependent, and the Mössbauer resonance will move to lower velocity. There is also a significant zero-point motion contribution at absolute zero.

Fortunately, the second-order Doppler shift is largely self-cancelling when δ values are compared in similar compounds, and is therefore conveniently ignored in most instances.

267

5.3.2 Magnetic hyperfine interactions

If the nucleus has a spin I and a magnetic moment μ, the Hamiltonian describing the interaction with a magnetic field of flux density B is given by:

$$\mathcal{H} = -\mu \cdot B = -g_N \mu_N I \cdot B \qquad (5.19)$$

where μ_N is the nuclear magneton $(eh/4\pi m_p = 5.04929 \times 10^{-27} \text{A m}^2 \text{ or J T}^{-1})$ and g is the nuclear g-factor $[=\mu/(I\mu_N)]$. The solution of this Hamiltonian gives the energy levels as:

$$E_m = -g_N \mu_N B m_z \qquad (5.20)$$

where m_z is the magnetic quantum number $m_z = I, I-1, \dots, -I$. Thus the magnetic field splits the nuclear level into $(2I + 1)$ non-degenerate equally spaced levels. For a Mössbauer transition from an excited state with $I_e = \frac{3}{2}$ to a ground state with $I_g = \frac{1}{2}$, one finds that both states are split according to the values of the magnetic moments μ_e and μ_g. Transitions can take place between these levels if $\Delta m_z = 0, \pm 1$, and are illustrated in Fig. 5.6, there being six allowed resonance lines. The relative intensities of these are determined by some complex vector-coupling equations which are angular dependent, but in this instance for a randomly oriented sample simplify to a strict $3 : 2 : 1 : 1 : 2 : 3$ ratio as shown in the stick diagram. Higher nuclear spin states result in more complex spectra.

The magnetic hyperfine interaction also is a product of a nuclear constant $(g_N \mu_N)$ and an external variable (the flux density B). The latter may be applied by an external magnet, but can also be intrinsic to the atom. Any unpaired electron in an atom induces a slight imbalance in the s-electron spin-density at the nucleus because it interacts differently with the parallel and antiparallel spins of the other electrons, and this results in a large local magnetic field of up to 100 T at the nucleus. In a magnetically ordered solid, the direction of the unpaired spin and thus of the field are effectively frozen, and the result is magnetic hyperfine splitting. In a paramagnetic solid, the direction of the spin is rapidly changing by electronic spin-relaxation, and the time average of B is usually zero within the lifetime of the Mössbauer excited state, so no splitting is seen. Exceptions can occur when the relaxation time is very long, but these generally embrace the intermediate time-dependent situation, and discussion of this is deferred until later in the chapter.

5.3.3 Electric quadrupole interactions

The electric quadrupole interaction in Mössbauer spectroscopy is analogous to that in n.q.r. spectroscopy. Either or both of the Mössbauer nuclear levels may have a spin of $I > \frac{1}{2}$ and thus a nuclear quadrupole moment eQ. The electronic charge distribution about the nucleus is usually not spherical, and an electric field gradient tensor may be defined at the nucleus in terms of the electrostatic potential V such that:

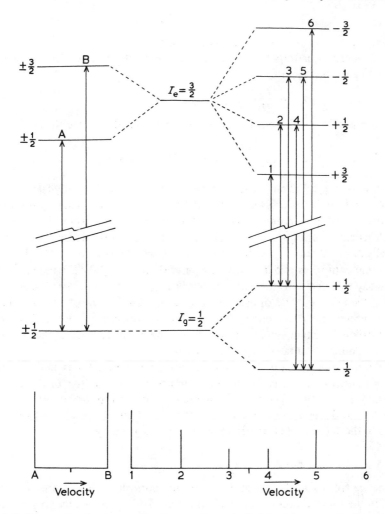

Fig. 5.6 Energy level schemes for the quadrupole (left) and magnetic hyperfine (right) splitting expected for a 3/2 → 1/2 Mössbauer transition. The relative inversion of the 3/2 and 1/2 multiplets in the latter case signifies a change in the sign of the magnetic moment, and the levels are scaled to be appropriate to ^{57}Fe.

$$E_{ij} = -V_{ij} = -(\partial^2 V/\partial x_i \partial x_j)$$

$$(x_i, x_j = x, y, z) \tag{5.21}$$

Because the Laplace equation demands that:

$$V_{xx} + V_{yy} + V_{zz} = 0 \tag{5.22}$$

it is only necessary to specify two parameters to define the tensor completely, and these are chosen as the maximum value of the field gradient ($V_{zz} \equiv eq$), and an asymmetry parameter defined as:

269

$$\eta = (V_{xx} - V_{yy})/V_{zz} \tag{5.23}$$

where $|V_{zz}| > |V_{yy}| \geqslant |V_{xx}|$ such that $0 \leqslant \eta \leqslant 1$.

The nuclear quadrupole moment interacts with the electric field gradient according to the Hamiltonian:

$$\mathcal{H} = \frac{eQ}{2I(2I-1)} (V_{zz}\hat{I}_z^2 + V_{yy}\hat{I}_y^2 + V_{xx}\hat{I}_x^2) \tag{5.24}$$

In the case of $I = \frac{3}{2}$ this equation may be solved exactly to give the eigenvalues:

$$E_Q = \frac{e^2 q Q}{4I(2I-1)} [3I_z^2 - I(I+1)] \left(1 + \frac{\eta^2}{3}\right)^{1/2} \tag{5.25}$$

where $I_z = \pm\frac{3}{2}, \pm\frac{1}{2}$. Thus there are two energy levels, at $+(e^2 qQ/4)(1 + \eta^2/3)^{1/2}$ for $I_z = \pm\frac{3}{2}$, and $-(e^2 qQ/4)(1 + \eta^2/3)^{1/2}$ for $I_z = \pm\frac{1}{2}$. For higher spin states, explicit equations can only be written if $\eta = 0$, but in any event the energy levels remain a series of Kramers' doublets. The selection rule for transitions is usually $\Delta m = 0, \pm 1$, so that for an $I_e = \frac{3}{2} \to I_g = \frac{1}{2}$ transition the Mössbauer spectrum comprises two lines with a separation of $(e^2 qQ/2)(1 + \eta^2/3)^{1/2}$ which is usually referred to as the *quadrupole splitting* and given the symbol Δ. When $\eta = 0$, then $\Delta = e^2 qQ/2$ or half the quadrupole coupling constant defined in n.q.r. spectroscopy. The energy level scheme is shown in Fig. 5.6, and is appropriate for example to ^{57}Fe and ^{119}Sn. The intensity ratio of the doublet becomes unity for a randomly oriented sample.

It is important to note that the spectrum is symmetrical only in the case of a $\frac{3}{2} \to \frac{1}{2}$ transition. There is thus no means of establishing the sign of V_{zz} or the magnitude of η. However, these parameters can be determined in other ways. The angular dependence of the ratio of the intensities of the $\pm\frac{3}{2} \to \pm\frac{1}{2}$ (π) transition to the $\pm\frac{1}{2} \to \pm\frac{1}{2}$ (σ) transition when $\eta = 0$ is given by:

$$\frac{I_\pi}{I_\sigma} = \frac{3(1 + \cos^2\theta)}{2 + 3\sin^2\theta} \tag{5.26}$$

so spectra from a single-crystal can determine both the direction and sign of V_{zz}. In the event that no single crystals are available, it is usually possible to determine the sign of V_{zz} from a spectrum measured with the sample in a large external magnetic field of the order of 3–10 T. In the case of ^{57}Fe, the $\pm\frac{1}{2} \to \pm\frac{1}{2}$ transition splits into an apparent triplet, while the $\pm\frac{3}{2} \to \pm\frac{1}{2}$ transition splits into a broadened doublet. However, combined magnetic/quadrupole interactions are too complex to treat in detail in this account.

The electric field gradient V_{zz} can originate in two ways. Each electron in the atom makes a contribution to the tensor V_{ij}, and if the orbital occupation is not spherical, the total value of V_{zz} will be non-zero because of this *valence* contribution. Thus, if there is an excess of electron density along the z axis (electrons in p_z, d_{z^2}, d_{xz}, or d_{yz}), V_{zz} will be negative in sign, and if the excess is in the xy plane (electrons in p_x, p_y, p_{xy}, $d_{x^2-y^2}$,), V_{zz} will be positive in sign. If the immediate chemical bonding dominates V_{zz}, we can now say something about the occupation of the orbitals.

An additional contribution originates from distant charge (perhaps ionic charges) which is therefore called the *lattice* contribution. This term is particularly important for an s-state ion such as high-spin Fe^{3+} with a d^5 configuration where the 'valence' contribution is formally absent. It must be realized of course that the distinction between 'valence' and 'lattice' is not rigorous since a lattice charge can induce an electric field gradient in the valence electrons by polarisation.

However, it is now clear that the quadrupole interaction is a measure of the asymmetry of the atomic environment and relates directly to the electron orbitals involved in chemical bonding.

5.3.4 Lattice dynamics

Both the probability of recoilless absorption and the second-order Doppler shift are governed by the vibrational characteristics of the absorbing nucleus. Therefore it follows that a study of these can provide information about the lattice dynamics of the solid matrix. The latter may be either a true compound of the resonant isotope or a non-resonant compound doped with a small quantity of a resonant impurity.

The recoilless fraction is determined by Equation (5.9), and is a function of the mean-square vibrational amplitude $\langle x^2 \rangle$, while the second-order Doppler shift is given by Equation (5.18), and is a function of the mean-square velocity $\langle v^2 \rangle$. This distinction is rather important because the mean value $\langle x^2 \rangle$ is weighted towards the low-frequency vibrations of large amplitude (the acoustical modes of the crystal) while $\langle v^2 \rangle$ is weighted towards the high-frequency vibrations (optical modes). The actual vibrational behaviour of any real solid is usually complex, and these two measurements are not strictly comparable except for the 'ideal' solid to which even pure metals only approximate.

The simplest mathematical treatment is to use the Debye model, which assumes a continuum of vibrational harmonic oscillator frequencies between zero and a maximum value of v_D, and following the distribution law $N(v) = $ const $\times v^2$ where $N(v)$ is the number of vibrational states of frequency v. A characteristic temperature θ_D called the Debye temperature is defined using the relationship $hv_D = k\theta_D$. The average frequency is then given by $hv = \frac{3}{4} k\theta_D$. The probability of exciting the different levels is temperature dependent; atoms vibrate more readily as the temperature increases, and the incorporation of this into the Debye model leads directly to the expression for f in Equation (5.10). If the Debye model is valid, f is a function of E_R, T, and θ_D. Only the latter is unknown, and can be directly determined from values for f. The low-temperature value of f is given by $f = \exp\left[-3E_R/(2k\theta_D)\right]$, and is independent of temperature, being effectively determined by the zero-point vibrations. The second-order Doppler shift is also temperature independent just above absolute zero, with a zero-point value of $\delta E_\gamma/E_\gamma = -(9/16)k\theta_D/(Mc^2)$, but at high temperatures it approaches the classical limit of $\delta E_\gamma/E_\gamma = -(3/2)RT/(Mc^2)$ which is linearly dependent on temperature.

Although these concepts have been applied with some success to metallic systems, the vibrational characteristics of molecular compounds are very poorly understood. However, by empirical reasoning it is sometimes possible to assess the relative importance of optical and acoustical vibrations. In particular, a polymeric structure usually leads to an abnormally high recoilless fraction at high temperatures when compared to a similar but monomeric compound.

The recoilless fraction is only strictly isotropic in a cubic crystal, and atoms in an asymmetric site can be expected to show anisotropy in f. The Mössbauer spectrum of a powder merely records an averaged value, but at the present time very few measurements of anisotropy in single crystals have been made. An important consequence of anisotropy is that the powder-averaged intensities of hyperfine spectra can deviate from the expected values. Thus the quadrupole spectrum from a $\frac{3}{2} \rightarrow \frac{1}{2}$ transition should be two lines of equal intensity, but with anisotropy of f, a small but detectable deviation may occur. This phenomenon is known as the Goldanskii–Karyagin effect, but although well substantiated it is difficult to distinguish from similar effects caused by partial orientation of microcrystallites on compacting.

5.3.5 Time-dependent phenomena

The time-scale for observation of a Mössbauer event is the lifetime of the nuclear excited state. This is long compared with the frequency of the atomic vibrations, so the resonant nucleus is only influenced by the time average of the atomic motion. However, in a ferromagnetic or antiferromagnetic material, the atomic spin S is oriented in a particular direction for long enough to appear static to the Mössbauer nucleus, while in a paramagnetic substance S changes its direction by relaxation very quickly so that the time-averaged value $\langle S \rangle$ is effectively zero within less than 0.1 μs. Thus, in the first case the atomic spin induces a static hyperfine field at the nucleus which results in magnetic hyperfine splitting, while in the paramagnet the time-averaged induced field is zero, and no hyperfine effect is seen.

However, it is possible to conceive of an intermediate situation where the spin direction changes in a time comparable to the Mössbauer lifetime, and this causes important effects on the Mössbauer spectrum. The relaxation which causes the atomic spin to change direction relative to the nuclear spin takes place by two main mechanisms. In spin–lattice relaxation there is a transfer of energy to the phonon modes of the lattice, so that the rate of relaxation becomes slower with decrease in temperature. In spin–spin relaxation the dipolar interactions between adjacent atomic spins results in a mutual spin-flip which conserves the total energy. This process is therefore independent of temperature, but slows rapidly as the internuclear distance increases.

The Mössbauer spectrum observed when the relaxation time is of the order of the excited state lifetime has to be calculated by taking the ensemble average using equations directly related to those for motional narrowing effects in n.m.r. spectroscopy. These calculations show that as the relaxation time decreases, the magnetic hyperfine components broaden inwards and eventually coalesce into the motionally narrowed single line (or quadrupole pattern) corresponding to a zero field. It is easiest to illustrate relaxation effects by

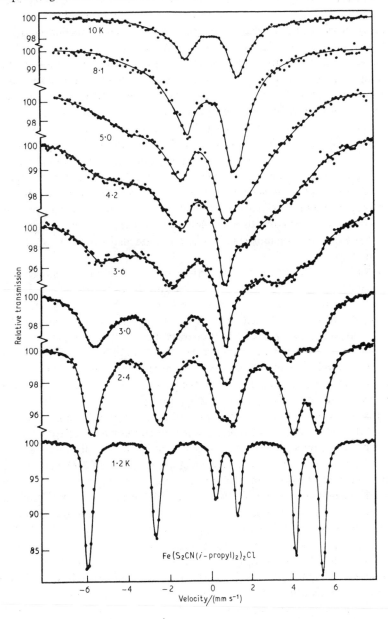

Fig. 5.7 Mössbauer spectra of $(Pr^i_2 NCS)_2$ FeCl in the temperature range 1.2–10 K showing the effects of a decrease in the relaxation time. [5.5]

273

example, and the experimental data for $(Pr_2^i NCS)_2 FeCl$ in Fig. 5.7 illustrate
the main features [5.5]. This unusual compound has an $S = \frac{3}{2}$ intermediate
spin-configuration, and the $S_z = \pm\frac{3}{2}$ electronic ground state has a long relaxation
time. However, the $S_z = \pm\frac{1}{2}$ excited state becomes thermally populated above
1 K, and can relax quickly. The observed magnetic hyperfine splitting at 1.2 K
is quickly broadened and collapses to a simple quadrupole splitting as thermal
excitation induces a drastic decrease in the relaxation time.

From the earlier remarks, it will be seen that relaxation in a paramagnetic
compound will be slow at lower temperatures and when the atomic spins are
comparatively dilute. The spin–lattice relaxation mechanism involves the spin–
orbit interaction $\lambda L \cdot S$, and is therefore dependent on the orbital angular momentum $\langle L \rangle$. For an S-state ion such as high-spin Fe^{3+} where $\langle L \rangle = 0$ one
expects a comparatively slow relaxation, and indeed any Fe^{3+} compound with
an Fe – Fe separation of greater than 6 Å gives a spectrum with very broad lines
which become increasingly asymmetric on cooling. Compounds of high-spin
Fe^{2+} and low-spin Fe^{III} which have a non-zero $\langle L \rangle$ rarely show these effects.
However, paramagnetic relaxation is frequently found in biological iron com-
pounds because of the large distances between the paramagnetic atoms.
Numerous examples of paramagnetic relaxation are also known in rare-earth
compounds.

For an $S = \frac{1}{2}$ ion, the observed hyperfine splitting is similar to a magnetic
hyperfine splitting in a ferromagnetic compound. However, when $S > \frac{1}{2}$ one
sees a superposition of the hyperfine spectra from the $S_z = \pm\frac{1}{2}, \pm\frac{3}{2}$, etc.
Kramers' doublets which usually have different relaxation times and therefore
show motional narrowing to different degrees. Paramagnetic relaxation spectra
in, for example, the $S = \frac{5}{2}$ Fe^{3+} ion are therefore complex.

A similar kind of relaxation known as superparamagnetism is sometimes found
in magnetically ordered materials when the particle size or effective magnetic
domain size becomes very small. The number of interacting spins is then
smaller than usual, and the probability of spin-relaxation is proportionately
greater. The net result is an apparent inward broadening and collapse of the
normal magnetic hyperfine splitting which is similar to paramagnetic relaxation,
and has been seen for example in ultra-small particles of α-$Fe_2 O_3$ and α-FeOOH.

Another process which can result in spectrum averaging is electron hopping
between adjacent atomic sites. Thus $Eu_3 S_4$, which is nominally $Eu^{2+}Eu_2^{3+}S_4$,
gives a spectrum at low temperature comprising two distinct resonance lines in
the intensity ratio of 1 : 2 as expected [5.6]. Between 210 K and 273 K the
spectrum broadens and coalesces into a single line centred at the centre of
gravity of the low-temperature spectrum. The extra electron on the Eu^{2+} site
can 'hop' to a neighbouring Eu^{3+} site in a process characterized by a relaxation
time τ and an activation energy E_A so that $\tau = \tau_0 \exp(-E_A/kT)$.

The presence of atomic diffusion is usually seen as a broadening of the
Mössbauer line, but the theoretical treatment is more difficult, and comparatively
few data are available.

5.4 MOLECULAR STRUCTURE

We have seen that the hyperfine interactions in the Mössbauer spectrum can provide information about the immediate electronic environment of the resonant nucleus. In some instances the arrangement of the nearest-neighbour atoms may be unknown, and one can hope to determine the atomic geometry. In other cases the geometry is known or can be inferred from other data, but information is lacking concerning the formal oxidation state, electronic configuration, or chemical binding. It is perhaps in this latter context of electronic structure that Mössbauer spectroscopy is most useful, but there are numerous examples in which molecular configurations have been determined.

5.4.1 Site symmetry and molecular geometry

Any individual application of Mössbauer spectroscopy to determine molecular geometry is decided by circumstance, but the following examples will serve to indicate the usefulness of the technique. As a simple example, the ^{129}I spectrum of $[IF_6]^+[AsF_6]^-$ is a single line, thereby confirming that the $[IF_6]^+$ cation has a regular octahedral geometry (cubic symmetry); in contrast, the $[IF_6]^-$ anion in $CsIF_6$ shows a large quadrupole splitting because the additional lone-pair of electrons is stereochemically active and results in a distorted 7-coordinate geometry [5.7].

Another good example from iodine chemistry concerns the geometry of the compound $I_2Cl_4Br_2$, which is derived from the planar bridged structure of I_2Cl_6. Accordingly, there are six possible structures (a)–(f), as shown in Fig. 5.8. The ^{129}I spectrum shows two superimposed quadrupole patterns [5.8], indicating that the two iodine atoms are not identical. This eliminates structures (b), (c), and (d). The quadrupole coupling constants $e^2q^{127}Q_g/h$ at 80 K are $+3040$ and $+2916\,MHz$, compared with $+3060\,MHz$ in I_2Cl_6. Thus one iodine in $I_2Cl_4Br_2$ has a geometry similar to that in I_2Cl_6, eliminating structures (e) and (f), and establishing structure (a) as the correct one.

One of the earliest successes of Mössbauer spectroscopy was in establishing a structure for the iron carbonyl $Fe_3(CO)_{12}$. The then accepted geometry, based on incomplete X-ray data, was an equilateral triangle of $Fe(CO)_2$ groups joined to each other by three pairs of bridging carbonyl groups so that each iron atom was equivalent. The observed Mössbauer spectrum as illustrated in Fig. 5.9 comprises three lines of equal intensity [5.9], the outer pair of which was assigned to a quadrupole splitting of two apparently identical iron atoms, and the centre line to a third atom in a different but more symmetrical environment. Clearly, this spectrum was incompatible with the proposed structure.

A key to the correct structure was provided by a study of the hydride derivative $[Fe_3(CO)_{11}H]^-$ which shows a very similar spectrum (see parameters in Table 5.2) and therefore has the hydrogen atom in a bridging position

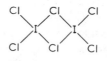

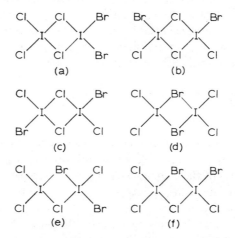

Fig. 5.8 The six possible planar structures for $I_2Cl_4Br_2$.

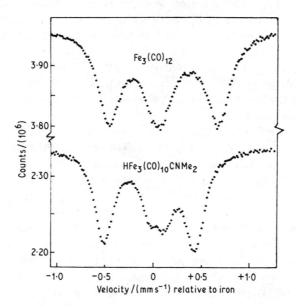

Fig. 5.9 The Mössbauer spectra of $Fe_3(CO)_{12}$ and $HFe_3(CO)_{10}CNMe_2$. [5.11]

Table 5.2 Mössbauer parameters[†] for derivatives of $Fe_3(CO)_{12}$ at 80 K

Compound	Outer doublet		Inner doublet	
	Δ	δ	Δ	δ
$Fe_3(CO)_{12}$	1.13	+0.11	~0	+0.05
$[Fe_3(CO)_{10}H]^-$	1.41	+0.04	0.16	+0.02
$HFe_3(CO)_{10}CNMe_2$	0.94	−0.04	0.16	+0.04

[†] Values in mm s^{-1}. δ is given relative to iron metal at room temperature.

replacing a carbonyl between the two equivalent iron atoms [5.10]. The proposed structure for this compound (Fig. 5.10) and the parent carbonyl have now been confirmed by X-ray analysis.

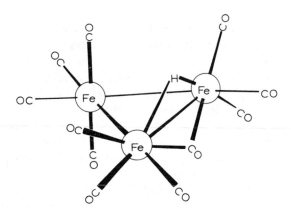

Fig. 5.10 The molecular structure of $[Fe_3(CO)_{11}H]^-$.

Similar success has been achieved in establishing a structure for the compound initially reported as $HFe_3(CO)_{11}CNMe_2$. The Mössbauer spectrum (Fig. 5.9) indicates that it is also a derivative of $Fe_3(CO)_{12}$ with any structural changes affecting the bridged iron atoms equally [5.11]. Mass-spectroscopy was used to confirm the proposed formulation of $HFe_3(CO)_{10}CNMe_2$, a novel structure featuring one hydride and a $CNMe_2^+$ bridge replacing the carbonyls between the two equivalent iron atoms.

5.4.2 Geometrical isomerism

There are many instances of pairs of compounds with the same chemical formulation but different three-dimensional configurations. Perhaps the most familiar example is *cis/trans* geometrical isomerism in octahedral coordination compounds of the type MA_2B_4, where the A groups are in either adjacent (*cis*) or diametrically opposed (*trans*) coordination positions. The Mössbauer spectra of *cis* and *trans* isomers have proved to be characteristically different. The electric field gradient at the central atom M reflects the asymmetry of the geometry. Regardless of the nature of A and B, and provided that the geometry is close to octahedral in both cases without a large *trans*-effect in the bonding, it can be shown that the values of V_{zz} in *cis*-MA_2B_4 and *trans*-MA_2B_4 are in the ratio 1 : 2 and of opposite sign. An example of this using the ^{57}Fe resonance is given by *cis*- and *trans*-$FeCl_2(p\text{-MeO}\cdot C_6H_4 \cdot NC)_4$ where $\Delta = -0.83$ and $+1.59$ mm s^{-1} respectively [5.12]. Similar results have also been found in 6-coordinate organotin compounds, even in the presence of chelating ligands which distort the geometry considerably. Thus in $Me_2Sn(acac)_2$ we find $|\Delta| = 3.93$ mm s^{-1}, and in $Ph_2Sn(acac)_2$ $|\Delta| = 2.14$ mm s^{-1}, so they may be assigned *trans* and *cis* stereochemistries respectively.

5.5   ELECTRONIC STRUCTURE

The hyperfine interactions in the Mössbauer spectrum find a ready application in the study of the electronic environment of the resonant nucleus. We have already seen that the chemical isomer shift is not only a direct function of the s-electron density at the nucleus, but is also influenced indirectly by inter-penetration shielding-effects from p-, d-, and f-electrons. The quadrupole splitting derives from an asymmetric occupation of the p-, d-, and f-orbitals, and its magnitude, sign, and direction can be easily related to the electron orbitals used in bonding. In those compounds which are magnetically ordered, the hyperfine field relates to the number and orbital angular momentum of the unpaired electrons. Thus the hyperfine interactions are sensitive to most changes in chemical binding in the valence shells of the atom.

   Paramagnetic compounds often show strongly temperature-dependent spectra because of changes in the orbital occupation by thermal excitation, but in contrast the electronic properties of diamagnetic compounds are almost temperature-independent. In the latter case the Mössbauer spectrum gives correspondingly less useful information, and the primary observable parameters are the chemical isomer shift and the quadrupole splitting. In isolation these may convey very little information, but a systematic study of large series of related compounds generally leads to a useful semi-empirical interpretation of the data.

5.5.1 Oxidation state and electron
configuration

An example of useful empiricism is in the determination of the formal oxidation state and electron configuration in a compound. For example, in the spectra of tin compounds one finds a large difference in chemical isomer shift between those with the formal oxidation state II and those which have lost the $5s^2$ electron-pair to give oxidation state IV (approximately $+2.4$ to $+4.1$ and

-0.5 to $+2.2$ mm s^{-1} relative to SnO$_2$ respectively). The lack of overlap between the two allows a unique assignment of the oxidation state. Thus in certain types of transition-metal complex such as $(Et_4N)_3 Pt(SnCl_3)_5$ which contain nominal SnCl$_3^-$ groups and are made from a tin(II) starting material, the observed shifts of ~ 1.6–1.8 mm s^{-1} show that the electronic configuration at the tin is more typical of the IV oxidation state.

More complicated in detail, but perhaps more useful in application, are the changes found for different oxidation states in a transition metal such as iron or ruthenium. The different 3d-electron configurations of iron have a marked shielding effect, particularly on the 3s-orbital, as shown by the calculated values [5.13] for $\psi_{ns}^2(0)$ given in Table 5.3. A progressive increase in the number of 3d-electrons causes a reduction in the total s-electron density. The sign of $\delta R/R$ is negative, so the chemical isomer shift increases; thus an Fe^{2+} compound (3d^6) is shifted to more positive velocity than an Fe^{3+} compound (3d^5). However, absolute calibration of the chemical isomer shift is very difficult because the pure ionic configurations are unknown in nature; all chemical compounds feature some degree of covalent bonding, particularly those containing cations with a large formal charge such as Fe^{3+}. The variations in covalent overlap with different ligands result in a range of chemical isomer shifts for any given formal configuration, and an empirically compiled summary of the experimental data is shown in Fig. 5.11. This figure also illustrates another important point. Some oxidation states such as Fe^{2+} (3d^6) can exist in more than one electronic configuration, i.e. high-spin ($S = 2$), intermediate-spin ($S = 1$), and low-spin ($S = 0$). These differ in electron correlation and shielding effects on the s-electrons, and therefore show different shifts.

Table 5.3 Electron densities at $r = 0$ for different configurations of an Fe atom

Electrons per cubic Bohr radius	3d^8	3d^7 Fe$^+$	3d^6 Fe^{2+}	3d^5 Fe^{3+}	3d^6 4s^2 free atom
$\psi_{1s}^2(0)$	5378.005	5377.973	5377.840	5377.625	5377.873
$\psi_{2s}^2(0)$	493.953	493.873	493.796	493.793	493.968
$\psi_{3s}^2(0)$	67.524	67.764	68.274	69.433	68.028
$\psi_{4s}^2(0)$					3.042
$2\sum_n \psi_{ns}^2(0)$	11,878.9	11,879.2	11,879.8	11,881.7	11,885.8

A particularly striking example [5.14] of the determination of electronic configuration is given by spectra from a compound of iron(II) with a hydro-tris-(1-pyrazolyl)borate chelate (Fig. 5.12) which is in the $^5T_2$ high-spin state at room temperature (quadrupole doublet with $\delta = +1.04$, $\Delta = 3.74$ mm s^{-1}), and in the $^1A_1$ low-spin state at low temperature (single line with $\delta = 0.47$ mm s^{-1} at 4.2 K). The intermediate spin-crossover region sees a mixture of the two forms, which probably adopt different crystal structures resulting in only a slow inter-

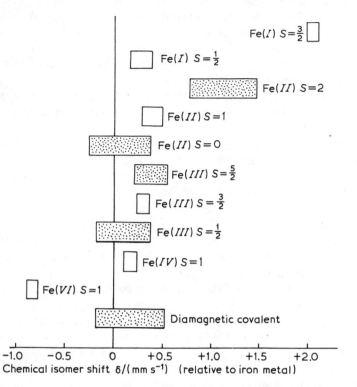

Fig. 5.11 An approximate representation of the ranges of chemical isomer shifts relative to iron metal at room temperature for different electron configurations of iron. The most common configurations are shown shaded, and as a general rule the more ionic compounds within each block have the higher shift. The shift usually decreases as spin-pairing takes place.

conversion. The temperature dependence of the spectrum in the intermediate region is a sensitive monitor of the system, and gives evidence for both particle-size effects and hysteresis behaviour in the interconversion.

Any increase in covalent character will result in an effective decrease in 3d-shielding because of the nephelauxetic effect of electron withdrawal towards the ligands. The result is a corresponding decrease in shift, so that for instance the shifts for the anhydrous iron(II) halides at 4.2 K are in the order FeF_2 ($\delta = 1.48$ mm s^{-1} relative to iron metal), $FeCl_2$ (1.16), $FeBr_2$ (1.12), and FeI_2 (1.04). Iron (II) compounds with a tetrahedral coordination are more covalent than corresponding octahedral compounds, and the effect of this is shown in Fig. 5.13 for some simple halogen and pseudohalogen derivatives. However, in interpreting small differences in shift, the presence of the small contributions from the second-order Doppler shift must not be overlooked.

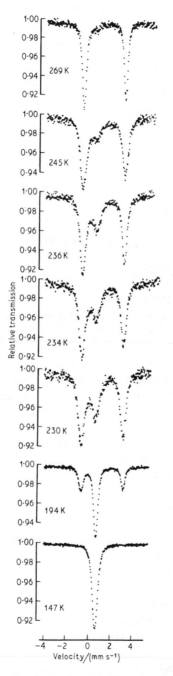

Fig. 5.12 Typical spectra for the hydro-tris-(1-pyrazolyl)borate chelate complex of iron(II) showing the $^5T_2 - {}^1A_1$ crossover. [5.14]

281

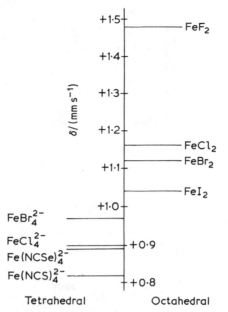

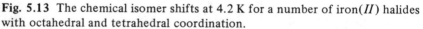

Fig. 5.13 The chemical isomer shifts at 4.2 K for a number of iron(*II*) halides with octahedral and tetrahedral coordination.

In compounds with a high degree of covalent bonding such as $[Fe^{II}(CN)_6]^{4-}$, it is often difficult to distinguish similar configurations such as $[Fe^{III}(CN)_6]^{3-}$, as both show a similar shift. The difference of ~ 0.08 mm s^{-1} should be compared with the much larger separation, for example, of ~ 0.79 mm s^{-1} between $FeBr_2$ and $FeBr_3$. However, the $S = 1/2$ configuration of low-spin Fe^{III} generally favours a more distorted molecular geometry with a temperature-dependent quadrupole splitting, and empirical criteria such as these can also be of value in ambiguous circumstances.

5.5.2   Covalent bonding

The interpretation of the hyperfine parameters in covalent compounds is a difficult problem, particularly for the main-group elements such as tin and antimony. However, in a transition metal such as iron the valence 3d-orbitals are often only partially involved in chemical bonding, and therefore retain a degree of identity which is particularly useful, especially in paramagnetic configurations. In a typical covalent compound such as $[Fe^{II}(CN)_6]^{4-}$ one has to consider the effects of π-overlap between the 3d-orbitals and the π^*-antibonding orbitals of the CN$^-$ ligands. The various contributions to the chemical isomer shift can be enumerated as follows: (1) a direction contribution from 4s-electrons; (2) indirect contributions from 3d-electrons via shielding, which are (a) the pure ionic or non-bonding shielding effect of the $3d^n$

configuration, (b) covalent overlap with the ligands in which filled ligand orbitals donate an effective total of n_2 electrons to the 3d-orbitals, and (c) covalent overlap with the ligands in which filled metal 3d-orbitals donate an effective total of n_3 electrons to the π^* ligand orbitals. The effective number of 3d-electrons is thus $n_{eff} = n + n_2 - n_3$. For a molecular orbital description in octahedral symmetry, this represents an overlap of the metal t_{2g} and ligand π and π^* orbitals, and of the metal e_g and ligand σ orbitals. Thus n_2 represents σ-donation from the ligands, and n_3 largely represents π-back-donation from metal.

Numerical estimation of these contributions in $[Fe^{II}(CN)_6]^{4-}$ and $[Fe^{III}(CN)_6]^{3-}$ has been made using molecular orbital methods [5.15]. Although the value of n is decreased by one electron, the back-donation via n_3 is decreased proportionately, while n_2 remains effectively constant, so the value of n_{eff} remains almost unchanged. Thus the increase in formal 3d-population in the hexacyanoferrate(*II*) is offset by additional delocalization towards the ligands so that the s-electron density and the chemical isomer shift are similar in both compounds. However, the two electron configurations remain orbitally distinct, as one is diamagnetic while the other is paramagnetic.

5.5.3 Magnetic ordering

Mössbauer spectroscopy is an excellent method of studying magnetically ordered materials. Below the ferromagnetic or antiferromagnetic ordering temperature, the Mössbauer resonance is split by a magnetic hyperfine interaction. The value of the magnetic field increases with decreasing temperature until it reaches a saturation value at 0 K. An approximate rule of thumb for ^{57}Fe spectra is that the value at 0 K is about 11 T per unpaired electron, so a field of 55 T can be expected for an Fe^{3+} (d^5) ion. This value is reduced by covalency, and deviations occur when $\langle L \rangle \neq 0$. The study of the magnitude and direction of this field is of considerable importance in the study of spin-exchange interactions. When a quadrupole interaction is also present, the directional properties of the two interactions ensure that, if the orientation of one of them with respect to the molecular axes is known, that of the other can be determined.

An excellent example of this is given by the compound $LiFe_2F_6$ in which the Mössbauer spectrum determines not only the relative orientation of the spin-axis and the electric field gradient tensor, but also allows a complete assignment of the magnetic structure [5.16]. This compound is antiferro-magnetically ordered below 105 K. Above this temperature the Mössbauer spectrum is comparatively simple with two quadrupole doublets of equal area which are characteristic of high-spin Fe^{2+} and high-spin Fe^{3+}. The formulation is thus $LiFe^{2+}Fe^{3+}F_6$. The parent compound FeF_2 has the simple rutile structure with the iron in a distorted octahedral geometry, and $LiFe_2F_6$ adopts the related tri-rutile structure by ordering of the Li^+ cations into every third

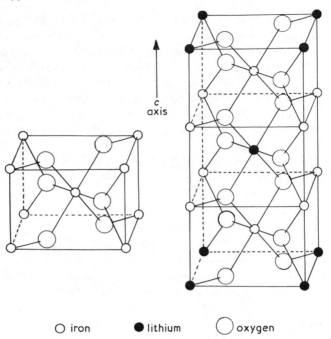

O iron   ● lithium   ◯ oxygen

Fig. 5.14 The rutile structure of FeF_2 and the tri-rutile structure of $LiFe_2F_6$.

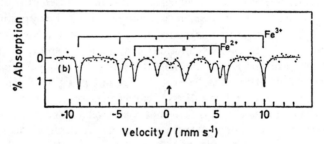

Fig. 5.15 The Mössbauer spectrum of $LiFe_2F_6$ at 4.2 K. The two component hyperfine fields are Fe^{2+} (24 T) and Fe^{3+} (59.5 T). The former is combined with a large quadrupole interaction ($\Delta = 3.1$ mm s^{-1} and $\eta = 0.4$) whose minor axis V_{xx} is along the spin direction, resulting in the line positions shown by the stick diagram.

layer. The structures are shown in Fig. 5.14. X-ray data cannot distinguish Fe^{2+} from Fe^{3+}, and it had been assumed that these cations were disordered in the antiferromagnetic phase.

The Mössbauer spectrum at 4.2 K is shown in Fig. 5.15, together with a calculated assignment of the Fe^{2+} and Fe^{3+} hyperfine patterns. The sharp resonance lines contrast with the broad spectra found in similar rutile phases such as $FeCoNiF_6$ which are disordered, and show that $LiFe_2F_6$ must have an

284

ordered cation arrangement. The interpretation of the hyperfine patterns is too complicated to detail here, but that from Fe^{2+} is similar to FeF_2 in which the magnetic spins are aligned along the crystal c axis and are perpendicular to V_{zz}. The principal values of the electric field gradient tensor are determined by the immediate environment of the Fe^{2+} site which is almost identical in the two compounds, so we can be confident that the spins are also aligned along the c axis in $LiFe_2F_6$.

The only magnetic ordering along the c axis which is antiferromagnetic is one involving strong ferromagnetic coupling between Fe^{2+} and Fe^{3+} across each Li^+ layer, and a weaker antiferromagnetic coupling through the fluoride ions between adjacent Fe^{2+} and Fe^{3+} layers as follows:

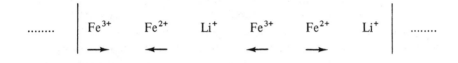

Thus the Mössbauer spectrum leads to a very detailed knowledge of the magnetic and crystal structure of this compound.

5.5.4 Biological applications

It is now recognized that many of the large organic protein molecules responsible for chemical reactions in living matter contain one or more metal atoms whose oxidation–reduction and coordination properties are of fundamental importance, and that the complicated protein chain acts largely as a template to ensure a high specificity for reaction. Thus the haemoglobin proteins, which act as an oxygen carrier, function by a direct bonding of oxygen molecules to an iron atom.

The widespread occurrence of iron-containing metalloproteins, and the specific attention of Mössbauer spectroscopy to the active centre, has led to extensive biological use of the technique. The major drawback is the need to cultivate organisms on an ^{57}Fe enriched diet to ensure adequate absorption intensity. However, the oxidation state of the iron can be determined in different chemically activated states of the protein, and paramagnetic electronic configurations frequently show relaxation broadening at low temperatures.

The haemoglobin derivatives provide an excellent example of this. The iron is coordinated by four nitrogens in a planar porphyrin unit, and is attached to the protein by the nitrogen of a histidine unit below the plane. The sixth coordination position is vacant and provides a site for coordination of small molecules such as oxygen. For example, the reduced form of haemoglobin gives the spectrum of a high-spin iron(*II*) compound with $S = 2$, while oxyhaemoglobin and haemoglobin-carbon monoxide show spectra for a low-spin iron(*II*) $S = 0$ configuration. However, the fluoride derivative has an iron(*III*)

$S = 5/2$ configuration, and the cyanide is a low-spin $S = 1/2$ iron(*III*) compound.

As a second example, the ferredoxins are enzymes which catalyse photo-chemical reactions in plants and photosynthetic bacteria. Ferredoxins from the higher plants have a molecular weight in the region of 12000–24000 and contain two atoms each of Fe and labile sulphide (i.e. not attached to the amino-acid cysteine), and accept one electron upon reduction. The Mössbauer spectrum at 195 K of the oxidized form of the ferredoxin from the green alga *Scenedesmus* (Fig. 5.16) shows a simple quadrupole doublet from two identical $S = 5/2$ high-spin Fe^{3+} cations [5.17]. From measurements in an external magnetic field it can be shown that they couple antiferromagnetically to give a diamagnetic ground-state and therefore must be close together in the protein. Upon reduction a second quadrupole doublet characteristic of $S = 2$ high-spin Fe^{2+} appears. The Fe^{2+} and Fe^{3+} pair also couple antiferromagnetically to give a net spin of $S = \frac{1}{2}$. E.s.r. measurements show the coordination of both irons to be tetrahedral. The data in combination lead to a model for the active centre which can be formulated as $(protein-S)_2 Fe(S)_2 Fe(S-protein)_2$. Detailed information such as this is not easily obtained by other methods.

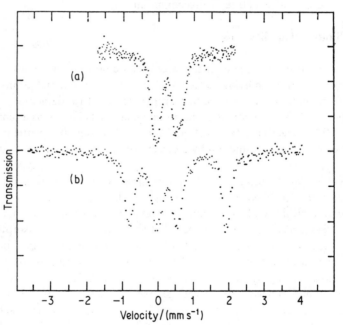

Fig. 5.16 Mössbauer spectra at 195 K of (a) the oxidized and (b) the reduced forms of *Scenedesmus* ferredoxin enriched with ^{57}Fe. [5.17]

5.6 SOLID-STATE CHEMISTRY

The Mössbauer spectrum of a pure stoichiometric compound may show hyperfine interactions from at most two or three unique nuclear environments. This will no longer necessarily be true in non-stoichiometric compounds which have a large range of composition. The most important classes of these materials are the ionic compounds such as oxides, sulphides, and silicate minerals, and the metallic alloy phases. The spectra recorded are often not only a function of the normal elemental composition, but also of order–disorder within the lattice. The spectrum of a particular sample may be very dependent on its mode of preparation, and subsequent thermal and mechanical treatment.

5.6.1 Non-stoichiometry in ionic lattices

Mössbauer spectroscopy is best suited to a study of grossly non-stoichiometric compounds, although in Section 5.6.3 we shall see that impurity doping studies can be used in special circumstances.

We have already mentioned the relationship between FeF_2 with the rutile structure and $LiFe_2F_6$ with an ordered tri-rutile structure. There are several more compounds of this type, and in particular compounds with simple divalent cation substitution such as $FeCoNiF_6$, Mg_2FeF_6, and $MgFe_2F_6$ all have the rutile structure but feature cation disorder. They are all magnetically ordered at 4.2 K, but the hyperfine patterns are broadened (Fig. 5.17) compared with those in FeF_2 and $LiFe_2F_6$ because random disorder of the cations causes small random variations in the local environment of the Fe^{2+} cations, and hence variations in the magnitude of the hyperfine field [5.16]. The Fe^{2+} ion is very sensitive to cation disorder.

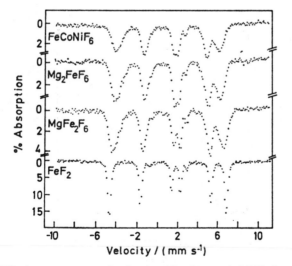

Fig. 5.17 Mössbauer spectra of mixed rutile phases at 4.2 K showing the line-broadening caused by disorder in comparison to FeF_2 which is stoichiometric.

287

Considerable interest attaches to the magnetic properties of non-stoichiometric iron oxides. As an example, one can consider the spinel oxides of the type $A[B_2]O_4$ in which the $M - O - M$ bond angles are such that the tetrahedral A-site/octahedral B-site exchange interactions far exceed the A$-$A and B$-$B interactions. Thus some magnetically concentrated oxides such as $Zn[Fe_2]O_4$ and $Ge[Fe_2]O_4$ are paramagnetic down to less than 20 K, while $MgFe_2O_4$ which has Fe^{3+} cations in both A and B sites is ferromagnetically ordered at room temperature. Many of these oxides can form a complete range of solid solutions, and the spectra in samples intermediate between magnetic and non-magnetic end-members show very complex spectra as a result of variations in site environment. A drastic reduction in the strength of the exchange interactions at some sites which are only weakly coupled to the rest of the lattice can lead to relaxation narrowing. Thus in the oxide $(Zn_{0.5}Fe_{0.5})$ $[Co_{0.5}Fe_{1.5}]O_4$, no less than 11% of the B-site iron atoms are not strongly coupled to the A-sites. In this kind of situation the magnetic hyperfine structure in the Mössbauer spectrum is generally broad and complex, and is highly temperature-dependent. A systematic study of the effects of compositional variation and temperature can be of considerable value in such circumstances.

5.6.2   Metals and alloys

It is difficult to characterize the electronic state of the atom in metallic solids where some of the electrons are in a collective band system of energy levels. The chemical isomer shift of the Mössbauer spectrum can be used as a sensitive probe in studying changes in the band-system across a range of alloy composition.

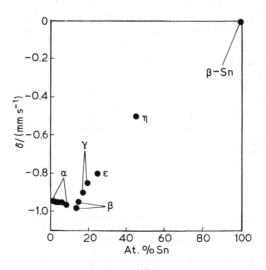

Fig. 5.18 The chemical isomer shift of ^{119}Sn in the Sn$-$Cu phases. [Ref. 5.18]

As an example we can consider the copper–tin system [5.18]. The shift is almost independent of composition within the face-centred cubic copper-rich α-phase (<10 at. % Sn) as shown in Fig. 5.18. On the other hand, there is an almost linear concentration dependence encompassing the β (~15%), γ(~16–20%), ε(Cu_3Sn), and η(Cu_6Sn_5) phases and β-tin. The conduction-electron system in copper ($3d^{10}4s^1$) comprises a collective 4s-band which is only half filled. Addition of tin ($5s^2 5p^2$) introduces additional electrons into the conduction band, and the large negative shift with respect to β-Sn indicates an unusually low s-electron density at the tin which is consistent with this transfer. The configuration of the tin remains essentially constant as the 4s-band is initially filling, and it is only at the β-phase that the tin takes a more active part in the band-structure of the metals.

Most interest attaches to those alloys which are magnetically ordered. In many iron alloys the magnetic field at a particular nucleus is sensitive to the number and nature of the nearest-neighbour atoms. Thus, if there are n nearest neighbour sites containing solute atoms, the magnetic field at the iron deviates from that in pure α-iron by $\Delta B = n\delta B_n$, i.e. the effect is linearly dependent on

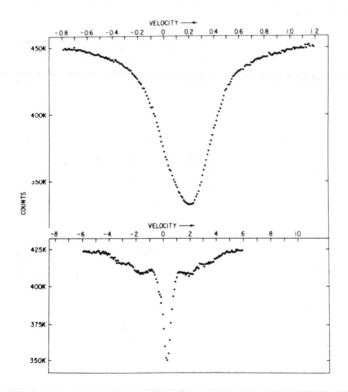

Fig. 5.19 Mössbauer spectra for the alloy $Fe_{0.604}Al_{0.396}$ showing the magnetic splitting introduced by crushing the ordered alloy. Note the different velocity scales for the two spectra. [Ref. 5.19]

the number of neighbours. This expression can be modified to include successive nearest neighbour shells, the effects being significant up to about the sixth shell.

In most iron alloys the field decreases with substitution (the exceptions are cobalt and nickel alloys), so alloying generally causes a reduction and inward broadening of the hyperfine pattern. However, when the solute is non-magnetic, there comes a point where increasing substitution causes a breakdown in the continuity of the magnetic exchange through the lattice and a progressive collapse to a paramagnetic spectrum. In this situation the presence of order-disorder phenomena are important. Thus the ordered alloy FeAl, which ideally has a body-centred cubic lattice (CsCl structure), has eight non-magnetic Al neighbours about each Fe. Crushing the alloy causes ferromagnetic behaviour to appear and this is clearly seen in Fig. 5.19 for an alloy of composition $Fe_{0.604}Al_{0.396}$ because plastic deformation creates large numbers of antiphase boundaries across which the number of Fe—Fe nearest neighbours is significantly greater than in the ordered alloy [5.19].

5.6.3   Impurity doping

There are many interesting compounds which do not contain a suitable Mössbauer isotope, but it may still be possible to obtain useful information by

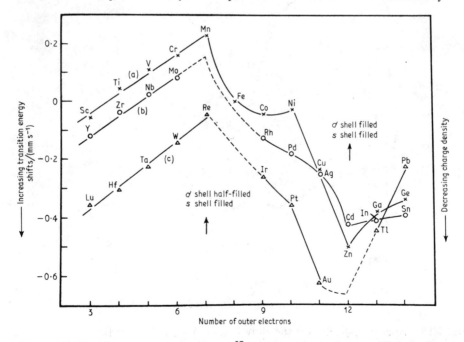

Fig. 5.20 The chemical isomer shift of ^{57}Co impurities in metals of (a) the 3d, (b) the 4d, and (c) the 5d transition-metal series. [5.21]

doping the compound with a Mössbauer precursor as an impurity probe, and using the compound as a source. The most common use of this technique is in ^{57}Co or ^{119}Sn doping of metal alloys. A daughter ^{57}Fe atom in a magnetic alloy will couple magnetically with the surrounding host atoms and thereby monitor the magnetic properties of the latter. For example, the spectrum of ^{57}Co in single crystals of nickel shows that the direction of the ^{57}Fe impurity spin and hence of the ordered Ni spins changes from the [111] axis to the [100] axis at 490 K [5.20].

The chemical isomer shift is also a monitor of the band-structure of the host. Figure 5.20 shows the shift for ^{57}Co impurity in the transition metals [5.21], which has a systematic dependence upon the electron configuration of the metal which is similar for the 3d, 4d, and 5d series. The maxima and minima at the half-filled and filled d-shell configurations show that the observed variations in shift derive in part from the d-band filling of the host metal.

5.6.4 Decay aftereffects

In most applications of Mössbauer spectroscopy the energetic nuclear events which precede the emission of the Mössbauer γ-ray can be ignored. The Auger cascade following K-electron capture in ^{57}Co can cause momentary ionization with charge states of up to +7. In metallic matrices the mobility of the electrons causes a return to 'normality' before the subsequent Mössbauer decay, and thus the nuclear processes remain unrecorded.

In many non-conducting compounds, however, the daughter ^{57}Fe atom is often found to exist in more than one oxidation state. It was originally thought that any Fe^{3+} charge state, for example, was still in the process of decay to Fe^{2+} if the latter was also present. The Fe^{2+} : Fe^{3+} ratio should alter when the spectrum is recorded in delayed coincidence with the preceding 123 keV γ-ray, but it is known that this ratio usually remains constant, and that the Fe^{3+} state is already in equilibrium in the source matrix.

In some molecular compounds such as Fe(acac)$_3$, the emission spectra from ^{57}Co doped samples resemble very closely the spectra of electron-irradiated undoped samples [5.22]. It seems highly likely that autoradiolysis occurs in many instances, and it therefore becomes difficult to decide whether the emission spectrum recorded is that of the original molecular complex, or of an unidentified radiolysis product.

Nevertheless, because the Mössbauer spectrum can be recorded using a relatively small number of decay events which are thereby effectively independent of each other, it has a valuable role to play in the study of the aftereffects of nuclear decays.

BIBLIOGRAPHY

The following texts may be of interest to the reader who wishes to enquire further:—

Wertheim, G.K., *Mössbauer Effect: Principles and Applications*, Academic Press, New York (1964).

An Introduction to Mössbauer Spectroscopy, Ed. May, L., Hilger, London (1971).

Chemical Applications of Mössbauer Spectroscopy, Ed. By Goldanskii, V.I. and Herber, R.H., Academic Press, New York (1968).

Greenwood, N.N. and Gibb, T.C., *Mössbauer Spectroscopy*, Chapman and Hall, London (1971).

Spectroscopic Properties of Inorganic and Organometallic Compounds, Vol. 1 (1967) and succeeding volumes, Ed. Greenwood, N.N., The Chemical Society, London.

Mössbauer Effect Methodology, Ed. Gruverman, I.J., Vol. 1 (1965) and succeeding volumes, Plenum Press, New York.

Gibb, T.C., *Principles of Mössbauer Spectroscopy*, Chapman and Hall, London (1976).

REFERENCES

5.1 Mössbauer, R.L., *Z. Physik*, **151**, 124, (1958); *Naturwissenschaften*, **45**, 538, (1958); *Z. Naturforsch.*, **14a**, 211 (1959).

5.2 Pound, R.V. and Rebka, G.A., *Phys. Rev. Letters*, **3**, 554 (1959).

5.3 Kistner, O.C. and Sunyar, A.W., *Phys. Rev. Letters*, **4**, 412 (1960).

5.4 Reddy, K.R., Barros, F. de S. and DeBenedetti, S., *Phys. Letters*, **20**, 297 (1966).

5.5 Wickman, H.H. and Wagner, C.F., *J. Chem. Phys.*, **51**, 435 (1969).

5.6 Berkooz, O., Malamud, M. and Shtrikman, S., *Solid State Comm.*, **6**, 185 (1968).

5.7 Bukshpan, S., Soriano, J. and Shamir, J., *Chem. Phys. Letters*, **4**, 241 (1969).

5.8 Pasternak, M. and Sonnino, T., *J. Chem. Phys.*, **48**, 1997 (1968).

5.9 Herber, R.H., Kingston, W.R. and Wertheim, G.K., *Inorg. Chem.*, **2**, 153 (1963).

5.10 Erickson, N.E. and Fairhall, A.W., *Inorg. Chem.*, **4**, 1320 (1965).

5.11 Greatrex, R., Greenwood, N.N., Rhee, I., Ryang, R. and Tsutsumi, S., *Chem. Comm.*, 1193 (1970).

5.12 Bancroft, G.M., Garrod, R.E.B., Maddock, A.G., Mays, M.J. and Prater, B.E., *Chem. Comm.*, 200 (1970).

5.13 Watson, R.E., *Phys. Rev.*, **118**, 1036; **119**, 1934 (1960).

5.14 Jesson, J.P., Weiher, J.F. and Trofimenko, S., *J. Chem. Phys.*, **48**, 2058 (1968).

5.15 Shulman, R.G. and Sugano, S., *J. Chem. Phys.*, **42**, 39 (1965).

5.16 Greenwood, N.N., Howe, A.T. and Menil, F., *J. Chem. Soc. (A)*, 2218 (1971).

5.17 K.K. Rao, R. Cammack, D.O. Hill and C.E. Johnson, *Biochem. J.*, **122**, 257 (1971).

5.18 Chekin, V.V. and Naumov, V.G., *Zhur. Eksp. Teor. Fiz.*, **50**, 534 (1966).

5.19 Huffman, G.P. and Fisher, R.M., *J. Appl. Phys.*, **38**, 735 (1967).

5.20 Chandra, G. and Radhakrishnan, T.S., *Phys. Letters*, **28A**, 323 (1968).

5.21 Quaim, S.M., *Proc. Phys. Soc.*, **90**, 1065 (1967).

5.22 Baggio-Saitovitch, E., Friedt, J.M. and Danon, J., *J. Chem. Phys.*, **56**, 1296 (1972).

Index